MULTI-DISCIPLINARY SUSTAINABLE ENGINEERING: CURRENT AND FUTURE TRENDS

MULTIDISCIPLINARY SUSTAINABILITY, NONFICTION & CURRENT AND FUTURE TRENDS

PROCEEDINGS OF 5TH NIRMA UNIVERSITY INTERNATIONAL CONFERENCE ON ENGINEERING (NUICONE 2015), AHMEDABAD, INDIA, 26–28 NOVEMBER, 2015

Multi-disciplinary Sustainable Engineering: Current and Future Trends

Editors

P.N. Tekwani, M. Bhavsar & B.A. Modi
Institute of Technology, Nirma University, Ahmedabad, India

CRC Press
Taylor & Francis Group
Boca Raton London New York Leiden

CRC Press is an imprint of the
Taylor & Francis Group, an **informa** business

A BALKEMA BOOK

Published by: CRC Press/Balkema
P.O. Box 11320, 2301 EH Leiden, The Netherlands
e-mail: Pub.NL@taylorandfrancis.com
www.crcpress.com – www.taylorandfrancis.com

First issued in paperback 2020

ISBN 13: 978-0-367-73710-8 (pbk)
ISBN 13: 978-1-138-02845-6 (hbk)

Visit the Taylor & Francis Web site at
http://www.taylorandfrancis.com

and the CRC Press Web site at
http://www.crcpress.com

Multi-disciplinary Sustainable Engineering: Current and Future Trends – Tekwani, Bhavsar & Modi (Eds)
© 2016 Taylor & Francis Group, London, ISBN 978-1-138-02845-6

Table of contents

Infrastructure project planning and management

Chemical process development and design

Technologies for green environment

Sustainable manufacturing processes

Design and analysis of machine and mechanism

Energy conservation and management

Preface

We are pleased to publish the proceedings of the 5th Nirma University International Conference on Engineering (NUiCONE 2015), entitled Multi-disciplinary Sustainable Engineering: Current and Future Trends. This volume of proceedings from the conference provides an opportunity for readers to engage with a selection of refereed papers that were presented during the conference. Researchers from industry and academia were invited to present their research work in the areas listed below. The research papers presented in these tracks have been published in this book with the support of CRC Press/Balkema, Taylor & Francis Group. This volume definitely will provide a platform to proliferate new findings among the researchers.

- Concrete and Structural Engineering
- Infrastructure Project Planning and Management
- Chemical Process Development and Design
- Technologies for Green Environment
- Sustainable Manufacturing Processes
- Design and Analysis of Machine and Mechanism
- Energy Conservation and Management

Editors
Dr. P.N. Tekwani
Dr. Madhuri Bhavsar
Dr. B.A. Modi

Multi-disciplinary Sustainable Engineering: Current and Future Trends – Tekwani, Bhavsar & Modi (Eds)
© 2016 Taylor & Francis Group, London, ISBN 978-1-138-02845-6

Introduction

Nirma University International Conference on Engineering, NUiCONE, is a flagship event of the Institute of Technology, Nirma University, Ahmedabad, Gujarat, India. NUiCONE 2015 was planned with major interesting changes in the mode of events/themes to make it a multi-conference with many inter-disciplinary technical themes, encompassing and enabling researchers from a broad range of disciplines. This conference follows the successful organization of four national conferences and four international conferences in previous years. More than 170 research papers were submitted, out of which 71 papers were selected after rigorous reviews, including plagiarism checks and blind-fold technical reviews by multiple reviewers. The selected papers were presented during the conference by the authors. The expert speakers and conference delegates were from all across India, covering researchers and eminent experts from academia, industry and government R&D organizations like IPR, ISRO, PRL etc. An exciting new set of events was organised, specially to enable and attract increased participation by practicing engineers, technologists and technopreneurs from the industry through special knowledge sharing sessions, involving applied technical papers by industry participants based on case-study applications, white-papers, panel discussions, industrial exhibitions of innovations and technology products etc.

Padmashri Dr. Vijay Bhatkar (Maharashtra Bhushan, Fellow IEEE, Fellow ACM, Fellow CSI, INAE, NAScI, IETE, MASc, GES), Chairman of ETH Research Lab, Founder Chancellor of India International Multiversity, Chancellor of D.Y. Patil Education Society (Deemed University), National President of Vijnana Bharati, has graced the inaugural function as the chief guest. An eminent scientist, Padmashri Dr. Avinash Chander (President, Sensors Research Society, India, Former Secretary, Department of Defence R&D, Former Director General DRDO and Scientific Advisor to Raksha Mantri) was the key note speaker of the conference.

Mr Devdatta Bhaskar Hambardikar (Head of R&D and Manufacturing, Division Manager-Mitsubishi Electric India Pvt Ltd.) was invited as the chief guest for the valedictory function. Several eminent speakers from India (IFFCO, Watermann Pumps, Linde, ORACLE, IBM, NVIDIA, GIFT City, Mehta Group, Stantec Consulting, AMC, Bhoomi Consultants) delivered plenary addresses during the conference.

Acknowledgement

Putting together NUiCONE 2015 was a team effort. We first thank the authors for providing the papers and presenting their research work. We are grateful to the program committee and the reviewers, who worked very hard in reviewing papers and providing feedback for authors. Due credit should be given to CRC Press/Balkema, Taylor & Francis Group for publishing the proceedings of the research papers presented in the conference. Finally, we thank the hosting organization Nirma University, technical co-sponsors (Institutions for Engineers India, GESIA, BISAG, IEEE Gujarat Section, ASHRAE western India chapter, ACDOS, ISA, ASCE India section), sponsors, international advisory committee members and organising committee members.

ORGANISING COMMITTEE

Chief Patrons:	Shri K.K. Patel, *Vice President, Nirma University*
	Dr. Anup Singh, *Director General, Nirma University*
Patron:	Dr. P.N. Tekwani, *Director, IT, NU*
Conference Chair:	Dr. Madhuri D. Bhavsar, *IT, NU*
Conference Co-Chairs:	Dr. B.A. Modi, *IT, NU*
	Prof. Arup Dasgupta, *Ex-officio Expert, IEEE Gujarat Section*
Members:	Shri D.P. Chhaya, *Director, A&GA, NU*
	Dr. G. Ramchandran Nair, *Executive Registrar, NU*
	Dr. R.N. Patel, *Head of Mechanical Deptartment, ITNU*
	Dr. Sanjay Garg, *Head of Computer Deptartment, ITNU*
	Dr. P.V. Patel, *Head of Civil Department, ITNU*
	Dr. S.S. Patel, *Head of Chemical Department, ITNU*

INTERNATIONAL ADVISORY COMMITTEE

1. Dr Devdas Shetty, *Ph.D. P.E., Dean, School of Engineering and Applied Science, University of the District of Columbia, USA*
2. Laurencas Raslavicius, *Vice Dean of the Faculty of Mechanical Engineering, Kaunas University of Technology, Lithuania*
3. Dr Rishi Gupta, *Victoria University, Canada*
4. Dr Ajai, Scientist, *SAC-ISRO, Ahmedabad, India*
5. Shri. S.K. Patel, *Chief Engineer, Road and Building Department, Government of Gujarat, India*
6. Ms Alpa Sheth, *Structural Consultant, VMS Consultant, Mumbai, India*
7. Prof Toshihiko Kuwabara, *Tokyo University of Agriculture and Technology, Japan*
8. Prof Raghu Echempati, *Kettering University, Flint, MI, USA*
9. Shri Sanjay Desai, *CEO, RBD Engineers, India*
10. Shri Rajesh Sampat, *CEO, Inspiron Engineering Pvt. Ltd., Ahmedabad, India*
11. Dr Borko Furht, *Professor, Florida Atlantic University, USA*
12. Dr Srinivas Sampalli, *Professor, Dalhousie University, Halifax, Nova Scotia, Canada*
13. Mrs Sutapa Ranjan, *Institute of Plasma Research, Gandhinagar, India*
14. Mr G.K. Panchal, *Adviser, Yadav Measurement (P) Ltd., Udaipur, India*

15. Dr P.S.V. Nataraj, *Professor, IIT Bombay, Mumbai, India*
16. Dr Sisil Kumarawadu, *Professor, University of Moratuwa, Sri Lanka*
17. Dr Ram B. Gupta, *Professor, Auburn University, USA*
18. Uriel R. Cukierman, *President, IFEES, Universidad Technologica Nacional, Argentina*
19. Prof Radhakant Padhi, *Aerospace Engineering Department, IISc-Bangalore, India*
20. Shri Bharat Patel, *Former Chairman, The Institution of Engineers (I), Gujarat Centre, India*
21. Dr Suparana Mukherji, *Professor, IIT Bombay, Mumbai, India*
22. Shri Jaimin R. Vasa, *President, Gujarat Chemical Association (GCA), India*
23. Prof (Dr) K. Gopakumar, *Chairman, Department of Electronic Systems Engineering (DESE) (formerly known as CEDT), IISc, Bangalore, India*
24. Dr Atif Iqbal, *Associate Professor, College of Engineering, Qatar University, Qatar*
25. Shri Varuneshkumar, *MD, Veeral Controls, India*
26. Dr K.S. Dasgupta, *Director, Indian Institute of Space Science and Technology Thiruvananthapuram, India*
27. Prof Lajos Hanzo, *Head of Wireless Group, ECS, Faculty of Physical Sciences and Engineering, University of Southampton, Southampton, UK*
28. Shri Jay Ruparel, *President, GESIA, Ahmedabad, India*
29. Shri Nilesh Ranpura, *Manager, ASIC Design, E-infochip, Ahmedabad, India*
30. Amrita Alpadhi, *Information Developer, IBM, India*
31. Shri Jagdish Shukla, *Ex District Vice President, ISA Gujarat, India*
32. Shri P.N. Parikh, *Consultant UPL and Sector Expert (Chlor-Alkali), BEE (Ministry of Power) Government of India*
33. Prof Ravindra Gudi, *Professor, Chemical Engineering Department, IIT Bombay, Mumbai, India*
34. Dr Neerja Gupta, *Principal, BAC College, Syndicate Member, Gujarat University, Ahmedabad, India*

TECHNICAL COMMITTEE

Dr R.N. Patel (Chair)
Prof Arup Dasgupta, Ex. Officio, IEEE Gujarat Section

Dr Jayesh Ruparelia	Dr R.K. Mewada
Dr S.P. Purohit	Dr Parul Patel
Prof Zunnun Narmawala	Dr Kamal Mehta
Dr Santosh Vora	Prof C.R. Mehta
Dr Yogesh Trivedi	Dr Amisha Naik
Dr Dipak Adhyaru	Prof S.A. Mehta
Prof S.J. Joshi	Prof A.M. Lakdawala
Dr Richa Mishra	Prof Kunal Pathak

PUBLICATION COMMITTEE

Dr P.V. Patel (Chair)	
Prof Rushbh Shah	Prof Leena Bora
Prof Payal Oza	Prof Vibha Patel
Prof Amitayu Chakraborty	Prof Vijay Savani
Prof B.A. Shah	

LIST OF REVIEWERS

Civil Engineering Track

1. Dr P.H. Shah, *Professor and Principal, U.V. Patel College of Engineering, Ganpat University, Mehsana*
2. Dr P.P. Lodha, *Associate Professor, Civil Engineering Department, GEC, Valsad*
3. Dr Anjana Vyas, *Professor and Acting Dean, Faculty of Technology, CEPT University, Ahmedabad*
4. Dr L.B. Zala, *Professor and Head, Civil Engineering Department, BVM, Vallabh Vidyanagar*
5. Dr P.R. Patel, *Professor, Institute of Technology, Nirma University, Ahmedabad*
6. Dr Gaurang Joshi, *Associate Professor, Civil Engineering Department, SVNIT, Surat*
7. Dr Srinivas Arkatkar, *Assistant Professor, Department of Civil Engineering, SVNIT, Surat*
8. Dr U.H. Pandya, *Sr. Scientist (R & D), Excel Crop Care, Bhavnagar*
9. Dr H.S. Patil, *Professor, Department of Applied Mechanics, SVNIT, Surat*
10. Dr B.S. Munjal, *Senior Scientist, Space Applications Centre, ISRO, Ahmedabad*
11. Dr B.J. Shah, *Professor, Department of Applied Mechanics, L.D. College of Engineering, Ahmedabad*
12. Dr D.P. Soni, *Associate Professor, Civil Engineering Department, SVIT, Vasad*
13. Shri Himat Solanki, *Visiting Faculty, Institute of Technology, Nirma University, Ahmedabad*
14. Dr P.V. Patel, *Professor and Head, Institute of Technology, Nirma University, Ahmedabad*
15. Dr U.V. Dave, *Professor, Institute of Technology, Nirma University, Ahmedabad*
16. Dr S.P. Purohit, *Professor, Institute of Technology, Nirma University, Ahmedabad*
17. Dr C.D. Modhera, *Professor, Department of Applied Mechanics, SVNIT, Surat*
18. Dr Yogesh Patil, *Assistant Professor, Department of Applied Mechanics, SVNIT, Surat*
19. Dr Chirag Patel, *Professor and Head, Civil Engineering Department, Sir P.S. University, Udaipur*

Chemical Engineering Track

1. Dr A.P. Vyas, *Principal, S.B. Patel College of Engineering, Mahesana*
2. Dr Nitin Bhate, *Associate Professor, Chemical Engineering Department, M.S. University, Baroda*
3. Dr P.N. Dave, *Professor, Department of Chemistry, KGSKV Katchchh University, Bhuj*
4. Dr S.S. Patel, *Professor and Head, Institute of Technology, Nirma University Ahmedabad*
5. Dr J.P. Ruparelia, *Professor, Institute of Technology, Nirma University Ahmedabad*
6. Dr R.K. Mewada, *Professor, Institute of Technology, Nirma University Ahmedabad*
7. Dr M.H. Joshipura, *Professor, Institute of Technology, Nirma University Ahmedabad*
8. Dr Vikas Lakhera (Mechanical), *Professor, Institute of Technology, Nirma University Ahmedabad*

Mechanical Engineering Track

1. Dr H.K. Raval, *Professor, Mechanical Engineering Department, SVNIT, Surat*
2. Dr H.J. Nagarsheth, *Professor, Mechanical Engineering Department, SVNIT, Surat*
3. Dr D.P. Vakharia, *Professor, Mechanical Engineering Department, SVNIT, Surat*
4. Dr D.V. Bhatt, *Professor, Mechanical Engineering Department, SVNIT, Surat*
5. Dr Jyotirmay Banerjee, *Professor, Mechanical Engineering Department, SVNIT, Surat*
6. Dr Amit Trivedi, *Head—Production Engineering Department, BVM, Vallabh Vidyanagar, Gujarat*
7. Dr Bharat Ramani, *Associate Professor, Marwadi Education Foundation's Group of Institutions, Rajkot*
8. Dr Jayesh Ratnadhariya, *Principal, Alpha Engineering College, Khatraj, Ahmedabad*
9. Dr Ragesh G. Kapadia, *Professor, Mechanical Engineering Department, SVMIT, Bharuch*

10. Prof P.D. Solanki, *Professor and Head, Mechanical and Automobile Engg., LDCE, Ahmedabad*
11. Dr Hetal N. Shah, *Assistant Professor, Mechanical Engineering Department, Charutar Institute of Technology, Changa*
12. Dr K.P. Desai, *Professor, Mechanical Engineering Department, SVNIT, Surat*
13. Dr N.M. Bhatt, *Director, Gandhinagar Institute of Technology, Vil. Moti Bhoyan, Khatraj, Gandhinagar*
14. Dr Vishal N. Singh, *Professor, A.D. Patel Institute of Technology, New Vallabh Vidyanagar*
15. Dr Dhaval A. Jani, *Professor, A.D. Patel Institute of Technology, New Vallabh Vidyanagar*
16. Dr Vishvesh J. Badheka, *Assistant Professor, Pandit Deendayal Petroleum University, Gandhinagar*
17. Dr Surendra Kumar Kachhwaha, *Professor, Pandit Deendayal Petroleum University, Gandhinagar*
18. Dr Sanket N. Bhavsar, *Professor, Mechatronics Department, G.H. Patel College of Engineering, Vallabh Vidhyanagar*
19. Dr Vinod N. Patel, *Professor, G.H. Patel College of Engineering, Vallabh Vidhyanagar*
20. Mr Manoj Gupta, *Engineer-SE (Cryogenics Division), Institute of Plasma Research, Bhat, Gandhinagar*
21. Dr R.N. Patel, *Professor and Head, Institute of Technology, Nirma University Ahmedabad*
22. Dr V.J. Lakhera, *Professor, Institute of Technology, Nirma University Ahmedabad*
23. Dr D.S. Sharma, *Professor & Head, Mechanical Engineering Department, MSU Baroda*
24. Dr K.M. Patel, *Professor, Institute of Technology, Nirma University Ahmedabad*
25. Dr B.A. Modi, *Professor, Institute of Technology, Nirma University Ahmedabad*
26. Prof Reena Trivedi, *Senior Associate Professor, Institute of Technology, Nirma University Ahmedabad*
27. Prof S.J. Joshi, *Associate Professor, Institute of Technology, Nirma University Ahmedabad*
28. Dr Manoj Chouksey, *Associate Professor, SGSITS, Indore*
29. Dr Manoj Gour, *Associate Professor, MITS Gwalior*
30. Dr B. Rawal, *Associate Professor, SGSITS Indore*
31. Dr Rahul Mulik, *Associate Professor, IIT Roorkee*
32. Dr Pankaj Jagad, *Associate Professor, Institute of Technology, Nirma University Ahmedabad*
33. Dr Chandrashekhar Sewatkar, *Associate Professor, College of Engineering Pune*
34. Dr Mukul Shrivastava, *Research Fellow, IIT Mumbai*
35. Mr Vijay Duryodhan, *Research Fellow, IIT Mumbai*
36. Dr Mitesh Shah, *Associate Professor, ADIT Anand*
37. Mr A.V. Pathak, *Scientist SG, SAC ISRO*
38. Dr Mayur Sutaria, *Associate Professor, CHARUSAT Changa*
39. Prof Chetan Mistry, *Associate Professor, Institute of Technology, Nirma University Ahmedabad*
40. Prof M.B. Panchal, *Associate Professor, Institute of Technology, Nirma University Ahmedabad*

About Nirma University

Nirma University is one of India's leading universities, based in Ahmedabad (Gujarat). The University was established in the year 2003 as a Statutory University under a special act passed by the Gujarat State Legislative Assembly. It is recognized by the University Grants Commission (UGC) under Section 2 (f) of the UGC Act. The University is duly accredited by the National Assessment and Accreditation Council (NAAC). The University is a member of the Association of Indian Universities (AIU) and the Association of Commonwealth Universities (ACU). Dr Karsanbhai K. Patel, Chairman of the Nirma Group of Companies and Chairman of NERF is the President of the University, functioning under the aegis of NERF. The University consists of a Faculty of Engineering and Technology, a Faculty of Management, a Faculty of Pharmacy, a Faculty of Law, a Faculty of Science, a Faculty of Architecture and a Faculty of Doctoral Studies and Research.

Concrete and structural engineering

Multi-disciplinary Sustainable Engineering: Current and Future Trends – Tekwani, Bhavsar & Modi (Eds)
© 2016 Taylor & Francis Group, London, ISBN 978-1-138-02845-6

Effect of gradation on pavement concrete mix with Blast Furnace Slag

B.G. Buddhdev
R.K. University, Rajkot, Gujarat, India
Government Polytechnic, Bhuj, Gujarat, India

H.R. Varia
Tatva Institute of Technological Studies, Modasa, Gujarat, India

ABSTRACT: Concrete is the second largest consumed material by humans worldwide. In this modern era, cement concrete pavements are in demand when compared with bituminous pavements in highway projects. In the recent era of green technology, waste utilization in the production of concrete, especially for pavements, has become the major concern and is the focus of the present study. Blast Furnace Slag (BFS) is one of the major wastes produced from steel processing plants around the globe. An enormous quantity of BFS is produced from these steel plants, but a very little quantity is utilized in different civil engineering applications. BFS can be utilized as a fine aggregate for the replacement of sand to produce concrete mainly for pavements. Sand, itself being a natural resource, has many technical limitations in terms of quality for the production of pavement concrete. In this paper, a comprehensive experimental program was considered to study the effect of gradation of BFS on pavement concrete when BFS was replaced with natural riverbed sand. In this regard, BFS was graded as per four zones of gradation of the Rajkot city area. The graded samples of BFS were utilized for the production of concrete to evaluate its different properties. Based on the results of the experiment, variations in the properties of pavement concrete were analyzed and the effect of the gradation of BFS on pavement concrete was studied. The results indicate that unlike sand, BFS provides more flexibility in terms of design mix when utilized as a fine aggregate in pavement concrete.

Keywords: gradation; blast furnace slag; pavement concrete; sand (fine aggregate); design mix

1 INTRODUCTION

Concrete is a vital and trusted material in the construction industry for many decades. It has too many applications and utilization in the construction industry including construction of the pavement. In the era of sustainable and green technology, concrete material has been exposed to new innovation and technological development in the form of varieties of waste material utilization to produce concrete. Blast Furnace Slag (BFS) is one of the many waste materials available. BFS, which is a by-product of the iron-making process, has an amorphous structure and pozzolanic properties.[3] It is a glassy material, typically with sand-to-gravel size particles, and provides superior skid characteristics in the pavement. Due to huge amounts of BFS production, stockpiling and land filling have become the major environmental issues for this waste material.[5] On the other hand, during the production of concrete, one of the important ingredients is the fine aggregate (sand). Generally, during the production of concrete, naturally available river bed sand is used as a fine aggregate. This ingredient is mainly responsible for the cohesion of concrete and enhances the compaction characteristics, and thus increases the workability of concrete. Due to rapid urbanization and increase in construction boom, concrete production is increased. This accounts for the increase in the requirement of basic raw materials for the production of concrete. As sand is a naturally available material from the river source, its availability against its demand is very much scarce. Again, due to environmental problems such as global warming, green house effect, and ozone depletion, the river capacity decreases on a daily basis. These factors lead to a short supply of natural sand available for the construction industry. This creates an imbalance in the situation of demand and supply. To find out the technical solution of this imbalanced situation, and keep the concrete material economical in this global scenario, an attempt has been made to utilize the waste material such as BFS in concrete production. This natural river bed sand can be fully replaced by BFS, which is purely a waste material, to economize the concrete production and at the same time to preserve and protect natural resources such as sand. Gradation plays an important role in the production of concrete and has attributes in the properties of fresh and hardened concrete. During the mix design of concrete, natural sand creates a limitation to the mix designer, i.e., it cannot be changed to a particular location, considering other ingredients as constant. Gradation of BFS is a controlled process, while it is used as a fine aggregate in concrete mix, provides flexibility to the designer to work out an optimum mix considering other ingredients from the same source. In this paper, the effect of gradation on the proposed concrete mix with BFS was investigated. Also, other properties such as slump, compressive and flexural strength, and density of concrete were evaluated to understand the behavior of concrete when it was prepared with BFS in place of natural sand.

2 REVIEW OF THE LITERATURE

Many researchers in the field of concrete technology have considered the importance of good grading in making quality concrete consistent with economy, and have directed their studies to achieve good grading of aggregate at construction sites. Fuller and Thompson (1907) concluded that grading for maximum density gives the highest strength and that the grading curve of the best mixture resembles a parabola.[4] Abrams et al. (1918) in the course of their investigations also found that the surface area of the aggregate varied widely without causing much appreciable difference in concrete strength. Therefore, they introduced a parameter known as "fineness modulus" for arriving at satisfactory grading.[1]

Edwards and Young (1919) proposed a method of proportioning based on the surface area of aggregates to be wetted. It was concluded that the concrete made from aggregate grading with the least surface area would require least amounts of water and would thus be the strongest.[13] Waymouth (1931) introduced his theory of satisfactory grading on the basis of "particle interference" considerations. He found the volume relationship between the successive size group of particles based on the assumption that particles of each group are distributed throughout the concrete mass in such a way that the distance between them is equal to the mean diameter of the particle of the next smaller size group plus the thickness of the cement film between them.[13]

Powers (1968) discussed various aggregate grading techniques including Fuller grading, the hypothesis that there was an ideal size gradation for concrete aggregate.[13] Dewar (1999), who developed a computer program that can predict aggregate packing, concluded that there is a very wide range of acceptable distributions, both continuous and gap-graded, which will result in economic concrete, provided the correct proportioning is achieved in each case.[2]

Many other methods have been suggested for arriving at an optimum grading. All these procedures, methods, and formulae point to the fact that none of them is satisfactory and reliable in field applications. At the site, a reliable satisfactory grading can only be decided by applying an actual trial and error method, which takes into consideration the characteristics of local materials with respect to size fraction, shape, surface texture, flakiness, and elongation index. The widely varying peculiar characteristics of course and fine aggregates cannot be solved by formulas and set procedures in practical applications.

One of the practical methods for arriving at the practical grading by the trial and error method is to mix aggregates of different size fractions at different percentages and to choose one sample, with a maximum weight or minimum void per unit volume, from all the alternative samples. Fractions that are, in fact, available in the field, or which could be made available in the field including that of fine aggregates (sand), will be used in producing samples.[12,13]

3 EXPERIMENTAL WORK

3.1 *Introduction*

Gradation is an important parameter to be considered for producing good workable concrete, which ultimately influences the strength and durability. This experimental work aimed at working out the optimum gradation in the case of BFS, as the gradation of BFS used as a fine aggregate is a controlled process, unlike the case of natural river bed sand. To perform the mix design, natural sand is classified into four zones as per its gradation according to the Indian Standard (IS) code provision. The courser gradation of sand attributes to zone I, and subsequently finer sand is classified as zone II, zone III, and zone IV, respectively. This gradation always plays an important role in mix design. Sand is a natural resource and cannot be changed to a particular location, considering other ingredients as constant, and thus causes a limitation to the mix designer. Gradation of BFS, used as a fine aggregate in concrete mix, is a controlled process and provides the designer a flexibility to work out an optimum mix while considering other ingredients from the same source.

3.2 *Gradation of BFS*[7]

To determine the effect of gradation on the proposed concrete mix, a trial was performed by grading the BFS into four zones of gradation as per the IS code provision (see Figure 1). The graded curves for the entire zone as per the guidelines of IS are shown in Figures 2–5. These

Figure 1. Graded sample of BFS of zone-I to zone-IV.

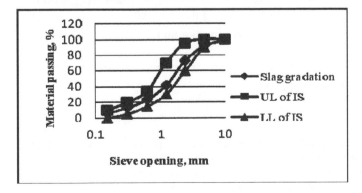

Figure 2. Gradation for zone-I.

5

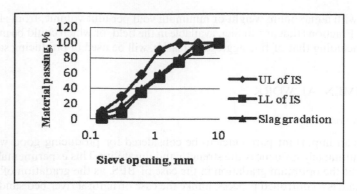

Figure 3. Gradation for zone-II.

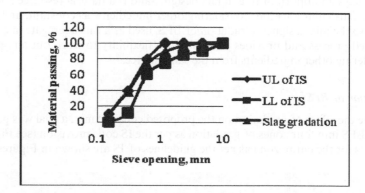

Figure 4. Gradation for zone-III.

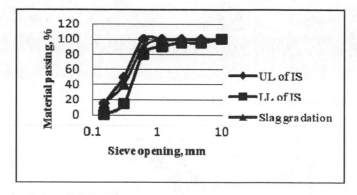

Figure 5. Gradation for zone-IV.

BFS samples were used in the proposed concrete mix of grade M40 (which is the most common grade for almost all types of pavement construction) for each of the zones, considering other concrete ingredients as constant. The effect of gradation on the concrete characteristics such as slump, density, compressive, and flexural strength of concrete was observed for 7 and 28 days of the curing period.

Table 1. Characterization of aggregates and BFS.

Test parameter	Coarse aggregate (20 mm to 10 mm)	Coarse aggregate (≤10 mm)	BFS
Specific gravity	2.84	2.77	2.72
Water absorption (%)	0.6	0.4	1.35

Table 2. Gradation for coarse aggregates.

IS sieve sizes (mm)	Analysis of coarse aggregate fraction (%) passing		Percentage Passing of different fractions		
	I 20 to 10 mm	II ≤10 mm	I 60 percent	II 40 percent	Combined 100 percent
20	95.34	100.00	57.20	40.00	97.20
10	2.52	65.32	1.51	26.12	27.64
4.75	Nil	5.76	–	2.30	2.30

Table 3. Mix proportions for 1 cubic meter of concrete.

Zone	Cement (kg)	BFS (kg)	Coarse aggregates (kg)		Water (kg)	Admixture (kg)
			≤10 mm	20 to 10 mm		
Zone-I	416	723	483	726	176	2.5
Zone-II	416	684	499	749	174	2.5
Zone-III	416	648	514	773	173	2.5
Zone-IV	416	609	530	796	172	2.5

3.3 *Characterization of the material used*[6,7,12]

A detailed characterization of the material used in this study was studied as per the current IS code provision. The physical properties of BFS are presented in Table 1. Raw materials required for the concrete mix were characterized for the aggregates used for concrete production. Properties such as specific gravity, water absorption, and gradation were evaluated experimentally and listed in Tables 1 and 2.

3.4 *Concrete mix design*[9,10,11]

Mix design for concrete of M40 grade was carried out as per the provision of IS: 10262:2009 & Indian Roads Congress (IRC): 44:2008. The stipulations for design mix considered were as follows: cement of 53 grade, maximum size of coarse aggregate 20 mm, minimum and maximum cement contents 325 kg/m³ and 425 kg/m³, respectively, maximum water-to-cement ratio 0.50, slump for ordinary concrete, and good degree of supervision. The mix design was performed based on the raw material selected and the results evaluated for their characterization. The design mix was prepared for all the zones classified as zone-I, zone-II, zone-III, and zone-IV. The design mix for 1 cubic meter of concrete produced is detailed in Table 3. The appropriate dosage of super plasticizer was also utilized for concrete mix to achieve the desired strength and workability that was required for concrete of higher grade such as M40.

4 RESULTS

Based on the experiment done to determine the effect of gradation on concrete with BFS, the results were evaluated for concrete properties such as density at different times, slump, and

Table 4. Density of concrete at different time periods.[14]

Zone	Avg. density at the time of testing (kg/m³)	Avg. density after 24 hr. (kg/m³)	Avg. density after 7 days (kg/m³)	Avg. density after 28 days (kg/m³)	% Decrease in water content after 24 hr.	% Increase in water ingress due to curing after 7 days	% Increase in water ingress due to curing after 28 days
Zone-I	2593.58	2541.24	2565.92	2612.25	2.02	0.96	2.72
Zone-II	2575.8	2535.3	2561.97	2588.94	1.57	1.04	2.07
Zone-III	2574.81	2531.35	2554.07	2583.47	1.69	0.89	2.02
Zone-IV	2546.17	2506.66	2529.38	2559.17	1.55	0.90	2.05

Table 5. Slump of concrete.[14]

Zone	Zone-I	Zone-II	Zone-III	Zone-IV
Slump (mm)	30	38	34	39

Table 6. Compressive and flexural strength of concrete.[8,14]

Zone	Avg. compressive strength (N/mm²)		Avg. flexural strength (N/mm²)	
	7 days	28 days	7 days	28 days
Zone-I	34.79	49.48	4.205	4.913
Zone-II	38.60	54.26	4.224	4.963
Zone-III	36.46	52.59	4.186	4.902
Zone-IV	35.10	50.69	4.152	4.894

the compressive and flexural strength for 7 and 28 days of the curing period. These results for density parameters are summarized in Table 4, for slump in Table 5, and for the compressive and flexural strength in Table 6 for 7 and 28 days of the curing period. The density of concrete at different time periods was evaluated and is indicative of good protection against moisture absorption. The slump value obtained was almost the same as per the assumed workability in design mix. Compressive strength and flexural strength of concrete at 7 days of the curing period were comparatively less than the normal value of concrete; however, it was at par at 28 days of the curing period, which was possibly due to the pozzolanic effect of small-sized particles present in the BFS used as a fine aggregate.

5 CONCLUSION

An experimental work was carried out to study the effect of gradation on the properties of pavement concrete mix with BFS used as a fine aggregate by the replacement of natural river bed sand. Based on the characterization of the raw materials used and design mix worked out for all the four zones, the results for the various properties of the pavement concrete produced were found to be satisfactory. The density of the concrete produced was superior when compared with the normal value of the concrete material. Compressive and flexural strength at 7 days of the curing period were comparatively less than the normal value of the concrete, but it became at par at 28 days of the curing period, indicative of showing some pozzolanic effect of BFS on the concrete produced. The results indicated that zone-II and zone-III types of gradation of BFS were the most appropriate concrete mix for the chosen coarse aggregates. Alternatively, coarse aggregates with other characteristics can be optimized with the available gradation of BFS from any of the four zones, providing a flexibility to achieve the best concrete mix without any compromise. The effect of gradation on concrete mix for grade M40 was established with BFS and the same can be utilized for other grades of concrete mix. Also, the use of BFS in place of natural sand for concrete production is a good indication, which definitely saves the natural resource because, currently in metro cities, artificial sand is utilized for the production of concrete.

REFERENCES

[1] Abrams, D.A. (1918). Design of Concrete Mixtures. Chicago: Structural Materials Research Laboratory, Lewis Institute.
[2] Dewar, J.D. (1999). Computer Modeling of Concrete Mixtures. London: E & FN Spon
[3] Emery, J.J. (1980). "Palletized Lightweight Slag Aggregate," Proceedings of Concrete International, Concrete Society.
[4] Fuller W. and Thompson S.E. (1907) The laws of proportioning concrete. American Society of Civil Engineers, 33, 67–143
[5] Jahangirnejad S., Dam T.V., Morian D., Smith K. (2013) "Use of Blast Furnace Slag as a Sustainable Material in Concrete Pavements" Paper submitted in the 92nd Annual Meeting of the Transportation Research Board, National Research Council, Washington
[6] Indian Standard IS: 383:1970 Coarse and Fine Aggregates from Natural Sources
[7] Indian Standard IS: 2386:1963 Test for Aggregates
[8] Indian Standard IS: 516:1959 Methods of Tests for Strength of Concrete
[9] Indian Standard IS: 456:2000 Plain and Reinforced Concrete-Code of Practice
[10] Indian Standard IS: 10262:2009 Concrete Mix Proportioning-Guidelines
[11] Indian Road Congress IRC: 44:2008 Cement Concrete Mix Design for Pavements
[12] Neville A.M. and Brooks, J.J., "Concrete Technology", John Wiley & Sons, Inc.
[13] Shetty M.S., "Concrete Technology Theory & Practices", S. Chand & Company Ltd., NewDelhi.
[14] Test Reports: Atmiya Institute of Technology & Science-Testing & consultancy Cell- July-2014.

Multi-disciplinary Sustainable Engineering: Current and Future Trends – Tekwani, Bhavsar & Modi (Eds)
© 2016 Taylor & Francis Group, London, ISBN 978-1-138-02845-6

Behavior of cellular reinforced sand-silt mixture under monotonic loading

S.A. Mulani
Department of Geotechnical Engineering, College of Engineering, Pune, India

R.S. Dalvi
Department of Civil Engineering, College of Engineering, Pune, India

ABSTRACT: Monotonic triaxial tests were carried out on a loose sand-silt mixture to investigate the effect of cellular reinforcement on deviator stress. Triaxial tests were conducted on an unreinforced and a reinforced fully saturated sand-silt mixture of sample size 75×150 mm with three different confining pressures of 50, 100, and 150 kPa. The cellular reinforcement is made up of locally available waste plastic bottles used for drinking water. The reinforcement used in the present study is of 60 mm diameter with 20 mm and 40 mm height to study the effect of height of reinforcement on deviator stress. The study indicated that deviator stress increased as the height of reinforcement increased. The results showed that the shear strength parameters of reinforced sand silt mixture increased significantly.

Keywords: monotonic triaxial test; cellular reinforcement; sand silt mixture; compression

1 INTRODUCTION AND LITERATURE REVIEW

A wide range of natural and manmade reinforcements have been used to improve soil performance. Short discrete fibers made of polymeric or natural material have been used to improve the shear strength of soil (Gray and Ohashi, 1983; Gray and Maher, 1989). Fuziah Ahmad (2010) has suggested that natural resources may provide superior materials for improving soil structure base with cost effectiveness.

Michaowski and Cermak (2002) conducted drained triaxial compression tests on specimens of fiber reinforced sand. It is concluded that a small amount of synthetic fibers increases shear strength substantially. Further, Ibraim et al. (2010) carried out a series of laboratory experiments for improving the monotonic undrained response of loose clean sand. They observed that the reinforcement inclusions reduce the potential for occurrence of liquefaction for both compression as well as tension and convert strain softening response to strain hardening response. Lal and Mandal (2013) performed monotonic triaxial tests on 35- and 50-mm diameter cellular reinforced dry fly ash. For both 35- and 50-mm diameter cellular reinforcements (geocells), the peak deviator stresses and the corresponding strains increased with increased reinforcement height. The maximum increase in shear strength parameters occurs when the cellular reinforcement was placed in two layers at one-third the height from the extreme top and bottom of the sample. The peak deviator stresses, corresponding strains, and the evaluated shear strength parameters for 50-mm diameter geocells were found to have higher values in comparison to the 35-mm diameter geocells.

From the literature survey, it is observed that monotonic triaxial tests were not performed on sand-silt mixture with cellular reinforcement to improve shear strength parameter. Therefore, in this paper, we performed a series of consolidated undrained triaxial tests on saturated sand-silt mixture to observe its shear strength using cellular reinforcement. The effect of height of reinforcement on the shear strength was also studied in this work.

2 MATERIALS AND METHODS

2.1 *Materials*

The sand used in these experiments was obtained locally and has been classified as SP according to the Unified Soil Classification System (USCS). The grain size distribution curve is as shown in Figure 1. The index properties of sand are $G = 2.58$, $\gamma_{max} = 15.55$ KN/m³, $\gamma_{min} = 14.31$ KN/m³, $e_{max} = 0.803$, $e_{min} = 0.659$, $D_{50} = 0.3$, and the silt used is nonplastic having $G = 2.79$, $\gamma_{max} = 14.67$ KN/m³, $\gamma_{min} = 11.65$ KN/m³, $e_{max} = 1.395$, $e_{min} = 0.902$, $D_{50} = 0.02$.

The tests are conducted for 30% relative density with 25% silt content and for three confining pressures of 50 kPa, 100 kPa, and 150 kPa. In total, nine tests were conducted on a 75×150 mm sample size. The detailed test program is as shown in Table 1.

2.2 *Cellular reinforcement*

The cellular reinforcement used in this study is as shown in Figure 2. It is made up of locally available used and wasted water bottles of 60 mm diameters. These bottles were cut across the length to form a geocell. Two different heights of the reinforcement 20 mm and 40 mm are used to study the effect of height of cellular reinforcement on sand silt mixture.

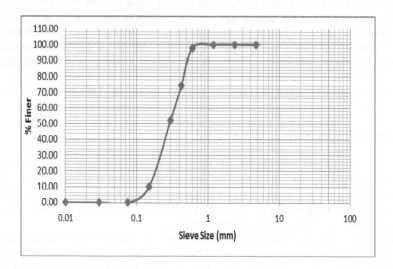

Figure 1.   Grain size distribution of sand.

Table 1.   Experimental program.

Sample size (mm)	RD (%)	Silt content (%)	Confining pressure (kPa)	Cellular height (mm)
				0
			50	20
				40
				0
75×150	30	25	100	20
				40
				0
			150	20
				40

Figure 2.   Cellular reinforcement.

Figure 3.   Triaxial test instrument.

2.3   *Experimental investigation*

A standard triaxial test apparatus used to conduct the consolidated undrained tests includes.

Load Frame: 50 kN capacity, Ram Speed: 0.00001–9.9999 mm/min, displacement transducer: $\pm$ 50 mm, pore pressure transducer ranging from 0.1 to 10 kg/cm^2, constant pressure system having a capacity of 20 Kg/cm^2, and data acquisition system for recording load, pore pressure, and displacement during the test. The details of the apparatus are as shown in Figure 3.

2.4   *Specimen preparation method*

All triaxial tests were performed on the sand with 25% of silt content on cylindrical specimens of 75×150 mm. Moist tamping method was used to prepare a sand silt sample. In this method, a known quantity of sand for achieving particular density was mixed with 25% of

Figure 4.   Placing of cellular reinforcement.

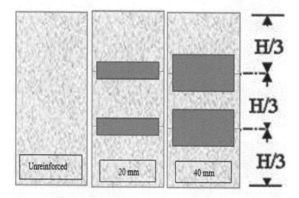

Figure 5.   (a) Unreinforced (b) 20 mm height of reinforcement (c) 40 mm height reinforcement.

silt. Then, 5% of water was added in the sand silt mixture and placed in a split mold in 6–7 layers. It was ensured that the placement of the cellular reinforcement is in such a way that the center of reinforcement coincides with the center of the sample (Figures 4 and 5). Tamping was done for each layer with the help of a hammer to achieve a uniform density.

2.5   *Test procedure*

After specimens were prepared, their caps were placed and sealed with "O" rings. A negative pressure of 10 kPa was applied to the specimens to reduce disturbance during removal of split mold and triaxial cell installation. When the cell was filled with water, a confining pressure of 35 kPa was applied to the samples. The saturation of the specimen was done by applying CO_2 and back pressure technique. The control of saturation is done using Skempton's pore pressure parameter B. The sample was considered to be fully saturated if the B value is at least equal or greater than 0.95. After completion of the saturation process, the confining pressure was slowly increased to provide the desired effective confining pressure. All the samples were isotropically consolidated and loaded at the strain rate of 1.2 mm/min. For each confining pressure load, displacement, pore pressure, and volume change were recorded using data acquisition system during the test after every 5 seconds. All the tests were conducted up to the maximum axial strain of 20%.

3 RESULTS AND DISCUSSION

3.1 *Unreinforced sand silt mixture*

Figure 6 shows the deviator stress-strain behavior of unreinforced soil for 50, 100, and 150 kPa confining pressures. It was observed that for unreinforced specimens, the stress-strain relation becomes curved at very small strains and achieve a peak at a strain of about 3%. The resistance of soil then gradually decreases until this test is arbitrarily stopped at a strain of 20%. It is also seen that as the confining pressure increases from 50 kPa to 150 kPa, the peak value of deviator stress increases from 13 to 45 kPa, and steady state is achieved at a large percentage of strain.

A typical graph of increase in excess pore pressure with axial strain during shearing is shown in Figure 7.

It is observed that the peak value of pore pressure was reached at a strain of about 2% and remained constant at large percentages of strain for all confining pressure. As all tests have been conducted at specific sand sizes (Hazirbaba. (2005)), the voids between sand particles are actually increasing due to addition of silt. Therefore, the sand-silt specimen becomes loose, which results in increased pore pressure.

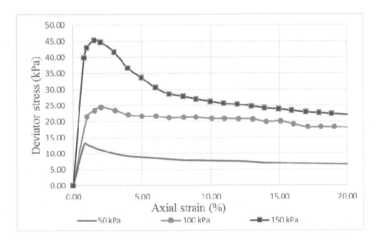

Figure 6.   Deviator stress vs strain for unreinforced specimens.

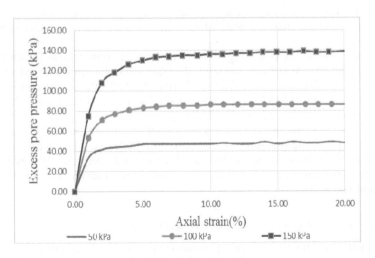

Figure 7.   Excess pore pressure vs strain of unreinforced specimen.

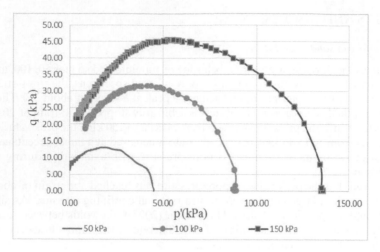

Figure 8.   p' vs q of unreinforced specimen.

Effective stress paths for unreinforced specimens for all confining pressures are shown in Figure 8. For all confining pressures, contractive behavior has been observed for all the tests conducted in the present work. Further, it is also seen that contractive behavior increased with increase in confining pressures.

3.2   *Reinforced sand-silt mixture*

Figure 9 shows deviator stress vs axial strain of the 40-mm tall cellular reinforced specimen.

For reinforced sand silt mixture, deviator stress increases as the confining pressure increases from 50 kPa to 150 kPa. However, the peak value of stress was observed at 20% strain from 52 kPa to 134 kPa.

Effective stress paths for the 40-mm tall reinforced specimen is as shown in Figure 10 for all confining pressures. It is observed that after initial contraction all the samples dilated.

3.3   *Effect of confining pressure on peak deviator stress*

The variations in peak deviator stress with increase in confining pressure for unreinforced and reinforced sand-silt mixture with 20 mm and 40 mm height of geocell is shown in Figure 11.

It is observed that as the confining pressure increases deviator stress for unreinforced as well as reinforced soil. Maximum value of deviator stress for unreinforced sand-silt mixture has been observed to increase from 6.52 kPa to 22.09 kPa, as the confining pressure increases from 50 kPa to 150 kPa. However, for the 20-mm reinforced sand-silt mixture, it is observed to be between 27.14 kPa to 58.95 kPa. However, pronounced increase in peak deviator stress has been observed for the 40-mm reinforced specimen from 52.15 kPa to 134.1 kPa with increase in confining pressure. This indicated that as the confining pressure increased, the peak value of deviator stress also increased for unreinforced and reinforced specimens. However, the maximum value of deviator stress has been observed for the 40-mm cellular reinforcement specimen.

3.4   *Effect of height of reinforcement on peak deviator stress*

Figure 12 shows the effect of height of reinforcement on peak deviator stress for unreinforced and reinforced specimens for all confining pressures. It is observed that for 50 kPa

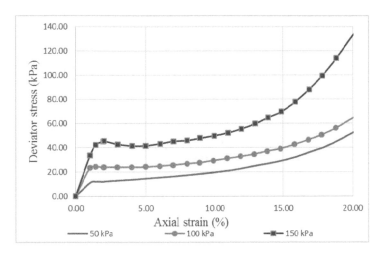

Figure 9. Deviator stress vs axial strain for the reinforcement of 40 mm height.

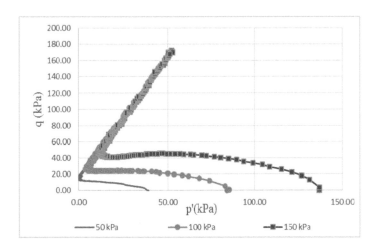

Figure 10. p' vs q for the 40-mm tall reinforced specimen.

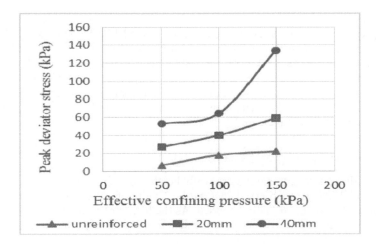

Figure 11. Effect of confining pressure on unreinforced and reinforced specimens.

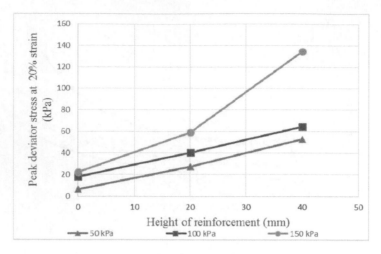

Figure 12.   Effect of height of reinforcement for unreinforced and reinforced specimens.

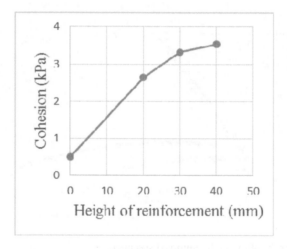

Figure 13.   Cohesion for unreinforced and reinforced specimens.

confining pressure, deviator stress is increased from 6.52 kPa to 52.65 kPa. However, for 150 kPa confining pressure, peak deviator stress for unreinforced soil is 22.69 kPa and as the height of reinforcement (geocell) increased to 40 mm, a higher value of peak deviator stress (134.11 kPa) has been observed. This indicated that compared to the unreinforced specimens, the cellular reinforced specimens showed higher deviator stress at large percentages of axial strain. The reason could be an increase in the surface area for 40-mm cellular specimens as the height of reinforcement increased.

3.5   *Variation in shear strength parameters*

Variation of shear strength parameters with height of reinforcement are shown in the Figures 13 and 14. It was observed that as the height of reinforcement increased, shear strength parameters (c and Φ) also increased. As the height of reinforcement increased, surface area in contact of soil also increased which leads to increase in the interface friction angle between the soil and reinforcement.

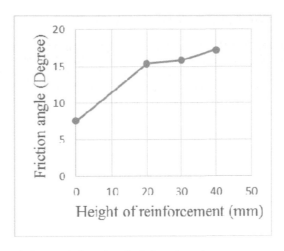

Figure 14.   Friction angle for unreinforced and reinforced specimens.

4   CONCLUSION

Consolidated undrained tests were conducted on unreinforced and reinforced (20 mm and 40 mm geocell) specimens to study the effect of height of reinforcement on peak deviator stress. The following conclusions are made:

1. As the confining pressure increased, peak deviator stress also increases for unreinforced as well as reinforced specimens. Maximum value of deviator stress has been observed for reinforced specimen due to increase in the interface friction between cellular reinforcement and sand silt mixture.
2. Contractive behavior has been observed for unreinforced specimens. However, dilative behavior increased with increase in confining pressure for reinforced specimens.
3. As the height of cellular reinforcement increased, deviator stress also increased for samples tested in the present work. The maximum value of deviator stress is observed for the 40-mm tall cellular reinforced specimen due to increase in the surface area of the reinforcement. Consequently, interface friction of cellular reinforcement and sand silt increased.
4. Shear strength parameters (c and Φ) increased as height of reinforcement increased.

REFERENCES

ASTM Standard D4767, 2007. Standard Test Method for Consolidated Undrained Triaxial Compression Test for Cohesive Soil. *ASTM International, West Conshohocken, PA.*
Fauziah, A. 2010. Performance evaluation of silty sand reinforced with fibers. *Geotextiles and Geomembranes* 28: 93–99.
Gray, D.H. and Ohashi, H. 1983. Mechanics of fiber reinforcing in sand. *Journal of Geotechnical Engineering Division,* 109 (3): 335–353.
Gray, D.H. and Maher, M.H. 1989. Admixture stabilization of sand with discrete randomly distributed fibers. *Proceedings of XII Inte rnational Conference on Soil Mechanics and Foundation Engineering Rio de Janeiro Brazil*, 2:1363–1366.
Hazirbaba, K. 2005. Pore Pressure generation characteristics of sands and silty sands: A strain approach. *PhD Thesis* The University of Taxas Austin, Austin.
Ibraim, E., Diambra, A., Muir, W.D. and Russell, A.R. 2010. Static liquefaction of fibre reinforced sand under monotonic loading. *Geotextiles and Geomembranes,* 28: 374–385.
Khedkar, M.S. and Mandal, J.N. 2010. Behavior of cellular reinforced soil wall under uniformly distributed load. *Proc., 9th Int. Conf. on Geosynthetics, Brazilian chapter of the international geosynthetics society* (IGS—Brazil): 1693–1698 Rua ribeirao claro, Saupaulo, Brazil.

Michalowski, R.L. and Cermak, J. 2002. Strength anisotropy of fiber-reinforced sand. *Computers and Geotechnics,* 29 (4): 279–299.

Lal, R.R. and Mandal, J.N. 2011.Cellular reinforced fly ash retaining wall—experimental and finite element analysis." *Proceeding Indian Geotechnical Conference—IGC-2011*: 667–670. Kochi, India.

Lal, R.R. and Mandal, J.N. 2012. Feasibility study on fly ash as backfill material in cellular reinforced walls. *Electronic Journal of Geotechnical Engineering,* 17: 1637–1658.

Lal, R.R. and Mandal, J.N. 2013. Study of cellular reinforced fly ash under monotonic loading. *International Journal of Geotechnical Engineering,* 7: 91–104.

Multi-disciplinary Sustainable Engineering: Current and Future Trends – Tekwani, Bhavsar & Modi (Eds)
© 2016 Taylor & Francis Group, London, ISBN 978-1-138-02845-6

Application of fuzzy-neuro based hybrid technique for damage detection of fixed-fixed beam

D.K. Agarwalla, A.K. Dash, H.C. Das & P.M. Tripathy
*Institute of Technical Education and Research, Siksha 'O' Anusandhan University,
Bhubaneswar, Odisha, India*

ABSTRACT: Hybridization of Neural Network (NN) and Fuzzy Logic (FL) have attracted researchers from various scientific and engineering domains to develop adaptive intelligent systems to address real-time applications. In the present analysis, fuzzy logic and neural network have been adopted to form a damage identification tool for structural health monitoring of the beam structure with made of steel with both ends fixed. The proposed methodology utilizes the modal characteristics of the fixed-fixed beam structure using numerical modeling techniques and anticipates the position and severities of the damage present in the system. The robustness of the proposed technique has been realized by conducting experiments on the steel fixed-fixed beam with different damage characteristics.

Keywords: Neural Networks; Fuzzy Logic; fixed-fixed beam; damage; experiment

1   INTRODUCTION

1.1   *Overview*

Most of the structural failures encountered are caused by material fatigue and presence of damages in structures. Therefore, damages of any form are to be diagnosed as early as possible to maintain the integrity of the structures. In spite of the existence of many traditional methods, presence of any damage can't be ensured without diagnosing the entire structure. Fuzzy-Neuro hybrid computing technique is a potential tool for solving complex problems. If the parameters representing a system can be expressed in terms of linguistic rules, a fuzzy inference system can be built up. A neural network can be built, if data required for training from simulations are available. From the analysis of NN and FL it is observed that the drawbacks of the two methods are complementary and, therefore, it is desirable to build an integrated system combining the two techniques. The learning capability is an advantage for NN, while the creation of a linguistic rule base is an advantage for fuzzy logic. Hence, the hybrid fuzzy-neuro technique can be used for identifying cracks present in a structural system using vibration data.

1.2   *Literature review*

Meesad and Yen [1] have proposed an innovative neuro-fuzzy for pattern classification application, especially for vibration monitoring. They have incorporated a fuzzy set into the network design to handle imprecise information. The neuro-fuzzy classifier proposed has been equipped with a one-pass, on-line, and incremental learning algorithm. They have classified the neuro-fuzzy network on the basis of Fisher's Iris data. The neuro-fuzzy network achieved 97.33% correct classification after they used the west-land data set, which consists of vibration data collected from a US Navy CH-46E helicopter for better classification. By using various torque levels, they have achieved 100% correct classification. Far et al. [2] have experimented on model-based fault detection and isolation of a U-tube steam generator in a nuclear power plant. They have used two types of neuro-fuzzy networks. They have considered the Takagi-Sugeno (TS) fuzzy model for residual generation and the Mamdani model for residual evaluation. They have used a Locally Linear Neuro-Fuzzy (LLNF) model which has been trained by the locally linear model tree (LOLIMOT) algorithm. From the experiment they have concluded that a qualitative description of faults has been extracted from the fuzzy rules obtained from the Mamdani model. Zhu et al. [3] have come up with an integrated approach for structural damage identification using the Wavelet neuro-fuzzy model. Using the Wavelet Transform (WT) algorithm's feature of filtering random noise (ANFIS) the structural behavior can be properly modeled and the interval modeling technique can quantify the damage index accurately. Finally, they concluded from the results and some other signal processing methods that the proposed method can be used to identify both the time and location of unexpected structural damage. Nguyen et al. [4] have presented a new Beam Damage Locating (BDL) method based on an algorithm. The algorithm has made by the combination of an Adaptive Fuzzy Neural Structure (AFNS) and an average quantity solution to achieve a wavelet transform coefficient (AQWTC) of beam vibration signal. AFNS has been used for remembering undamaged beam dynamic properties and AQWTC has been used for signal analysis. The experiment has been done by dividing the beam into two elements. From the experiment they have concluded that the effectiveness of the approach which combined fuzzy neural structure and wavelet transform method has been demonstrated. Ayoubi and Isermann [5] have described knowledge based fault detection and diagnosis from analytic and heuristic symptom generation due to diagnostic reasoning. They have investigated undetermined parameters just as membership functions, relevance weights of antecedents, and priority factor of rules. Finally, they have come up with an application of the neuro-fuzzy system for online monitoring of air pressure in vehicle wheels. Subbaraj and Kannapiran [6] have discussed the design and development of Adaptive Neuro-fuzzy Inference System (ANFIS) based fault detection and diagnosis of pneumatic valve used in cooled water spray systems in the cement industry. The performance of the developed ANFIS model has been compared with Multilayer Feed Forward Neural Network (MLFFNN) trained by the back propagation algorithm. From the simulation result, they have found that ANFIS performed better than ANN.

2 NUMERICAL MODELING

Figure 1 illustrates the fixed-fixed beam, subjected to axial load (P_1) and bending moment (P_2), which effectuate a combining effect in terms of longitudinal and transverse motion of

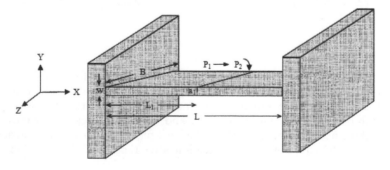

Figure 1.   Fixed—fixed beam with axial and bending load.

the beam, respectively. The beams contain damage in the transverse direction of depth "a_1" having width "B" and thickness "W". The existence of damage in the beam structure modifies the localized flexibility square matrix of two dimensions.

At the damaged portion, strain energy release rate can be explained as [7];

$$J = \frac{1}{E'}(K_{I1}+K_{I2})^2, \text{ where } \frac{1}{E'}=\frac{1-v^2}{E}\text{(for plane strain condition);}$$
$$= \frac{1}{E}\text{(for plane stress condition)}$$

K_{I1}, K_{I2} are stress intensity factors for the 1st mode of vibration for load P_1 and P_2, respectively. The values of stress intensity factors from the referred article [7] are:

$$\frac{P_1}{WB}\sqrt{\pi a}\left(F_1\left(\frac{a}{W}\right)\right) = K_{I1}, \frac{6P_2}{W^2 B}\sqrt{\pi a}\left(F_2\left(\frac{a}{W}\right)\right) = K_{I2}$$

The expressions for F_1 and F_2 are as follows:

$$F_1\left(\frac{a}{W}\right) = \left(\frac{2W}{\pi a}\tan\left(\frac{\pi a}{2W}\right)\right)^{0.5}\left\{\frac{0.752 + 2.02(a/W) + 0.37\,(1-\sin(a\pi/2W))^3}{\cos(a\pi/2W)}\right\}$$

$$F_2\left(\frac{a}{W}\right) = \left(\frac{2W}{\pi a}\tan\left(\frac{\pi a}{2W}\right)\right)^{0.5}\left\{\frac{0.923 + 0.199\,(1-\sin(a\pi/2W))^4}{\cos(a\pi/2W)}\right\}$$

According to Castigliano's theorem (assuming strain energy due to the damage as U_t) the extra extension along the force P_i is:

$$\frac{\partial U_t}{\partial P_i} = u_i$$

The form of strain energy will have,

$$U_t = \int_0^{a_1} J\,da = \int_0^{a_1}\frac{\partial U_t}{\partial a}\,da$$

where $J = \frac{\partial U_t}{\partial a}$ the strain energy density function.

Hence, from the equations mentioned above we can have:

$$\frac{\partial}{\partial P_i}\left[\int_0^{a_1} J(a)\,da\right] = u_i$$

C_{ij} the flexibility influence coefficient by definition is:

$$\frac{\partial u_i}{\partial P_j} = \frac{\partial^2}{\partial P_j \partial P_i}\int_0^{a_1} J(a)\,da = C_{ij}$$

and can be expressed as,

$$\frac{WB}{E'}\frac{\partial^2}{\partial P_j \partial P_i}\int_0^{a_1}(K_{I2}+K_{I1})^2\,d\xi = C_{ij}$$

Using the above equation, the compliance C_{11}, C_{22}, C_{12} ($=C_{21}$) is as follows:

23

$$C_{11} = \frac{BW}{E'} \int_0^{\xi} \frac{\pi a}{B^2 W^2} 2(F_1(\xi))^2 \, d\xi$$

$$= \frac{2\pi}{BE'} \int_0^{\xi} \xi(F_1(\xi))^2 \, d\xi$$

$$C_{12} = C_{21} = \frac{12\pi}{E'BW} \int_0^{\xi} \xi F_1(\xi) F_2(\xi) \, d\xi$$

$$C_{22} = \frac{72\pi}{E'BW^2} \int_0^{\xi} \xi F_2(\xi) F_2(\xi) \, d\xi$$

The dimensionless form of the influence co-efficient will be:

$$\overline{C_{11}} = C_{11} \frac{BE'}{2\pi}; \quad \overline{C_{12}} = C_{12} \frac{E'BW}{12\pi} = \overline{C_{21}}; \quad \overline{C_{22}} = C_{22} \frac{E'BW^2}{72\pi}$$

The inversion of the compliance matrix will lead to the formation of local stiffness matrix and can be written as:

$$K = \begin{bmatrix} C_{11} & C_{12} \\ C_{21} & C_{22} \end{bmatrix}^{-1} = \begin{bmatrix} K_{11} & K_{12} \\ K_{21} & K_{22} \end{bmatrix}$$

The stiffness matrix for the damage position can be obtained as follows:

$$K' = \begin{bmatrix} k'_{11} & k'_{12} \\ k'_{21} & k'_{22} \end{bmatrix} = \begin{bmatrix} C'_{11} & C'_{12} \\ C'_{21} & C'_{22} \end{bmatrix}^{-1}$$

The stiffness matrix obtained from the above analysis has been used to estimate the modal parameters of the fixed-fixed beam and subsequently used as the input to the fuzzy-neuro hybrid system for damage detection. The material properties of steel, such as Young's modulus (200 GPa), density (7850 m³/Kg), and Poisson's ratio (0.30), have been introduced to solve the stiffness matrix.

3 FUZZY-NEURO HYBRID CONTROLLER FOR DAMAGE DETECTION

This section introduces a hybrid intelligent method for prediction of damage positions and their severities in a fixed-fixed beam structure having a transverse damage using inverse analysis. As the presence of damage alters the dynamic behavior of the beam, the first three relative natural frequencies and first three average relative mode shape differences of the damaged and intact beam for different damage positions and severities are calculated using numerical modeling. The calculated modal frequencies, mode shapes, Relative Damage Positions (RDP), and Relative Damage Severities (RDS) from the numerical modeling are used to design the fuzzy neural controller. The measured vibration signatures are used as inputs for the fuzzy segment of the hybrid controller, and initial relative damage position and severity are the output parameters. The first three relative natural frequencies, first three average relative mode shape differences, and the output from the fuzzy controller are used as inputs for the neural part of the hybrid controller with final damage position and severity as the output parameters. The measured vibration signatures are used to formulate a series of fuzzy rules and training patterns for the fuzzy and neural controller. Out of numerous damage characteristics, such as RDP and RDS, only ten damage scenarios have been shown in the Table 1. Finally, the validation of the proposed method is carried out dynamically by means of experimental results from the developed experimental set-up with the same damage characteristics as that of the proposed model. The fuzzy segment of the hybrid controller for damage detection has been developed using the Gaussian membership function. The Gaussian membership function based hybrid controller is shown in Figure 2. The results obtained from the proposed technique have been presented in Table 1.

$$RDS = \frac{\text{Depth of the damage in the beam}}{\text{Thickness of the beam}}$$

$$RDP = \frac{\text{Position of the damage from one fixed end}}{\text{Length of the beam}}$$

Table 1. Comparison of results obtained from Gaussian Fuzzy-Neuro controller and experimental analysis.

Relative First Natural Frequency "FNF"	Relative Second Natural Frequency "SNF"	Relative Third Natural Frequency "TNF"	Average Relative First Mode Shape Difference "FMD"	Average Relative Second Mode Shape Difference "SMD"	Average Relative Third Mode Shape Difference "TMD"	Gaussian Fuzzy-Neuro Result		Experimental Result	
						RDS	RDP	RDS	RDP
0.9979	0.9985	0.9993	0.0087	0.0036	0.0042	0.460	0.122	0.465	0.123
0.9962	0.9989	0.9991	0.0036	0.9729	0.2263	0.413	0.121	0.423	0.127
0.9936	0.9976	0.9987	0.0138	0.014	0.0832	0.167	0.373	0.167	0.378
0.9976	0.9991	0.9988	0.0014	0.0041	0.0812	0.335	0.125	0.342	0.127
0.9978	0.9983	0.9878	0.0036	0.0329	0.0141	0.174	0.243	0.171	0.250
0.9987	0.9972	0.9981	0.2936	0.3428	0.2623	0.223	0.270	0.230	0.280
0.9849	0.9982	0.9869	0.0134	0.0211	0.0119	0.414	0.373	0.421	0.372
0.9989	0.9973	0.9974	0.0017	0.0025	0.0079	0.467	0.248	0.482	0.270
0.9977	0.9847	0.9881	0.0079	0.0077	0.0292	0.337	0.374	0.341	0.378
0.9988	0.9974	0.9991	0.0057	0.0023	0.0155	0.423	0.283	0.419	0.290

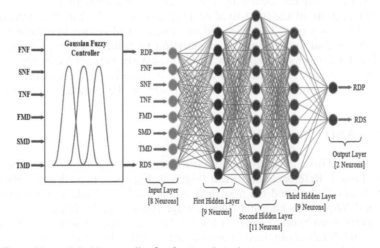

Figure 2. Fuzzy-Neuro hybrid controller for damage detection.

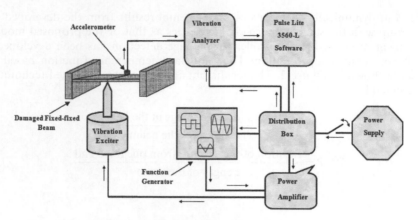

Figure 3(a). Schematic diagram experimental set-up with a fixed-fixed beam.

1. Vibration exciter
2. Delta Tron accelerometer
3. Composite cantilever beam platform
4. Vibration Analyzer
5. Vibration Monitor
6. Function Generator
7. Power amplifier
8. Power supply

Figure 3(b). Schematic diagram experimental set-up with a fixed-fixed beam.

4 EXPERIMENTAL SET-UP

Ample number of experiments have been conducted using the experimental set-up as shown in Figure 3 (a) and (b) for measuring the vibration signatures (natural frequencies and amplitude of vibration) of the fixed-fixed beam specimens made from steel with dimensions 1000 mm × 50 mm × 8 mm. The damaged and undamaged beams have been subjected to vibration with 1st, 2nd and 3rd modes of vibration by utilizing an exciter and a function generator. The dynamic characteristics of the beams have been recorded by placing the accelerometer along the length of the beams. The output obtained from the accelerometer in terms of natural frequencies and mode shapes are monitored on the vibration indicator.

5 RESULTS AND DISCUSSION

The results obtained from the fuzzy-neuro controller and the experimental analyses have been presented in Table 1. References in the text: Figure 1, Figs. 2–4, 6, 8a, b (not abbreviated)

The first three columns and the next three columns of the Table 1 depict the first three relative natural frequencies and first three average relative mode shape differences, respectively. The fuzzy-neuro hybrid controller for the damage detection is illustrated in Figure 2. The RDS and RDP values obtained from fuzzy-neuro hybrid controller are very close to the values obtained from experimental analysis. The percentages of deviations in the RDS and RDP are 3.4% and 3.5%, respectively.

6 CONCLUSION

The damage diagnostic tool, i.e., Fuzzy-Neuro Hybrid Controller with Gaussian membership function developed for the steel fixed-fixed beam has been found to be a significant tool for inherent damage detection. The results found from this controller have been validated by the experimental results with remarkable convergence. This damage detection technique can be used for various mechanical beam structures with different boundary conditions. However, the damage detection tool can also be used effectively for beams of different material.

NOMENCLATURE

a_1	= depth of damage
A	= cross-sectional area of the beam
$A_{i\,(i=1\text{ to }18)}$	= unknown coefficients of matrix A
B	= width of the beam
C_{11}	= Axial compliance
$C_{12} = C_{21}$	= Coupled axial and bending compliance
C_{22}	= Bending compliance
$\bar{C}_{11}$	= Dimensionless form of C11
$\bar{C}_{12} = \bar{C}_{21}$	= Dimensionless form of C12 = C21
$\bar{C}_{22}$	= Dimensionless form of C22
C'_{12}	= Axial compliance for damage position
$C'_{12} = C'_{21}$	= Coupled axial and bending compliance for damage position
C'_{22}	= Bending compliance for damage position
E	= Young's modulus of elasticity of the beam material
$F_{i\,(i=1,2)}$	= experimentally determined function
i, j	= variables
J	= strain-energy release rate
$K_{1,i\,(i=1,2)}$	= stress intensity factors for P_i loads
K_{ij}	= local flexibility matrix elements
K'	= Stiffness matrix for damage position
L	= length of the beam
L_1	= location (length) of the damage from fixed end
$P_{i\,(i=1,2)}$	= axial force (i = 1), bending moment (i = 2)
$u_{i\,(i=1,2)}$	= normal functions (longitudinal) $u_i(x)$
x	= co-ordinate of the beam
y	= co-ordinate of the beam
$y_{i\,(i=1,2)}$	= normal functions (transverse) $y_i(x)$
W	= depth of the beam
ω	= natural circular frequency
β_1	= relative damage location (L_1/L)
ρ	= mass-density of the beam

REFERENCES

[1] Meesad, P., Yen, G.G. 2009. Pattern Classification by a Neuro fuzzy Network: Application to Vibration Monitoring. *ISA Transactions*, 39(3): 293–308.
[2] Lucas, C., Far, R.R., Davilu, H., Palade, V. 2009. Model-Based Fault Detection and Isolation of a Steam Generator Using Neuro-Fuzzy Networks: *Neuro computing*, 72(13–15): 2939–2951.
[3] Zhu, F., Deng, Z., Zhang, J. 2013. An Integrated Approach for Structural Damage Identification Using Wavelet Neuro-Fuzzy Model: Expert System with Applications. 40(18): 7415–7427.
[4] Nguyen, S.D., Ngo, K.N., Tran, Q.T., Choi, S.B. 2013. A New Method for Beam-Damage-Diagnosis Using Adaptive Fuzzy Neural Structure And Wavelet Analysis. *Mechanical Systems and Signal Processing*, 39(1–2): 181–194.

[5] Ayoub, M., Isermann, R. 1997. Neuro-Fuzzy Systems for Diagnosis. *Fuzzy Sets and Systems*, 89(3): 289–307.

[6] Subbaraj, P., Kannapiran, B. 2014. Fault Detection and Diagnosis of Pneumatic Valve Using Adaptive Neuro-Fuzzy Inference System Approach. *Applied Soft Computing*, 19: 362–371.

[7] Tada, H., Paris, P.C., Irwin, G.R.. 1973. 'The stress analysis of cracks hand book', Del Research Corp. Hellertown, Pennsylvania.

Multi-disciplinary Sustainable Engineering: Current and Future Trends – Tekwani, Bhavsar & Modi (Eds)
© 2016 Taylor & Francis Group, London, ISBN 978-1-138-02845-6

Sloshing displacement in 2-D tank under dynamic excitation using Coupled Eulerian-Lagrangian approach

Aruna Rawat, Vasant Matsagar & A.K. Nagpal

Department of Civil Engineering, Indian Institute of Technology, Delhi, India

ABSTRACT: A two-dimensional (2-D) ground-supported rectangular rigid liquid storage tank filled with water and subjected to harmonic excitation is investigated using Coupled Eulerian-Lagrangian (CEL) Finite Element (FE) method. The Fluid-Structure Interaction (FSI) behavior of the tanks in model using CEL approach. In this approach the tank is modeled using Lagrangian elements and liquid with Eulerian elements. Sloshing displacement time history responses are investigated for rectangular tank subjected to uni-directional ground motions for different liquid height of the tank, amplitudes and frequencies of excitation. It is observed from the results that as amplitude of excitation increases the sloshing height increases. Further, the sloshing displacement in tank becomes significant in the case of excitation frequency coinciding with the natural frequency of the tank.

1 INTRODUCTION

Sloshing waves are related with various engineering problems, such as the liquid oscillations in liquid storage tanks caused by earthquakes, the motions of liquid fuel in aircraft and spacecraft propeller, the liquid motions in containers in ships. The wave forces can cause structural damage and loss of the motion stability of the structures. Sloshing in tank develops stresses in the tank walls depend on the distribution of the internal hydrodynamic pressure. Also, liquid sloshing amplitude depends on the nature, amplitude and frequency of the motion of the storage tank, tank geometry, properties and liquid height in the tank (Akyildiz & Unal 2006). Therefore, sloshing should be considered as one of the major aspects in the seismic design of the liquid storage tanks (Haroun & Housner 1981). For this reason, numerous analytical, numerical and experimental analyses of sloshing displacement in liquid storage tanks subjected to seismic ground motion have been studied during the last decades.

Many researchers have investigated the sloshing phenomenon using analytical approaches based on linear wave theory for evaluating the wave amplitude of liquid in the tanks, Abramson (1966), Veletsos (1984), Wu et al. (2001). Researchers have also applied numerical methods, such as Finite Element Method (FEM), Finite Difference Method (FDM) and Boundary Element Method (BEM), for evaluating the non-linear sloshing waves of large amplitude in liquid storage tanks. Faltinsen (1974, 1978) studied non-linear sloshing problems in rectangular tanks using numerical and analytical approaches based on perturbation theory. Ramaswamy & Kawahara (1987) used Arbitrary Lagrangian-Eulerian (ALE) finite element technique for analyzing the free surface flow problems. Chen et al. (1996) used finite difference method to simulate large-amplitude sloshing under harmonic and seismic excitations for two-dimensional (2-D) rectangular tanks. Ushijima (1998) used Arbitrary Lagrangian-Eulerian (ALE) formulation for the non-linear free surface oscillations in 3-D tanks. Mackerle (1999) presents a bibliography of references on fluid-structure interaction problems analyzed using finite element and boundary element approaches. Faltinsen et al. (2000) conducted a modal analysis of non-linear sloshing in a rectangular tank using the Bateman-Lake variational principle and also verified it with the experimental results. Ibrahim (2005) reviewed various researches on sloshing of liquid motion in different applications such as storage tanks, ship carriers, space vehicles and road tankers. Frandsen (2004) developed a Finite Difference Method (FDM) to analyze the non-linear sloshing motions of liquid subjected to vertical, horizontal and combined motions of tanks. Virella et al. (2008) studied the linear and non-linear wave theory for a 2-D rectangular tank using Abaqus where liquid was a modeled with acoustic elements for linear theory and ALE technique was used for non-linear theory.

Among the other numerical approaches, Tippmann et al. (2009) used Coupled Lagrangian-Eulerian (CEL) capability for simulation of sloshing in 2-D rectangular rigid and flexible tanks subjected to sinusoidal excitation. Mittal et al. (2014) used CEL approach for analyzing ground-supported 3-D liquid storage tank subjected to blast loading.

The objectives of the present investigation are (i) to study 2-D fixed base ground-supported rectangular rigid liquid storage tanks subjected to uni-directional harmonic excitation, using Coupled Eulerian-Lagrangian (CEL) finite element method; (ii) to evaluate the sloshing displacement time history responses in tanks subjected to uni-directional motions; and (iii) to study sloshing response for different liquid heights, excitation frequencies and amplitudes. The FE analyses are carried out using Abaqus (Abaqus user manual 2011), in which Eulerian elements are used to model the liquid in the tank in the CEL approach and acoustic elements are used in the CAS approach.

2   GOVERNING EQUATIONS OF LIQUID MOTION

A rectangular tank, L is the length and B is the width, H_L is the liquid height in the tank and h is the sloshing amplitude of the free liquid surface. The Laplace equation, which can be written in general three-dimensional space (x, y, z) and given as

$$\frac{\partial^2 \varphi}{\partial x^2} + \frac{\partial^2 \varphi}{\partial y^2} + \frac{\partial^2 \varphi}{\partial z^2} = 0 \text{ or } \nabla^2 \varphi \tag{1}$$

where ∇^2 is the Laplacian operator, gradient of the velocity potential function $\varphi(x, y, z, t)$, satisfying. The boundary conditions at the interface of liquid with the base and wall of the tank, respectively, are given as

$$\frac{\partial \varphi}{\partial n} = v_n(x, y, z, t) \tag{2}$$

where $\partial \varphi/\partial n$ is the outward normal derivative and $v_n(x, y, z)$ is the outward normal velocity component along the direction vector for an outward pointing normal n, to the liquid region. In the present study, both base and wall of the tank are considered rigid. Therefore, under horizontal ground motion, the normal vertical velocity, v_n of the liquid is equal to zero. On the free surface if the liquid forms amplitude of wave h (along z-direction) relative to the mean surface, the boundary condition is given as

$$g\frac{\partial \varphi}{\partial z} + \frac{\partial^2 \varphi}{\partial t^2} = 0 \tag{3}$$

where g is the acceleration due to gravity.

3   COUPLED EULERIAN-LAGRANGIAN (CEL)

The Coupled Eulerian-Lagrangian (CEL) approach in the finite element software Abaqus is used in the present study. The Eulerian elements in which the material flows through the mesh are used to model fluid flow problems because the Eulerian mesh does not deform. Generally, the Lagrangian elements fail in simulating the flow problems because of severe

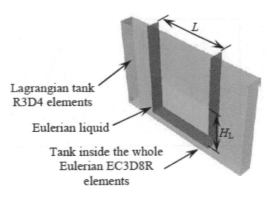

Figure 1. Coupled Eulerian-Lagrangian (CEL).

mesh distortion. In CEL approach, the free surface of liquid is traced when it flows through the mesh, using a volume fraction tool in each Eulerian element. The Eulerian Volume Fraction (EVF) is specified as a decimal between zero (EVF = 0) and one (EVF = 1) where EVF = 0 means that the elements are completely filled with void and EVF = 1 means that the elements are completely filled with liquid material. The Eulerian domain is modeled larger as compared to the actual dimensions of the tank and liquid in order to prevent any loss of liquid.

In the present study, the water is modeled using 3-D eight-node Eulerian continuum element with reduced integration (EC3D8R). The rigid tank is modeled with 3-D rigid four-node bi-linear quadrilateral discrete rigid elements (R3D4) as shown in Figure 1. A general contact algorithm is defined in CEL approach to model the interaction between liquid and the tank wall. The liquid is modeled herein using linear U_s-U_p Hugoniot form of the Mie-Grüneisen Equation-Of-State (EOS), which is described as a linear relationship between the shock and particle velocities given as

$$U_s = C_0 + U_p \tag{4}$$

where U_s and U_p are the velocities of shock and liquid particles, respectively. The parameter C_0 defines the velocity of sound through the liquid and s is a material constant. The pressure-density relation in the Mie-Gruneisen equation-of-state is given as

$$p = \frac{\rho_0 C_0^2 \eta}{(1 - s\eta)^2}\left(1 - \frac{\Gamma_0 \eta}{2}\right) + \Gamma_0 \rho_0 E_m \tag{5}$$

where $\eta = 1 - \rho_0/\rho$ is the volumetric compressive strain as a function of initial density ρ_0 and the current density ρ. Parameters Γ_0 and s are the material constants and E_m is the internal energy per unit mass. A small value of dynamic viscosity, v is also defined for the liquid for numerical reasons. The EOS parameters to model water in CEL approach are density $\rho = 1000$ kg/m^3, dynamic viscosity $v = 0.001$ N sec/m^2, velocity of sound through water $C_0 = 1500$ m/sec, material constants Γ_0 and s are taken as zero.

4 VALIDATION OF THE PRESENT APPROACH

In order to ensure the accuracy of the present approach for the dynamic analysis of liquid storage tanks, the results of sloshing displacement is evaluated using CEL approach and the result is compared with the numerical results obtained by Virella et al. (2008) using Arbitrary Lagrangian-Eulerian (ALE) formulation. For the 2-D rectangular tank, length, $L = 30.675$ m. and height of water, $H_L = 10.73$ m. subjected to external excitation of 0.01 g sin

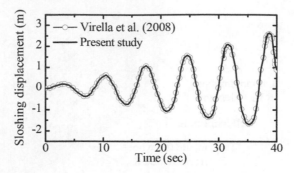

Figure 2. Coupled Eulerian-Lagrangian (CEL).

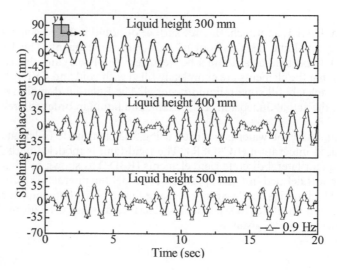

Figure 3. Sloshing displacement time history response subjected to amplitude of 5 mm. and external excitation frequency 0.9 Hz. for various liquid heights.

(ωt), where ω is the first fundamental frequency of tank. The density of water, $\rho = 983$ kg/m³ used by Virella et al. (2008) are considered herein for the validation analysis. Figure 2 shows the sloshing displacement response at the extreme top edge of the top water surface using the present approach and Virella et al. (2008).

5 NUMERICAL STUDY

The geometrical dimensions of 2-D rectangular tank considered in the present analyses is having a length, $L = 70$ cm. and liquid height $H_L = 30$ cm., 40 cm. and 50 cm. The tank is subjected to uni-directional displacement sinusoidal motion at the base of the tank, $x = A \sin(2\pi ft)$, where A is the amplitude of motion considered as 2 mm. to 8 mm. and f is the frequency Hz taken as 0.7 Hz., 0.8 Hz. and 0.9 Hz.. Figure 3 shows the time history response of sloshing displacement for liquid height 30 cm., 40 cm. and 50 cm. subjected to 5 mm amplitude and excitation frequency 0.9 Hz. The maximum sloshing is 58.4 mm., 41.68 mm. and 37.02 mm. for liquid height 30 cm., 40 cm. and 50 cm., respectively. It can be observed that as the liquid height increases the sloshing displacement reduces since the volume of liquid mass increases.

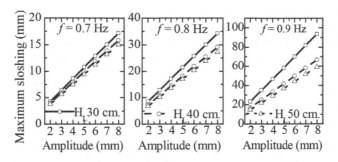

Figure 4. Maximum sloshing displacement for different amplitudes and excitation frequencies.

The sloshing displacement is studied for varying excitation frequencies and amplitudes from 2 to 8 mm. Figure 4 shows the effect of amplitudes and excitation frequencies on the sloshing displacement. It is observed that as the amplitude increases the sloshing displacement increases with increase in the excitation frequency.

6 CONCLUSION

A two-dimensional (2-D) ground-supported rectangular rigid liquid storage tank filled with water and subjected to harmonic excitation is investigated using Coupled Eulerian-Lagrangian (CEL) FE method. In this approach the tank is modeled using Lagrangian elements and liquid with Eulerian elements. Sloshing displacement time history responses are investigated for rectangular tanks subjected to uni-directional ground motions for different liquid height of the tank, amplitudes and frequencies of excitation. It is observed from the results that as amplitude and frequency of excitation increases the sloshing height increases. The sloshing height decreases with increase in the liquid height in the tank. The CEL approach of FE can be used for modeling the sloshing in the tank by avoiding the mesh distortion in liquid elements.

REFERENCES

Abaqus/Explicit User's Manual, Version 6.11. 2011. Dassault Systemes Simulia Corporation, Rhode Island, USA.
Abramson, H.N. 1966. The dynamics of liquids in moving containers, *Report SP 106*, NASA.
Akyildız, H., Unal, N.E. 2006. Sloshing in a three-dimensional rectangular tank: numerical simulation and experimental validation. *Ocean Engineering*, 33:2135–2149.
Chen, W., Haroun, M.A., Liu, F. (1996). Large amplitude liquid sloshing in seismically excited tanks. *Journal of Earthquake Engineering and Structural Dynamics*, 25:653–669.
Faltinsen, O.M. 1974. A non-linear theory of sloshing in rectangular tanks. *Journal of Ship Research*, 18:224–241.
Faltinsen, O.M. 1978. A numerical non-linear method of sloshing in tanks with two-dimensional flow. *Journal of Ship Research*, 22(3):193–202.
Faltinsen, O.M., Rognebakke, O.F., Lukovsky, I.A., Timokha A.N. 2000. Multidimensional modal analysis of nonlinear sloshing in a rectangular tank with finite water depth. *Journal of Fluid Mechanics*, 407:201–234.
Frandsen, J.B. 2004. Sloshing motions in the excited tanks. *Journal of Computational Physics*, 196(1):53–87.
Haroun, M.A., Housner, G.W. 1981. Earthquake response of deformable liquid storage tanks. *Journal of Applied Mechanics, ASME*, 48(1):411–418.
Ibrahim, R.A. 2005. Liquid sloshing dynamics: Theory and applications, *Cambridge University press*, UK.

Mackerel, J. 1999. Fluid-structure problems, finite element and boundary element approaches: A bibliography (1995–1998). *Finite Elements in Analysis and Design*, 31:231–240.

Mittal, V., Chakraborty, T., Matsagar, V. 2014. Dynamic analysis of liquid storage tank under blast using coupled Euler-Lagrange formulation, *Thin-Walled Structures*, 84:91–111.

Ramaswamy, B., Kawahara, M. 1987. Arbitrary Lagrangian-Eulerian finite element method for unsteady, convective incompressible viscous free surface fluid flow. *International Journal for Numerical Methods in Fluids*, 7(10):1053–1075.

Tippmann, J.D., Prasad, S.C., Shah, P.N. 2009. 2-D tank sloshing using the coupled Eulerian-LaGrangian (CEL) capability of Abaqus/Explicit. *Proc of SIMULIA Customer Conference*, London, England:1–11.

Ushijima, S. 1998. Three-dimensional arbitrary Lagrangian-Eulerian numerical prediction method for non-linear free surface oscillation. *International Journal for Numerical Methods in Fluids*, 26:605–623.

Veletsos, A.S. 1984. Seismic response and design of liquid storage tanks. Guidelines for the seismic design of oil and gas pipeline systems. *Technical Council on Lifeline Earthquake Engineering, ASCE*, New York, 255–370 & 443–461.

Virella, J.C., Prato C.A., Godoy L.A. 2008. Linear and nonlinear 2-D finite element analysis of sloshing modes and pressures in rectangular tanks subject to horizontal harmonic motions. *Journal of Sound and Vibration*, 312:442–460.

Wu, G.X, Eatock, T.R., Greaves, D.M. 2001. The effect of viscosity on the transient free-surface waves in a two-dimensional tank. *Journal of Engineering Mathematics*, 40:77–90.

Multi-disciplinary Sustainable Engineering: Current and Future Trends – Tekwani, Bhavsar & Modi (Eds)
© 2016 Taylor & Francis Group, London, ISBN 978-1-138-02845-6

Response of Elevated Water Tank subjected to near-fault and far-field earthquakes

Shailja Upadhyay
NFKTOR Engineering Project Consultants, Ahmedabad, Gujarat, India

Chirag Patel
AMD, GEC Modasa, Gujarat, India

ABSTRACT: Elevated Water Tanks (EWT) are widely used for water storage as they provide certain distinct advantages over other storage systems. Near-fault earthquakes exhibit seismic characteristics highly different from those of far field earthquakes. Due to this variation, higher damages observed in similar type of structures subjected to the near-fault. In this paper an attempt has been made to study the behavior of various EWT structures and structural parameters under the effect of various far field and near-fault earthquakes. For this purpose EWT having intz type container, various storage capacities and staging profiles are analyzed for near fault and far field earthquakes performing Time History Analysis. Comparison of the results shows that near fault earthquakes amplify the seismic response of the tanks and make them more vulnerable.

1 INTRODUCTION

1.1 *Near-fault and far-field earthquakes*

Recent earthquakes in Northridge (1994); Kobe (1995); Kocaeli, Turkey (1999); and Chi-Chi, Taiwan (1999) had caused devastated effects. These all earthquakes had epicenter nearby the well-developed urban areas which increased the severity of the damage. Researchers have noticed that the near-field ground motions are quite different from the usual far-field ones. On the contrary, the seismic codes deal with problems and design procedures related to far-field or intermediate epicentral distances. And hence in cases of near fault earthquakes, damage arose also when both design and detailing have been performed in perfect accordance with the code provisions. According to Mohraz, if the distance between site and fault is less than 20 km it can be called Near-fault earthquake. Near fault ground motion has high frequency change in short time duration, due to which a high magnitude pulse in the beginning of velocity and acceleration time history is seen (as shown in Figure 1). This pulse sends out the maximum domain of magnitude in a very small period. This result into increase in virtual stiffness, base shear, ductility demand and reduction in damping. Generally 2/3 is the ratio of

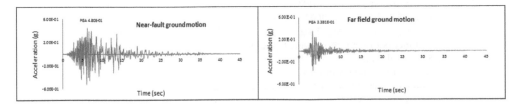

Figure 1. Near-fault and far-field time history record, Loma Prieta 1989.

vertical to horizontal spectrum for acceleration prescribed by many codes. This ratio becomes as high as 2 for near-fault ground motions.

1.2 *Near-fault earthquake and elevated water tanks*

Efforts are made by many researchers to study the characteristics of near-fault earthquake and its effect on various structures. Talking about Indian subcontinent 59% of its geographical area is vulnerable to seismic disturbance of varying intensities including the capital. Active fault mapping of India done by GSI (Geological Survey of India) showed that around 67 faults of regional extent exist in the country, many passing from the metro cities and highly populated urban areas. Even a small earthquake event in these regions can highly damage the EWTs located in those areas which cannot be afforded. Thus it is important to study the seismic behavior of EWTs during near fault earthquakes to improve the performance.

2 MODEL PROVISION AND GROUND MOTION SIMULATION

2.1 *Water tank models*

In this study, Intz shape tanks of 5 Lac, 10 Lac, 15 Lac and 20 Lac litters storage capacity have been modeled. Variations in staging height and pattern are made to understand its effect on structural behavior. Adapted staging heights are 12 m, 16 m, 20 m and 24 m and for each height two staging patterns, frame and shaft are taken under consideration. All the structures are made of M25 concrete. Tanks are designed in perfect accordance with the Indian Standard criteria for liquid retaining structure located in seismic zone IV. Table 1 briefs the general design criteria applied for all the EWT models, while structural configuration regarding the models are detailed in Table 2.

All the models of EWTs are created in STAAD.pro v8i. Top dome, cylindrical wall, conical dome and bottom domes are modeled using 3 and 4 noded plate elements. Staging are given fixed supports at the bottom ends. For all the domes radial and circumference meshing is done maintaining unit aspect ratio. Node to node connection was maintained to achieve monolithic behavior of structure. Tanks are considered as fully filled with water for all cases and water mass is added to the container walls according to alternate added mass approach given by Algreane Gareane A. I.et al.

2.2 *Ground motion simulation*

To study the seismic behavior of EWTs during near-fault and far-field earthquake time history analysis is performed, using near-fault and far-field time histories of three different earthquake events. Acceleration vs time records for Loma Prieta (1989), Northridge (1994) and Kobe earthquake (1995) earthquakes were collected from PEER NGA database. Seismic characteristics of the considered ground motions, like epicentral distance, Peak Ground

Table1.   General design data.

General design data	Value	Unit
Soil Bearing Capacity (S.B.C.)	200	N/mm^2
Type of Soil—IS 1893:2002	Medium	Type
Seismic Zone—IS 1893:2002	IV	No.
Distance between intermediate bracing	4	m
No. of columns	8	No.
Grade of concrete container	M-30	N/mm^2
Grade of concrete staging	M-25	N/mm^2
Grade of steel reinforcement	HYSD Fe-415	N/mm^2
Density of water	10	kN/mm^2

Table 2. Seismic parameters of time history records.

Earthquake	Station	Epicentral distance (km)	Hypocentral distance (km)	PGA (g)	PGV (cm/sec)	PGD (cm)
Loma Prieta (1989)	CDMG47006	28.98	33.84	0.334	26.940	5.330
	CDMG 47125	9.78	20.03	0.480	34.510	7.130
Northridge (1994)	USC 90059	23.18	29.05	0.140	9.090	2.050
	CDMG 24087	11.10	20.72	0.329	30.900	12.800
Kobe (1995)	Nishi-Akashi	8.70	19.90	0.486	35.73	10.75
	Kakogawa	24.20	30.10	0.266	21.66	7.60

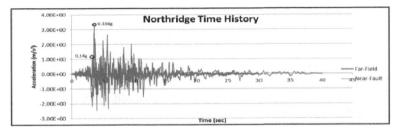

Figure 2. Time history record Northridge (1994).

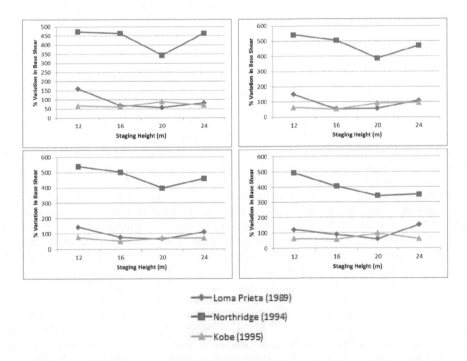

Figure 3. Percentage variation in base shear for frame supported tank of 5 lac, 10 lac, 15 lac and 20 lac litter (clockwise) storage capacities respectively.

Acceleration (PGA), Peak Ground Velocity (PGV) are illustrated in Table 2. For each earthquake one far-field record and one near-fault record has been considered for study. The difference of far-field and near-fault ground motion can be seen from the time history record shown in Figure 4. Figure 4 represents acceleration vs time record for Northridge (1994)

37

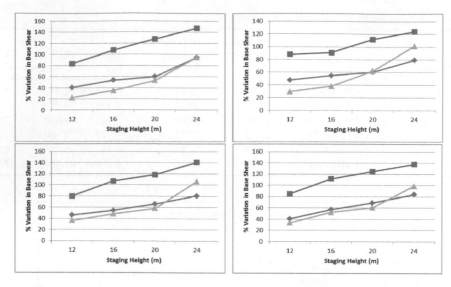

Figure 4. Percentage variation in base shear for shaft supported tank of 5 lac, 10 lac, 15 lac and 20 lac litter storage capacities respectively.

earthquake, where difference in PGA between two records and existence of pulse like motion is clearly visible.

3 RESULTS AND CONCLUSION

Here an attempt is made to understand the seismic behaviour of the elevated water tanks subjected to far-field and near-fault earthquake motions. The conclusions are made based on the comparison of the results generated by time history analysis for three different types of earth-quake motions. STAAD.pro analysis results are compared to do a parametric study of the structural responses of water tanks with change in storage capacity, staging type and staging height. Figures 5 and 6 shows the percentage variation in base shear for tanks with frame and shaft supported staging of various capacities respectively.

Graphs of base shear variation clears the idea of high effect of near fault ground motion on water tanks. Percentage variation in top level deflection in horizontal direction at the top crown level of EWT are shown in Figures 7 and 8 for frame and shaft supported tanks respectively. The graphs show that, the water tanks subjected to near-fault ground motions experience large horizontal deflection. For many cases this deflection is about 2 to 4 times higher than the deflection caused by the far-field ground motions. Response of the elevated water tank subjected to Near-Fault ground motion is very high as compared to Far-field ground motion and this difference is the highest for Northridge (1994) earthquake. Variation in base shear is comparatively higher for 12 m and 24 m staging height in case of frame supported water tank. Minimum variation is noted at 20 m staging height in case of Northridge (1994) and Kobe (1995) earthquake. While for Loma-Prieta earthquake variation is least for 16 m staging height. For shaft supported water tanks variation in base shear increases with increase in staging height.

Minimum variation is noted at 12 m staging height for every storage capacity and ground motion. The study also reflects that percentage variation in base shear is quite low for shaft supported tanks as compared to frame supported tanks. Base moment also follows the same pattern as base shear for every case of analysis. For deflection at various height levels we can notice huge increment in displacement values when tanks are subjected to Near-Fault ground motions. Deflections are very high for frame supported tanks than shaft supported tanks. For variation in deflection we can note that increase in deflection is very high for shaft

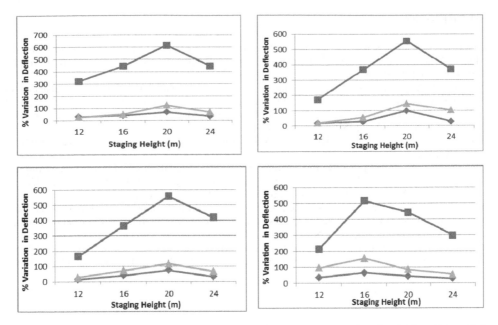

Figure 5. Percentage variation in top displacement for frame supported tank of 5 lac, 10 lac, 15 lac and 20 lac litter storage capacities respectively.

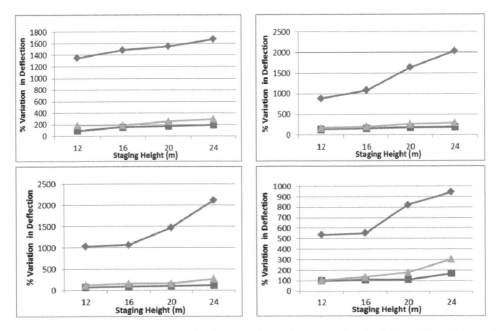

Figure 6. Percentage variation in top displacement for shaft supported tank of 5 lac, 10 lac, 15 lac and 20 lac litter storage capacities respectively.

supported tanks as compared to frame supported tanks. For frame supported water tanks percentage variation in deflection increase with staging height up to 20 m staging height and then decreases at the height of 24 m staging For shaft supported water tank percentage variation in deflection increase with the increment in staging height. With increase in storage capacity base shear and base moment values are increasing while deflection is decreasing.

REFERENCES

Dora Foti, 2014, "The Seismic Response of Protected and Unprotected Middle-Rise Steel Frames in Far-Field and Near-Field Areas", Hindawi Publishing Corporation Shock and Vibration. Volume 2014, Article ID 393870.

Gareane A.I. Algreane, S.A. Osman, Othman A. Karim, and Anuar Kasa, "Dynamic Behaviour of Elevated Concrete Water Tank With Alternate Impulsive Mass Configurations", Proceedings of The 2nd Wseas International Conference On Engineering Mechanics, Structures And Engineering Geology, Issn: 1790–2769, ISBN: 978-960-474-101-4.

Housner, (1963) "The Dynamic Behavior of Water Tank" Bulletin of the Seismological Society of America. Vol. 53, No. 2, pp. 381–387.

Maniatakis Ch.A., Taflampas I.M. and Spyrakos C.C., 2008 "Identification of near-fault earthquake record characteristics", World Conference on Earthquake Engineering, China, Pavel F, Aldea A, Vacareanu R," Near-field strong ground motion records from Vrancea earthquakes".

Tavakoli. H.R, Naeej. M, Salari. A, "Response of RC structures subjected to near-fault and far-fault, earthquake motions considering soil-structure interaction", international journal of civil and structural engineering Volume 1, No 4, 2011.

Tehrani Zade. M, Haj Najafi. L, "Assessing seismic behavior of eccentrically braced frames (ebfs) due to near-field ground motions". The 14th World Conference on Earthquake Engineering, China, 2008.

Multi-disciplinary Sustainable Engineering: Current and Future Trends – Tekwani, Bhavsar & Modi (Eds)
© 2016 Taylor & Francis Group, London, ISBN 978-1-138-02845-6

Nonlinear static analysis of an elevated RCC water tank

Prutha Vyas & Jahanvi Suthar
Department of Civil Engineering, Nirma University, Ahmedabad, Gujarat, India

ABSTRACT: Elevated water tanks are systems for storing huge masses of water at a certain height. Elevated tanks should be safe and functional during and after earthquakes for providing the necessary water supply for drinking and for firefighting. Hence, it is necessary to understand the seismic behavior of elevated tanks. Nonlinear static analysis is useful for understanding the post-elastic behavior of structures. Pushover analysis is a tool to perform nonlinear static analysis. In the present study, an elevated tank is designed with the limit state method as per IS:3370–2009 (part 1 and 2). A finite element model for the tank is generated in SAP2000 as per the design. Pushover analysis is performed on the developed tank finite element model to obtain the pushover curve, i.e., base shear v/s roof displacement and obtain the performance point and performance level of the elevated tank.

1 INTRODUCTION

Earthquake is the most destructive force that causes collapse and damage of structures and loss of lives. Due to the unpredictable nature of earthquake, it is vital that the design and construction of structures ensure endurance against earthquakes. Engineers cannot design the structures for the actual intensity of earthquake because the cost of construction would be inhibitive. The seismic design should be such that it prevents loss of life, ensures continuity, and minimum damage to property.

Water tanks are used to store water. Water tanks are designed as efficient and economical units for commercial as well as residential use. Seismic safety of liquid tanks is important. Water tanks should remain functional during and after an earthquake to ensure efficient supply of water to earthquake-affected regions and also for firefighting purposes.

Elevated water tank is a water storage facility supported by a tower. The design of such structures is a function of the hydrostatic pressure required to be produced by elevation of water and hence supply of water can be managed by gravity. This feature of elevated tank becomes advantageous in case of power outages after earthquakes in which the pumping system may not able to work without electricity.

2 NONLINEAR STATIC ANALYSIS

Structural engineers generally design the structure as per code based method. This method requires that structural components are evaluated for strength and serviceability in the elastic range. Linear elastic methods can predict the elastic capacity of the structure and indicates only where the first yield point will occur. But this method does not show failure mechanisms. During strong earthquakes, structures suffer significant inelastic deformations. Hence, to understand the nonlinear behavior of the structures, nonlinear analysis remains the option of choice. Nonlinear analysis basically includes nonlinear static analysis, which also known as pushover analysis and nonlinear dynamic analysis, i.e., nonlinear time history analysis.

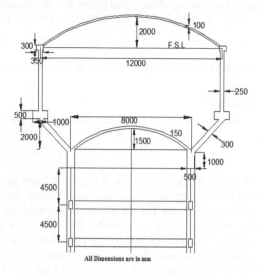

Figure 1. Sizes of various components of Elevated Intze Tank.

3 DESCRIPTION OF ELEVATED WATER TANK

Intze type water tank is considered here for designing the RCC elevated water tank. In the present study an Intze tank having a capacity of 6 lakh liters, 18 m height of frame type staging, eight circular columns, and three levels of bracings. M30 grade of concrete and Fe415 grade steel have been considered here as materials of choice. The tank is assumed to be in zone IV. As per IS:1893–2002 (part-1)[3] importance factor I = 1.5 and as per IS:1893–2007 (part-2) draft code [4], the response reduction factor R = 2.5 have been considered. Intze tank is designed as per IS:3370–2009 (part-1 & 2)[5].

The sizes of various components of Intze tank are as shown in Figure 1.

4 CRACK WIDTH CHECK

As per IS:3370–2009 (Part-2)[5] Appendix-B, B-6 crack width in concrete due to direct tension in walls should be checked. For limiting design, surface crack width should not be more than 0.2 mm.

5 STIFFNESS OF STAGING

For any supporting system of a water tank, the mass is assumed to be concentrated at the center of gravity (C.G) of the container, which is found out separately for the Tank Full and Tank Empty conditions. Staging is assumed to be having only stiffness and one-third of its mass is assumed to be lumped at the C.G. of the container. For evaluating the stiffness of a structure of given geometry, an arbitrary unit load (Say 10 kN) is applied at the C.G. of the container (which is at distance of "h" from top of staging, value of h is different for Tank Full and Tank Empty Conditions). For a particular tank, the model of staging is generated in SAP2000 [7] and at a height of CG of container, a master node is created on which an arbitrary load of 10 kN is applied on the master node as shown in Figure 2. The center of gravity of the tank was found to be at 4.42 m (full condition) and 4.02 m (empty condition) from the bottom of the container. Stiffness of the system is obtained by dividing the value of arbitrary force with the displacement obtained. Displacement at the

Figure 2. Stiffness of staging.

top of staging was found to be 0.996 mm. Thus, the stiffness of staging was calculated to be 10040.16 kN/m.

6 MODEL PROVISION

Stored water generates hydrostatic pressure on tank walls. Due to seismic forces, hydrodynamic forces are also exerted on the walls in addition to hydrostatic forces. These hydrodynamic forces are evaluated with the help of spring mass model of tanks.

As per Indian Standard code IS: 1893–2007 (part-2) draft code [4], the dynamic analysis of a liquid-containing tank is evaluated by spring mass model of tank. When a tank containing liquid vibrates, the liquid exerts impulsive and convective hydrodynamic pressure on tank wall. The parameters of this model depend on geometry of the tank and its flexibility.

The liquid in the lower region of the tank behaves like a mass that is rigidly connected to the tank wall. This mass is termed as impulsive liquid mass which accelerates along with the wall and induces impulsive hydrodynamic pressure on tank wall and similarly on base. Liquid mass in the upper region of tank undergoes sloshing motion. This mass is termed as convective liquid mass and it exerts convective hydrodynamic pressure on tank wall and base. Most elevated tanks are never completely filled with liquid. Hence a two-mass idealization of the tank is more appropriate as compared to a one-mass idealization. For elevated tanks with circular container, parameters mi, mc, hi, hc, and Kc shall be obtained with empirical formulas given in IS:1893–2007 (part-2) draft code[4]. After calculation with empirical formulas, the values of parameters are: hi = 2.44 m, hc = 4.04 m (from top of staging), Base Shear = 366.6 kN, and Base Moment = 8277.8 kN-m.

7 PUSHOVER ANALYSIS

Pushover analysis is one of the methods for performing nonlinear static analysis. It gives an idea about the post yield behavior of the structure, plastic hinge generation, and energy dissipation through plastic hinge formation. Pushover analysis is an approximate analysis method in which the given structure is subjected to monotonically increasing lateral forces. The roof displacement is plotted with base shear to get the global capacity curve as shown

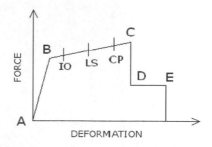

Figure 3. Global pushover curve.

in Figure 3. Five points labeled A, B, C, D, and E on the curve show the performance levels of the structure.

8 ANALYSIS IN SAP2000

Finite element software SAP2000 is used for modeling of the elevated tank. Components of tank, i.e., top dome, cylindrical wall, conical dome, and bottom spherical dome are modeled as shell elements. Top ring beam, middle ring beam, lower ring beam, columns of staging, and bracing are modeled as frame elements. Then, static load cases are defined, i.e., dead load, live load, earthquake load as per IS:1893 (part-1)-2002 [3]. As provided in SAP2000 default hinges to frame elements, i.e., PMM hinges are assigned for columns and M3 hinges are assigned for beam elements as described in ATC-40 [1]. After applying static load, nonlinear static loads have been defined. Generally, two static nonlinear cases are defined, one for gravity load and other for lateral load. In the case of lateral load, the static nonlinear analysis starts from the previous case of gravity nonlinear case. For particular case, pushover case is incrementally increasing load at the CG of tank of container.

9 RESULTS

Pushover analysis has been performed on an elevated Intze type water tank. The results obtained after nonlinear static analysis are Pushover curve (Base Shear Vs Roof Displacement), Capacity Spectrum Curve (ADRS Format), and Performance Point.

Pushover curve is obtained as shown in Figure 4. The ultimate base shear of the structure can be evaluated before failure and was obtained to be approximately 700 kN which is 1.9 times more than elastic base shear and the corresponding roof displacement is 180 mm. For obtaining the response spectrum curve as per IS:1893–2002 (part 1) [3], the value of Ca and Cv were calculated and assigned to the software. The values of Ca and Cv for all type of soils were taken as for medium soil and Zone IV as Ca = 0.24 and Cv = 0.33.

In Figure 5, the gray curve is the capacity spectrum curve and red curve is the Single Demand Spectra. The intersection point of Single Demand Spectra with the Capacity Spectrum Curve is the performance point. At performance point base shear was computed to be 487.75 kN and displacement is 68 mm.

At performance point of the structure, three hinges were formed at the column and of stage IO-LS; overall performance of the structure is of life safety stage. Hence, the structure has good capacity to resist future earthquakes as demand is observed to be less. Generated hinges at performance point of the tank are shown in Figure 6.

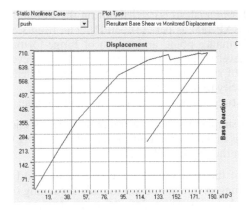

Figure 4. Pushover curve for elevated tank.

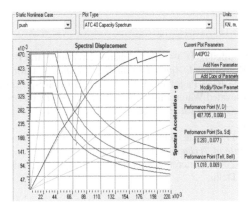

Figure 5. Performance point for elevated tank.

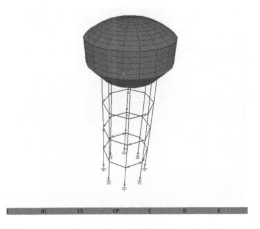

Figure 6. Generation of Hinges at performance point in elevated tank.

10 CONCLUSION

During earthquakes, elevated tanks are at a serious risk of failure. Pushover analysis is a tool for performing nonlinear static analysis. With pushover analysis, base shear v/s roof displacement curves have been obtained. In practice, ATC 40 [1] documents are referred for RCC structures with default hinge properties due to convenience and simplicity.

45

Nonlinear static pushover analysis is carried out on the elevated tank considering default hinges as generated by SAP2000 as per ATC-40 [1]. The results are presented in terms of capacity, demand, performance point, andpattern of hinge formation. The results help in identifying the damage level of the structure during earthquake events. For particular tanks, the overall performance of structure is of life safety stage, and hence, the structure has been found to have good capacity to resist future earthquakes.

Based on the current study, following conclusions can be made:

- Pushover analysis is a simple tool for performing nonlinear static analysis for structures.
- First hinge formation is in the base column, and after increasing the load, more number of hinges get generated in the above level of columns.

Overall performance of the structure is on life safety stage, and hence, the non-structural elements are severely damaged, but should not include the falling or collapse of heavy structural elements leading to less possibility of threat on life. It may cause injury but not loss of life.

REFERENCES

[1] Applied Technology Council (ATC)-40.1996, Seismic Evolution and Retrofit of Concrete Building, Report No. SSC 96–01, Volume I, *Applied Technology Council*, California.
[2] Ghateh, R., Kainosh, M.R. & Pagorzelski, W. 2013. Seismic Response Factors of Reinforcement Concrete pedestal in Elevated Water Tanks. *Journal of Engineering Structures* vol- 87.
[3] IS 1893–2002 (part I) .Criteria for Earthquake Resistant Design of Structures. *Bureau of Indian Standards*, New Delhi.
[4] IS 1893–2007 (part-II) Draft code. IITK-GSDMA Guidelines for Seismic Design of Liquid Storage Tank. *Indian Institute of Technology*,Kanpur, India.
[5] IS 3370 (part 1 and 2) -2009, Concrete Structures for storage of Liquids- code practice" *Bureau of Indian Standards*, New Delhi.

Multi-disciplinary Sustainable Engineering: Current and Future Trends – Tekwani, Bhavsar & Modi (Eds)
© 2016 Taylor & Francis Group, London, ISBN 978-1-138-02845-6

Evaluation of mechanical properties of pervious concrete with changing cement content

Nikhil S. Patel & Tejas M. Joshi
Institute of Technology, Nirma University, Ahmedabad, Gujarat, India

ABSTRACT: Pervious concrete was first used in the nineteenth century, but it has only recently begun to increase in popularity. As urban areas expand, the problems associated with runoff management have become more challenging. Generally in pervious concrete sand is not used but to increase the strength of pervious concrete, coarse aggregates will be replaced by various percentages of sand. The main properties to be studied are void ratio and compressive strength. Pervious concrete is produced using 10–20 mm aggregates. For the investigation, cement contents are used in the range from 250–450 kg/m^3 at an interval of 50 kg/m^3. The aggregate/cement ratio is kept constant as 4. Also, partial replacement of aggregates by sand is used, ranging from 0–40% at an interval of 10%. 3 w/c ratios are used as 0.3, 0.35 and 0.4. For this research paper 450 cubes are cast with different mix design. The paper is mainly focused on compressive strength and void ratio. From this investigation various graphs are produced. These graphs may be helpful for the mix design of pervious concrete.

1 INTRODUCTION

As urbanization increases in India and many parts of the world the problem of water logging and requirement of drainage is also increased. This is partly due to the impervious nature of the bituminous and concrete pavements. Pervious concrete which has an open cell helps significantly to provide high permeability due to its interconnected pores. It is a special type of concrete with a high porosity used for concrete flatwork applications that allows water from precipitation and other sources to pass directly through, thereby reducing the runoff from a site and allowing groundwater recharge. In pervious concrete, carefully controlled amounts of water and cementitious materials are used to create a paste that forms a thick coating around aggregate particles. Typically, between 15% and 30% voids are achieved in the hardened concrete, and flow rates for water through pervious concrete typically are around 0.34 cm/s.

2 EXPERIMENTAL PROGRAMME

This section provides information about concrete mix proportions and testing of concrete.

2.1 *Concrete mix proportioning*

The mix proportion includes 5 types of cement content between 250–450 kg/m^3 at 50 kg/m^3 interval. For each cement content, 3 w/c ratios are used as 0.3, 0.35 and 0.4. For each w/c ratio, 5 sand replacements are used from 0–40% at an interval of 10%. For example, a cement content of 250 kg/m^3 and the aggregate/cement ratio of 4:1 is used. So 1000 kg/m^3 of aggregates are used. Further, these aggregates are replaced by sand as shown in the Table 1. Similarly for other cement content is done.

Table 1 Mix design.

Sr. No.	Cement (kg/m³)	Aggregates (%)	Sand (%)	w/c ratio
1				0.3
2		100	0	0.35
3				0.4
4				0.3
5		90	10	0.35
6				0.4
7				0.3
8	250–450	80	20	0.35
9				0.4
10				0.3
11		70	30	0.35
12				0.4
13				0.3
14		60	40	0.35
15				0.4

All these mixes are made for single size of aggregates i.e. 75 mixes for 10–20 mm size aggregates. For each mix 6 cubes are cast i.e. 3 cubes for 7 days strength and 3 cubes for 28 days strength. So the research includes a total of 450 cubes.

2.2 Testing of concrete

For compressive strength test, 150 mm × 150 mm × 150 mm cubes of concrete are cast and tested to determine compressive strength. For the compressive strength test, cubes are tested as per IS 516 [2].

3 RESULTS AND DISCUSSION

3.1 Compressive strength

Void content

The void content is measured as per ASTM C1688 [1] (Standard Test Method for Density and Void Content of Freshly Mixed Pervious Concrete).

$$Void\,Content\,(\%) = \frac{T-D}{T} \times 100$$

$$where, D = \frac{Mc - Mm}{Vm}$$

M_c = mass of measure filled with concrete
M_m = net mass of concrete by subtracting mass of measure
V_m = volume of measure

$$T = \frac{Ms}{Vs}$$

Ms = total mass of materials batched
Vs = total absolute volume of materials

Sample calculation

For making 3 cubes by using 250 kg/m³ cement content with 0% sand and 0.3 w/c, following are the requirements.

      Cement = 3.725 kg.
      Aggregates = 14.9 kg.
      Sand = 0 kg.
      Water = 1.12 kg.
      After 7 days, the average density comes out to be 1880 kg/m³.
      D = 1880 kg/m³

$$T = \frac{3.275 + 14.9 + 0 + 1012}{\dfrac{3.725}{3.15 \times 1000} + \dfrac{14.9}{2.71 \times 1000} + \dfrac{0}{2.66 \times 1000} + \dfrac{1.12}{1 \times 1000}}$$

$$T = 2531 \text{ kg/m}^3.$$

$$Void\ content(\%) = \frac{2531 - 1880}{2531} \times 1000 = 25.72\%.$$

Compressive strength

Following are the results of 7 days and 28 days Compressive Strength (MPa) v/s Void Ratio (%) for various cement content with different sand content and w/c ratio.

    Figure No. 1,3,5,7 and 9 shows the 7 days compressive strength of 250,300,350,400,450 kg/m³ cement content respectively, with 0–40% sand and 0.3, 0.35 and 0.4 w/c. The 7 days compressive strength lies between 7.78–15.78 MPa and void content lies between 10.87–23.44%.

    Figure No. 2,4,6,8 and 10 shows the 28 days compressive strength for 250,300,350,400,450 kg/m³ cement content respectively with 0–40% sand and 0.3, 0.35 and 0.4 w/c. The 28 days compressive strength lies between 10.23–20.21 MPa and void content lies between 10.34–22.89%.

    Figure 11 shows the 28 days compressive strength for all 75 mixes including different cement content, sand content and w/c ratio. The compressive strength decreases as the void content increases. This is because of more voids present, which leads to less bonding between the aggregates resulting in the reduction of compressive strength.

    Figure 12 shows the 28 days compressive strength of different cement content. As the cement content increases the compressive strength increases. This is because of more cement paste available to bind the aggregates together which results into increased compressive strength.

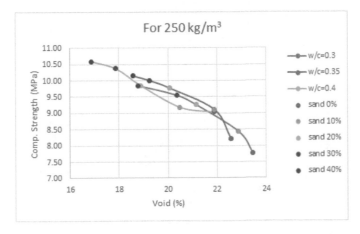

Figure 1.   7 days compressive strength.

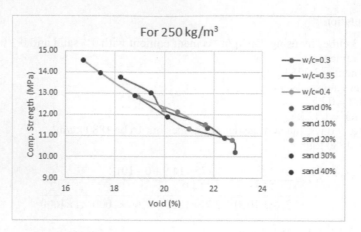

Figure 2.   28 days compressive strength.

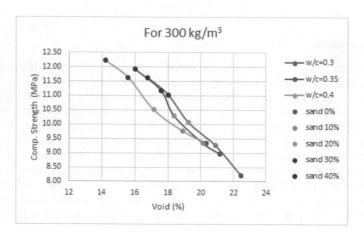

Figure 3.   7 days compressive strength.

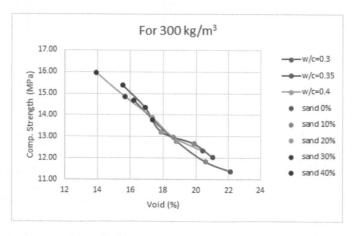

Figure 4.   28 days compressive strength.

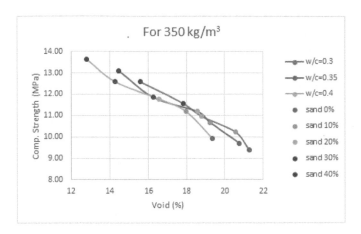

Figure 5. 7 days compressive strength.

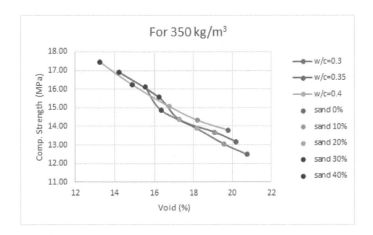

Figure 6. 28 days compressive strength.

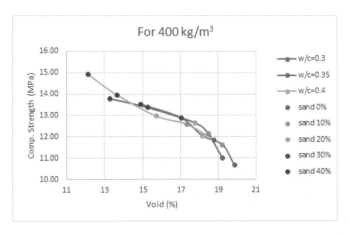

Figure 7. 7 days compressive strength.

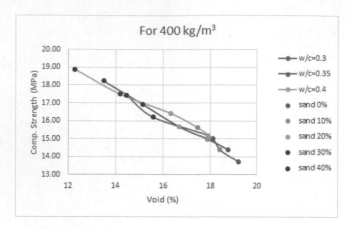

Figure 8.   28 days compressive strength.

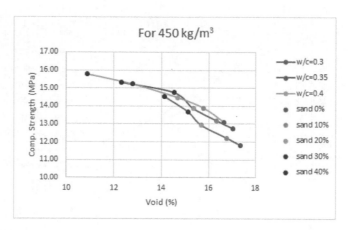

Figure 9.   7 days compressive strength.

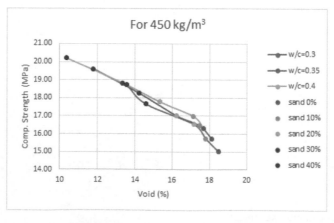

Figure 10.   28 days compressive strength.

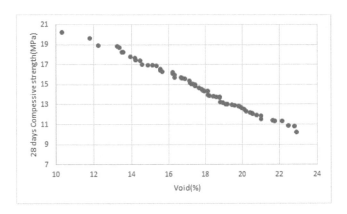

Figure 11.   28 days compressive strength.

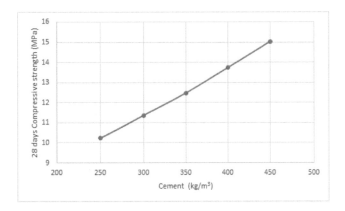

Figure 12.   28 days compressive strength.

4   CONCLUSIONS

The following conclusions can be drawn from the present investigation.

- With the increase in cement content, the compressive strength increases. This is because of more contact of cement paste with aggregates which helps to bind the materials.
- With the increase in w/c ratio, the compressive strength increases. However the void ratio decreases.
- With the increase in replacement of coarse aggregates with sand, the compressive strength increases. This is because of the filling of the voids with sand. However the void ratio decreases.
- This research focuses on the number of variations possible in the mix design of pervious concrete. So if cement content ranging from 250–450 kg/m³, w/c ranging from 0.3-0.4 and sand ranging from 0–40% is considered then such graphical representation may be helpful for the mix design of pervious concrete.

REFERENCES

[1]   ASTM C1688 Standard Test Method for Density and Void Content of Freshly Mixed Pervious Concrete.
[2]   IS: 516–1959 Methods of tests for strength of concrete, Bureau of Indian Standards, New Delhi, 1959.

Multi-disciplinary Sustainable Engineering: Current and Future Trends – Tekwani, Bhavsar & Modi (Eds)
© 2016 Taylor & Francis Group, London, ISBN 978-1-138-02845-6

Partial replacement of sand with pond ash in medium grade concrete

Akshar Gandhi & Urmil Dave

Department of Civil Engineering, Institute of Technology, Nirma University, Ahmedabad, Gujarat, India

ABSTRACT: The increase in demand for concrete and its constituents has resulted in the scarcity of natural resources. This has encouraged an interest in the research for alternative materials which could satisfy both the strength and performance criteria of concrete constructions. Pond ash, a waste product of thermal power plants is one of the materials that can be adopted as fine aggregate in concrete, replacing natural sand partially or completely. Encouraging the usage of such a waste material as a constituent in concrete can address the issues related to its disposal, environmental and ecological problems. Experimental study is conducted to explore the feasibility of using pond ash as replacement of fine aggregate with varying percentages such as 20%, 40% and 60%, in M40 grade concrete. Pond ash concrete is produced, tested and compared in terms of the mechanical properties such as compressive strength, split tensile strength and flexure strength with control concrete. Results of mechanical properties of pond ash concrete are at par with that of control concrete. This suggests that pond ash can be used as a replacement of fine aggregate in concrete.

1 INTRODUCTION

Ash is the residue left after combustion of coal in the thermal power plant. Particle size of the coal ash varies from about one micron to around 600 microns. The very fine particles (fly ash) collected from this ash generated by electrostatic precipitators are being used in the manufacturing of blended cements. Unused fly ash and bottom ash (residue collected at the bottom of furnace) are mixed in slurry form and deposited in ponds which are known as pond ash. Most of the thermal power plants in India adopt wet methods of disposal and storage of the ash in large ponds and dykes. In the wet method, both the fly ash collected from electrostatic precipitators and the bottom and grate ash are mixed with water and transported to the ponds in a slurry form. Pond ash is being produced at an alarming rate and efforts are required to safely dispose it and if possible find ways of utilizing it. Fly ash collected through hoppers has been widely accepted as pozzolana and is being used by the construction industry. Pond ash being coarser and less pozzolanic is not being used. More importantly in places where the fine aggregate is contaminated with harmful chemicals pond ash can act as a possible solution, as pond ash accumulation is posing environmental problems.

2 LITERATURE REVIEW

Siddique [1] investigated on the effect of fine aggregate replacement with Class F fly ash on the mechanical properties of concrete. Fine aggregates were replaced by 10%, 20%, 30%, 40%, 50% fly ash (by weight). The mixes were compared with control concrete in terms of compressive strength, split tensile strength and flexure strength at ages of 7, 14, 28, 56, 91, 365 days of curing. Substantial increase was observed in compressive strength and other mechanical properties with increase in percentage of fly ash gradually. This increase in strength was due to the pozzolanic action of fly ash. Singh and Siddique [2] investigated the

Table 1. Concrete mix proportioning.

Mix name	Cement (kg)	C.A20 mm (kg)	C.A 10 mm (kg)	F.A (kg)	P.A (kg)	Water (Lit.)	Admixture (kg)
PA00	433	809	436	663	–	158	3.46
PA20	433	809	436	531	132	158	3.46
PA40	433	809	436	398	265	158	3.46
PA50	433	809	436	265	398	158	3.46

effect of coal bottom ash as partial replacement of sand on properties of concrete. Natural sand was replaced with coal bottom ash by 10%, 20%, 30%, 40% and 50% of bottom ash. The strength development pattern of bottom ash concrete is similar to that of control concrete but there is decrease in strength at all the curing ages. The decrease in strength is mainly due to higher porosity and higher water demand on use of bottom ash in concrete. Bhangale and Nemade [3] studied the use of pond ash as a fine aggregate in cement concrete. Four mixes were prepared with varying percentages of replacement 0% (M1), 20% (M2), 30% (M3), and 40% (M4). The study showed that the compressive strength of concrete with pond ash increased with the increase in curing period.

3 EXPERIMENTAL PROGRAMME

This section gives brief information about the concrete mix proportioning and testing of pond ash concrete mixes and control concrete mix.

3.1 *Concrete mix proportioning*

The concrete mix design is based on provisions of IS 10262 [4] and is designed for grade M40. Water cement ratio is taken as 0.36 and Cement used is OPC 53 grade. Table 1 presents details of the constituents and their subsequent weights in accordance with the replacement of fine aggregates with pond ash for different concrete mixes.

4 RESULTS AND DISCUSSION

4.1 *Compressive strength*

Table 2 presents the results of compressive strength of various concrete mixes PA00, PA20, PA40 and PA60 at 7, 28, 56 and 90 days, respectively. Results in Table 2 indicates that the control concrete gains greater strength at the age of 7 days as compared to other pond ash concrete mixes. The compressive strength of pond ash concrete mixes, except PA60 is at par with control concrete mix at 28 days age. Except PA60 all other pond ash concrete mixes show notable increase in the compressive strength as compared to control concrete mix at 56 days. Substantial enhancement is observed in compressive strength for all pond ash concrete mixes, except PA60 as compared to that of control concrete at 90 days.

Comparison of percentage change in compressive strength of PA20, PA40 and PA60 with respect to PA00 at the age of 28 days, 56 days and 90 days, respectively is presented in Figure 1. Reduction of 0.6%, 1.6% and 22.5% has been observed in the compressive strength for PA20, PA40 and PA60, respectively as compared to that of PA00 at the age of 28 days as presented in Figure 1(a). Decrement of 0.3% and 15.7% is observed for PA20 and PA60, respectively and increment of 12.2% in compressive strength has been observed for PA40 at the age of 56 days as presented in Figure 1(b). Enhancement of 6.3% and 13.8% in compressive strength is observed for PA20 and PA40, respectively, while reduction of 19.2% has been observed for PA60 at the age of 90 days as presented in Figure 1(c).

Table 2. Results of compressive strength of varying percentages of pond ash.

| Mix* | Compressive Strength in N/mm² | | | | | | | |
	7 days	Avg.	28 days	Avg.	56 days	Avg.	90 days	Avg.
PA00	34.67	34.52	45.70	45.30	48.89	50.81	56.89	58.96
	35.11		48.00		51.11		58.67	
	33.78		42.20		52.44		61.33	
PA20	31.11	31.26	43.11	45.03	49.77	50.67	60.00	62.66
	32.00		46.67		51.56		63.11	
	30.67		45.30		50.67		64.88	
PA40	28.89	29.62	41.78	44.59	57.33	57.03	66.67	67.11
	30.22		45.33		58.67		68.89	
	29.78		46.67		55.11		65.78	
PA60	25.78	26.51	34.22	35.11	42.22	42.85	48.00	47.62
	27.55		34.67		43.56		46.00	
	26.22		36.44		42.67		48.88	

*PA00 is control concrete, PA20, PA40 and PA60 is concrete incorporating pond ash as 20, 40 and 60% replacement of fine aggregate.

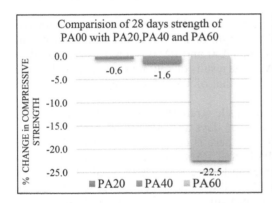

(a)

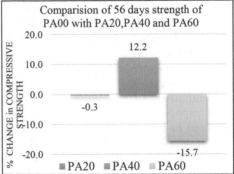

(b)

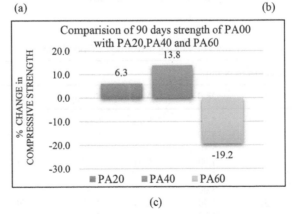

(c)

Figure 1. Comparison of % change in compressive strength of PA20, PA40, PA50 and PA60 wrt. PA00 at the age of at the age of (a) 28 days, (b) 56 days and (c) 90 days.

4.2 Flexure strength

Flexure strength of control concrete and pond ash concrete is evaluated at the ages of 7, 28, 56 and 90 days, respectively. Table 3 presents the results of flexure strength of various concrete mixes PA00, PA20, PA40 and PA60 at different ages.

Results in Table 3 indicate that 7 day flexure strength of pond ash concrete is at par with control concrete. The flexure strength of PA20 is at par with PA00, on the other hand, the flexure strength of PA40 and PA60 is lower as compared to that of PA00 at the age of 28 days. Except PA60, the results of flexure strength of pond ash concrete are at par with that of PA00 for the age of 56 days and 90 days, respectively.

Comparison of percentage change in flexure strength of PA20, PA40 and PA60 with respect to PA00 at the ages of 28 days, 56 days and 90 days, respectively is presented in Figure 2. Decrement of 6.9%, 13.0% and 47.6% has been observed in the flexure strength for PA20, PA40 and PA60, respectively as compared to that of PA00 at the age of 28 days as presented in Figure 2(a). Reduction of 8.1%, 14.9% and 42.4% has been observed for PA20, PA40 and PA60, respectively as compared to that of PA00 at the age of 56 days as presented in Figure 2(b). Decrement of 6.2%, 10.7% and 33.3% in flexure strength is observed for PA20, PA40 and PA60 respectively as compared to that of PA00 at the age of 90 days as presented in Figure 2(c).

4.3 *Split tensile strength*

Split tensile test is carried out on control concrete and pond ash concrete at the ages 7, 28, 56 and 90 days respectively. Table 4 presents the results of split tensile strength of concrete mixes PA00, PA20, PA40 and PA60 at different ages.

Results indicate that 7 day split tensile strength of control concrete is higher as compared to that of pond ash concrete. The split tensile strength of PA20 is at par with control concrete, on the other hand, split tensile strength of PA40 and PA60 is lower as compared to that of PA00 at the age of 28 days. Except PA60, the results of split tensile strength of all the pond ash concrete mixes is at par as compared to that of PA00 for the age of 56 days and 90 days, respectively.

Comparison of percentage change in split tensile strength of PA20, PA40 and PA60 with respect to that of PA00 at ages of 28 days, 56 days and 90 days, respectively is represented in Figure 3. It has been observed that there is decrement of 6.5%, 17.6% and 40.2% in split tensile strength for PA20, PA40 and PA60, respectively as compared to that of PA00 at the age of 28 days as presented in Figure 3(a). There is reduction in split tensile strength of 3.9%, 1% and 33.7% for PA20, PA40and PA60, respectively at the age of 56 days as presented in Figure 3(b). Enhancement of 2.6% in split tensile strength is

Table 3.   Results of flexure strength of varying percentage of pond ash.

Mix*	Flexure Strength in N/mm^2							
	7 days	Avg.	28 days	Avg.	56 days	Avg.	90 days	Avg.
PA00	3.20	3.40	5.70	5.40	6.00	5.80	6.40	6.00
	3.70		5.30		5.80		6.00	
	3.40		5.20		5.70		5.60	
PA20	3.20	3.36	4.90	5.03	5.20	5.36	5.60	5.63
	3.40		5.20		5.60		5.90	
	3.50		5.00		5.30		5.40	
PA40	3.20	3.33	4.90	4.70	4.90	4.96	5.20	5.36
	3.40		4.70		5.00		5.40	
	3.40		4.50		5.00		5.50	
PA60	2.00	2.00	3.20	2.83	3.20	3.36	4.00	4.00
	1.80		2.80		3.40		3.70	
	2.20		2.50		3.50		4.40	

*PA00 is control concrete, PA20, PA40 and PA60 is concrete incorporating pond ash as 20, 40 and 60% replacement of fine aggregate.

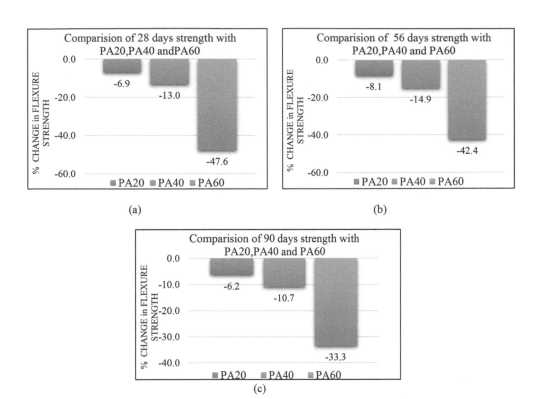

(a)

(b)

(c)

Figure 2.   Comparison of % change in flexure strength of PA20, PA40 and PA60 wrt. PA00 at the age of (a) 28 days, (b) 56 days and (c) 90 days.

Table 4.   Results of split tensile strength of varying percentage of pond ash.

| Mix* | Split Tensile Strength in N/mm² | | | | | | | |
	7 days	Avg.	28 days	Avg.	56 days	Avg.	90 days	Avg.
PA00	3.53	3.57	4.24	4.33	4.95	4.9	5.37	5.37
	3.37		4.24		4.81		5.23	
	3.53		4.52		4.95		5.51	
PA20	3.53	3.29	3.96	4.05	4.66	4.71	5.37	5.23
	3.25		4.38		4.52		5.09	
	3.11		3.82		4.95		5.23	
PA40	2.82	2.87	3.53	3.57	4.67	4.85	5.65	5.51
	2.97		3.67		4.81		5.51	
	2.82		3.53		5.09		5.37	
PA60	1.83	2.01	2.68	2.59	3.53	3.25	3.96	4.24
	2.26		2.83		3.25		4.24	
	1.96		2.26		2.97		4.52	

*PA00 is control concrete, PA20, PA40 and PA60 is concrete incorporating pond ash as 20, 40 and 60% replacement of fine aggregate.

observed for PA40, while reduction of 2.6% and 21.0% in split tensile strength is observed for PA20 and PA60, respectively as compared to that of PA00, at the age of 90 days as presented in Figure 3(c).

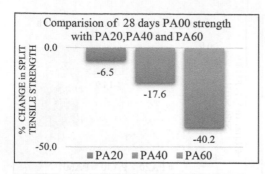

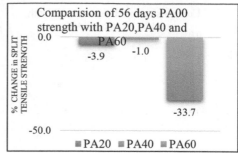

(a)   (b)

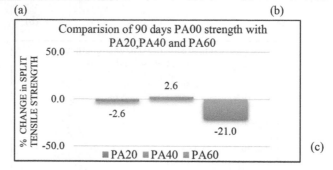

(c)

Figure 3.   Comparison of % change in split tensile strength of PA20, PA40 and PA60 wrt. PA00 at ages of (a) 28 days, (b) 56 days and (c) 90 days.

5   CONCLUSIONS

Following conclusions can be drawn from the present investigation:

1. Pond ash concrete mixes except PA60 exhibit superior performance as compared to that of control concrete at 90 days. Such behavior suggests that replacement of fine aggregate up to the extent of 40% is resulting into increment in compressive strength of pond ash mixes as compared to that of control concrete.
2. Flexure strength of PA20 and PA40 mixes exhibit at par performance as compared to that of PA00 mix at 90 days. Such behavior suggests that replacement of fine aggregate up to the extent of 40% is resulting in at par performance in flexure strength of pond ash mixes as compared to that of control concrete.
3. Split tensile strength of PA20 and PA40 exhibit at par performance with that of control concrete at 56 days. Pond ash concrete mixes PA20 and PA40 exhibit at par performance as compared to that of control concrete at 90 days. Such behavior suggests that replacement of fine aggregate up to the extent of 40% is resulting into increment in split tensile strength.

Based on the results of present investigation it is verified that it is possible to use pond ash as a replacement of fine aggregate without compromising the mechanical properties of concrete. This study opens up a major avenue for utilization of pond ash as potential viable material that can be used as the fine aggregate to produce concrete which is better than control concrete. It is believed that the proposed use of pond ash in concrete will help in alleviating the potential problem of dwindling of natural resources.

ACKNOWLEDGEMENT

Authors are thankful to Management and Director, Institute of Technology, Nirma University for the permission for conducting work and publication of the work.

REFERENCES

[1] Rafat Siddique, "Effect of fine aggregate replacement with Class F fly ash on the mechanical properties of concrete", Cement and Concrete Research 33 (2003), pp. 539–547.

[2] Malkit singh, Rafat Siddique, "Effect of coal bottom ash as partial replacement of sand on properties of concrete", Resources, Conservation and Recycling 72 (2013), pp. 20–32.

[3] P.P. Bhangale, P.M. Nemade (March 2013), "Study of Pond Ash Use as Fine Aggregate in Cement Concrete-Case Study", International Journal of Latest Trends in Engineering and Technology, Vol. 2, issue 2, pp. 292–297.

[4] IS: 10262–2009, Concrete mix proportion guidelines, Bureau of Indian Standards, New Delhi, 2009.

[5] IS: 516–1959 Methods of tests for strength of concrete, Bureau of Indian Standards, New Delhi, 1959.

[6] IS: 5816–1999 Splitting tensile strength of concrete method of test Bureau of Indian Standards, New Delhi, 1999.

Multi-disciplinary Sustainable Engineering: Current and Future Trends – Tekwani, Bhavsar & Modi (Eds)
© 2016 Taylor & Francis Group, London, ISBN 978-1-138-02845-6

Investigation on the pull-out behavior of BFRP bars

Husain S. Patrawala
SMV Engineers, Surat, India

Smitha Gopinath & A. Ramachandra Murthy
CSIR-Structural Engineering Research Centre, Taramani, Chennai, India

Paresh V. Patel
Department of Civil Engineering, Nirma University, Ahmedabad, India

ABSTRACT: The objective of the present study is to investigate the bonding behavior of an NSM (Near Surface Mounted) BFRP (Basalt Fiber Reinforced Polymer) bar and its interaction with the surrounding groove filler material and concrete. To simulate the real practical approach for NSM technique and to perform a direct pull-out test on BFRP bars, C-block concrete configuration has been used. Parameters, such as embedment length, have been considered to investigate their influence on the pull-out behavior. A BFRP bar of 10 mm diameter was inserted in a groove of size of 15 mm × 15 mm provided in the C-block. Different embedment lengths of 12 and 18 times the diameter of BFRP bars have been considered for the investigation. A novel analytical model is proposed and validated with the experimentally obtained bond stress-slip relation and found to be in good agreement. The proposed study intends to provide an insight for strengthening reinforced concrete beams using the NSM-BFRP technique.

Keywords: bond strength; NSM technique; BFRP bars; bond slip

1 INTRODUCTION

In construction, steel reinforced concrete is the most widely used structural material in the world. However, it is well known that in certain environments corrosion of steel reinforcement causes deterioration of structural elements which may lead to repair and strengthening of structure that turns into costly affair. In order to prevent these expenses, Fiber Reinforced Polymer (FRP) which incorporates non-corrosive properties has obtained generalized acceptance as a feasible reinforcement [1]. In this context, Near Surface Mounted (NSM) FRP bars are now emerging as a promising technique. Grooves are formed on the surface of the member, and FRP bars are installed by placing bars in grooves with an appropriate filling material.

Bonds between the BFRP bars and the existing structural elements is an important factor for the NSM technique to work effectively. For that purpose, investigations related to direct pull-out tests for BFRP bars can be carried out to know the bond slip relation and bond strength of BFRP bars. A modified direct pullout specimen was proposed keeping the practical advantages pertaining to direct pullout tests, such as possibility to conduct test, in slip control mode. It measured both loaded and free end slips and visual access to test zone [2]. By using this type of specimen, experiments were performed to observe the influence of parameters, such as embedment length, on the pull out behavior. This paper deals with the test results obtained and focusing on the specimen behavior, modes of failure, and the parameters which affect the results. Also, an attempt is made to develop a new bond-slip model pertaining to BFRP bars.

2   RESEARCH SIGNIFICANCE

The NSM technique has certain advantages as compared to external bonded reinforcement. This technique can be applied faster as no surface preparation is required besides the formation of the necessary groove and is less prone to accidental impact and mechanical damage, fire, and vandalism. The aesthetics of strengthened structural elements remain unchanged. The main aim is to study the bonding behavior between BFRP bars and concrete that corresponds to stress transfer from concrete into BFRP bars and to analyze the effect of the most critical parameters on bond performance.

3   EXPERIMENTAL INVESTIGATION

3.1   Test program

The specimen used for this investigation [3] is shown in Fig. 1. It consists of a C-shaped concrete block with a square groove in the middle of the block for embedment of the BFRP bar. The applied pull force is resisted by means of eight steel threaded bars inserted into an I-section which is placed on the top of C-block. The test variable embedment length ranges from 12 to 18 times the diameter of BFRP bar. The groove size has been kept as 15 mm × 15 mm and the diameter of BFRP used is 10 mm.

As the BFRP bar has a smooth configuration; in order to get a proper grip with groove filling material, a sand coating is done on the surface of the BFRP bar.

3.2   Material properties

C-shaped concrete blocks of M-25 grade has been cast. The mix proportion used for making concrete is given in Table 1.

The maximum size of aggregate used was 12 mm. The epoxy paste Sikadur-330 is used to adhere the BFRP bar to the concrete surface in the specified groove size upto a certain embedment length. The specifications of the adhesive are shown in Table 2.

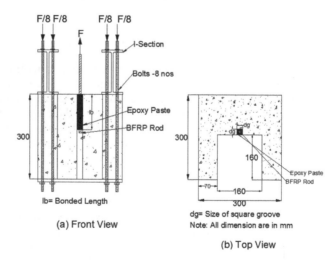

Figure 1.   Test specimen.

Table 1.   Mix proportion of M-25 concrete.

Grade of Concrete	w/c	Cement	Sand	C.A
M-25	0.5	1	1.8	2.8

Here, the symbol R indicated resin and H indicates hardener. The BFRP bar used for the experiment has a tensile strength of 660 MPa, modulus of elasticity of 55 GPa, and ultimate strain of 1.2%.

3.3 *Specimen preparation and testing*

In order to prepare a C-shaped concrete block, certain dimensions need to be extruded from the solid cube. So for that purpose a 160 mm × 160 mm × 300 mm thermocol was placed in a 300 mm × 300 mm × 300 mm steel mould. As per the information provided in literature [2], one square groove should be provided at the C.G of the block, using a wooden strip of square cross-section of specified dimension adhered at the center of the thermocol. After hardening, the concrete square groove was cleaned by means of air blasting. Then, epoxy paste was prepared by mixing the two components in 4(Resin):1(Hardener) proportion by volume. Epoxy was then poured into the groove up to half the depth; some mild compaction was given at some intervals so that the epoxy gets filled properly into the square groove. In all specimens, thermocol spacers were used to control the positioning of the bar and ensure that they are situated at the center of the groove. Fig. 2 shows a C-shaped concrete specimen preparation and also shows the procedure of adhering the BFRP bar in a square groove. Testing was conducted in displacement-control mode on a 50-kN universal testing machine with a 1-mm/min displacement rate. Fig. 3 shows the grip part of BFRP bar and experimental set-up.

Table 2. Material specification of adhesive (Sikadur-330).

Adhesive	Compressive strength (MPa)	Tensile strength (MPa)	Modulus of Elasticity (MPa)	Mixing Ratio
Sikadur-330	82	34	3489	4(R):1(H)

(a) (b) (c) (d)

Figure 2. (a) Use of thermocol in casting (b) Concrete C-block (c) Placing epoxy in the groove (d) Epoxy poured up to the bonded length.

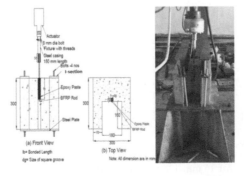

Figure 3. Schematic diagram of specimen and test set-up.

4 TEST RESULTS

The test results are reported in Table 3. Different failure modes were encountered such as crack propagation on epoxy, bar slipping out from the epoxy interface, and pull out at the interface between concrete and epoxy. The bond stress is calculated by dividing the pull out force with the contact area of the FRP bars.

$$\tau = \frac{P_u}{\pi . d . L_b} \tag{1}$$

τ = Bond strength, P_u = Pull out load at failure, d = diameter of FRP bar, L_b = Bonded length of FRP bar

Fig. 4 shows the bond slip relation for the BFRP bar. In the first stage, the bond stress was increasing linearly with increase in the slip value. After reaching the peak, the crack formation took place at the epoxy face at 7.2 kN which was the maximum peak load. In second stage, the yielding took place beyond the ultimate pull-out load with a large slip followed by decrement in slip values. The slip value of the BFRP bar was noted at the loaded end.

4.1 *Modes of failure*

The common mode of failure for most of the specimens using epoxy as an adhesive material was the crack propagation on the face of epoxy layer and the failure takes place at concrete epoxy interface as shown in Fig. 5. Once the crack initiation took place at epoxy face it gets propagated up-to free end. These cracks suggest the path of tensile force in the bar, which was transferred to epoxy. As the tensile strength of epoxy is much higher as compared to concrete, major failure should occur in concrete part, but this type of failure was not observed in any specimen. BFRP bar was able to take load up-to certain extent but after that slip started taking place between steel casing and the BFRP bar. The epoxy which was filled in steel casing did not get proper bond with BFRP bar.

Table 3. Test results for specimen.

Spec ID (Groove width-Bonded length)	P_m(kN)	τ_m(MPa)	s_m(mm)	Failure mode
15 mm-12db-(10 mm dia)	7.7	2.04	1.845	Failure at epoxy-concrete interface
15 mm-18db-(10 mm dia)	21.7	3.83	2.25	Cracks formed at epoxy layer

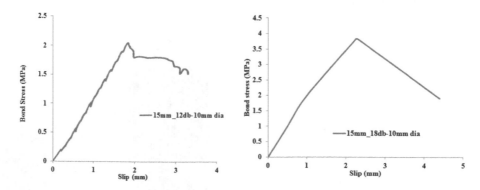

Figure 4. Bond-slip curve for 15 mm-12db (10 mm dia) and 15 mm-18db (10 mm dia).

(a)   (b)   (c)

Figure 5.   (a) Crack propagation at free end (b) Failure at epoxy-concrete interface (c) Crack propagation.

5   DEVELOPMENT OF ANALYTICAL MODEL

Fig. 6 shows a typical bond-slip curve for FRP bars. Based on **BPE** and **m-BPE** (Bertero–Popov–Eligehausen) model [4] one can divide this curve in two zones. The pre-cracking zone (ascending branch) and post-cracking zone (descending branch).

5.1   *Pre-cracking zone (ascending branch)*

At first steep ascending branch is formed where the bond between **FRP** bar and concrete remain intact due to adhesion till peak, where bond stress and slip values are maximum.

The bond–slip law follows the same ascending branch as discussed by BPE bond–slip relationship:

$$\left(\frac{\tau}{\tau_m}\right) = \left(\frac{s}{s_m}\right)^{\alpha} \qquad 0 \leq s \leq s_m \tag{2}$$

Here value of α should be less than 1 and it is obtained from curve fitting.
τ_m = Maximum bond stress (MPa), s_m = Maximum slip (mm).

5.2   *Post-cracking (descending branch)*

A descending branch is formed after reaching peak. When the tensile stress reaches the tensile strength of adhesive, cracks forms as a result bond stress tends to decrease with the increase in slip due to loss of adhesion between epoxy and concrete.

$$\left(\frac{\tau}{\tau_m}\right) = \left(\frac{s}{s_m}\right)^{\alpha'} \qquad s \geq s_m \tag{3}$$

For descending part equation remains same but in order to achieve decrement one parameter α' is included whose value lies in between 0 and −1. As a result bond stress decreases with increase in slip value and goes on increasing till bond stress value get constant.

5.3   *Analytical versus experimental results*

Based on the available experimental data the main parameters of the new developed model that is (α^+ and α^-) have been obtained by the using **MATLAB** tool in which linear regression analysis is performed and the value of parameter was found whose $R^2 \approx 1$.

Fig 7: Shows the analytical behavior of bond slip curve and it is compared with the experimental results.

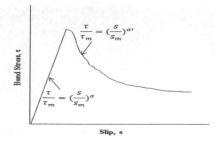

Figure 6.   Bond slip model.

Table 4.   Values of parameters.

Type of outer surface	α^+	α^-	τ_m(Mpa)	s_m(mm)
Sand Coating	0.725	0.85	3.84	2.25

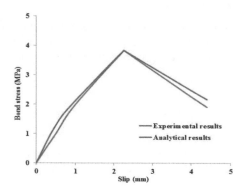

Figure 7.   Bond slip curve for 15 mm-18db (10 mm dia).

6   CONCLUSION

The experimental program has provided an understanding of the bond behavior of the NSM-FRP bar technique in concrete elements. Based on the experiment following conclusions can be drawn:

1. The BFRP bar and epoxy Sikadur-330 has shown good performance. There was no major crack or rupture of bar took place at the time of loading.
2. The C-block test method seems to be efficient.
3. The main mode of failure observed was the crack formation at the face of epoxy cover and the failure at epoxy-concrete interface.
4. Analytical model is developed which is useful to predict bond stress-slip relation from the experimental data.
5. The value of parameters such as α^+ (0.725) is useful to form plot the curve till cracking phenomenon takes place and α^- (0.85) which is useful to form curve in post cracking stage.
6. Developed model shows good agreement with the experimental results.

ACKNOWLEDGMENT

Authors thank the staff of Computational structural Mechanics Group for the support provided during this investigation. This paper is being published with the kind permission of the Director, CSIR-SERC.

REFERENCES

[1] Lorenzis L., Teng J.G., "Near-surface mounted FRP reinforcement: An emerging technique for strengthening structures", Composites: Part B 38 (2007), pp. 119–143.

[2] Lorenzis L., Rizzo A., Tegola A.L., "A modified pull-out test for bond of near-surface mounted FRP rods in concrete", Composites: Part B 33 (2002), pp. 589–603.

[3] Lorenzis L., Lundgren K. and Rizzo A., "Anchorage Length of Near-Surface Mounted Fiber-Reinforced Polymer Bars for Concrete Strengthening Experimental Investigation and Numerical Modeling", ACI Structural Journal, V. 101, No. 2, April, 2004, pp. 269–278.

[4] Cosenza E., Manfredi G. and Realfonzo R. "Behavior and modeling of bond of FRP rebars to concrete", Journal of Composites for Construction, Vol. 1. No.2, May, 1997, pp. 40–51.

Multi-disciplinary Sustainable Engineering: Current and Future Trends – Tekwani, Bhavsar & Modi (Eds)
© 2016 Taylor & Francis Group, London, ISBN 978-1-138-02845-6

Dynamic characterization of Shock Table

Kashyap M. Patel & Sharadkumar P. Purohit
Department of Civil Engineering, Institute of Technology, Nirma University, Ahmedabad, Gujarat, India

ABSTRACT: During earthquakes, structures experience base motion has the potential to damage the structure to different degrees. Understanding the dynamic response of the structure to the base motion is quite important. This is achieved by mounting a scaled model of structures on the Shake Table capable of simulating an earthquake scenario in the laboratory. However, such a facility is quite costly from both a development and operational point of view. The dynamic response of a structure can alternatively be determined using the Shock Table wherein impacts of various degrees are applied. The present study focuses on dynamic characterization of Shock Table in terms of acceleration, force, and dominant frequency. The experimental results extracted are compared with the analytical results derived using the Finite Element (FE) model of Shock Table through ANSYS. A good agreement between experimental and analytical results is observed. Shock Table was subjected to a force of about 1500 kN and has a dominant frequency of about 1 Hz.

1 INTRODUCTION

Earthquakes have a potential to damage most all manmade structures. Determining the response of structures during an earthquake is an important part of structure dynamics and earthquake engineering (Chopra 2009). It is usually obtained by testing scaled models of structures on Shock Table. In the absence of a sophisticated Shake Table, the study on the behavior of structures and damage caused due to earthquakes were attempted by using a non-sophisticated Shock Table. Shock Tables of various forms were used by researchers to carry out dynamic testing of scaled models of structures (Ersubasi & Korkmaz 2010). The Shock Table is a much cheaper alternative as compared to a costly Shake Table but gives a good enough insight to the dynamic problem. One such Shock Table testing facility as shown in Figure 1 was developed at the Civil Engineering Department, Institute of Technology, Nirma University, Ahmedabad, where the table was mounted on a roller and given shocks through a freely falling pendulum.

The Shock Table is 3.6 m wide and 6 m long and is mounted on four rollers that move on a reinforced cement concrete column of 350 mm × 300 mm fitted with a base plate. The table has two big cutouts of size 700 mm × 1700 mm placed symmetrically on the table to provide ease in construction of testing models. A pendulum (weight 1500 kg) made up of a steel box with a steel plate casing contributes to the pendulum, pivoted from top, to allow it to fall freely. On the other side of the shock table a reaction beam 250 mm wide by 400 mm depth and 3440 mm long is kept that helps to rebound the table and simulate an earthquake type of motion (Purohit & Patel 2005).

1.1 *Impact force estimation*

Shock Table facility consists of a pendulum pivoted at the top and weighs about 1500 kg. It is made up of a steel box having steel plates. The pendulum is pulled by a pulley system to different degrees of height and then allows it to fall freely. The free fall of pendulum produces an impact on the Shock Table and thus imparts an energy to it.

Chain Link

Pendulum

Stopper   Wheel Mounted Table

Figure 1. Shock Table facility and its components.

Table 1. Impact force and contact time of pendulum for different angle of swing.

Angle of swing	Force (kN)	Contact time (s)
10	178	0.0065
20	460	0.005
30	860	0.004
40	1514	0.003

Estimate of impact load is crucial in order to calculate acceleration produced in the Shock Table. In the present study, the impact load is estimated using impulse momentum principle. An important variable is duration of impact, which is obtained experimentally. Equation (1) states the impulse momentum principle.

$$F = \frac{m \times v}{\Delta t} \quad v = \sqrt{2 \times g \times h} \tag{1}$$

where m = mass of pendulum; v = velocity of pendulum; Δt = contact time between pendulum and table, and h = height of pendulum from mean position; g = gravitational acceleration.

Table 1 shows impact force and contact duration of pendulum with Shock Table for various degrees of angle. Note that maximum impact force is produced with maximum rise of pendulum.

2 FINITE ELEMENT (FE) ANALYSIS

FE analysis of Shock Table is done using ANSYS 14.5 software. Transient dynamic analysis is carried out to determine response of the Shock Table. Shock Table is modeled using SOLID187 element which is a 10-node tetrahedral element having three degrees of freedom at each node. Material properties used in FE Modeling are for concrete—M15 grade of concrete and Fe 250 grade for steel section used.

A mesh size of 75 mm is considered based on convergence analysis. Dynamic analysis is carried out for 0.1 s and acceleration (in "g") is extracted at a time interval of 0.001 s for various locations as shown in Figure 2. The location of external acceleration is in line with experimental instrumental setup.

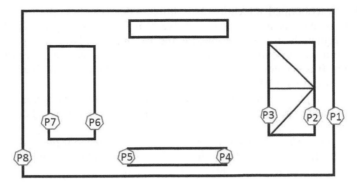

Figure 2. Point of measurement for acceleration on Shock Table.

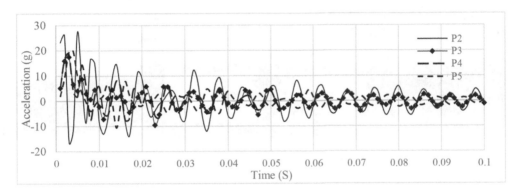

Figure 3. Numerical acceleration time history at various locations of Shock Table for pendulum swing at 20°.

Figure 3 shows the acceleration time history response of Shock Table for pendulum swing of 20° at location number 2, 3, 4, and 5.

It is seen from Figure 3 that peak acceleration of 28 g is produced at location 2.

3 EXPERIMENTAL PROGRAM

An experimental program is evolved for Shock Table facility available at Institute of Technology, Nirma University. Experimental program consists of application of shocks to the Shock Table through a pendulum weighing 1500 kg freely falling from varying degrees of swing at 10°, 20°, 30°, and 40°. Six Uniaxial and one Triaxial accelerometers are used to capture acceleration at various locations as shown in Figure 2. Acceleration time history is captured through data acquisition system and Lab VIEW software.

Figure 4 shows the complete experimental setup with instrumentation used for characterization of Shock Table.

Note that Figure 5 includes acceleration time history as given in Figure 2 for comparison. It is evident that acceleration time history shows a good agreement between experiment and FE solutions. The predominant frequency content of Shock Table subjected to impact is also determined through FFT analysis. Figure 6 shows FFT analysis of acceleration time history captured for all locations on Shock Table. It is found that predominant frequency of Shock Table is 1 Hz.

Peak acceleration obtained at location 2 for different degrees of pendulum swing by experimental and FE analyses is summarized in Table 2. It is clear that peak acceleration value shows good agreement between experimental and FE analyses. Note that peak acceleration has not been captured for pendulum swing of 40° as it has crossed the upper threshold value of accelerometer. It is evident from Table 2 that peak acceleration obtained experimentally shows good agreement with FE analysis value.

Table 2 shows comparison of maximum acceleration for different degrees of shocks.

Response obtained experimentally as well as analytically for various shocks is plotted as shown in Figure 7.

It is seen from Figure 7 that peak value of acceleration obtained is in good agreement with FE analysis results. It is also evident that the time history plot obtained experimentally shows fair agreement with the plot derived through FE analysis.

Figure 4.   Experimental setup with instrumentation for measurement.

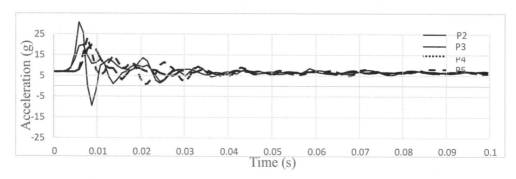

Figure 5.   Experimental acceleration time history at various locations of Shock Table for pendulum swing at 20°.

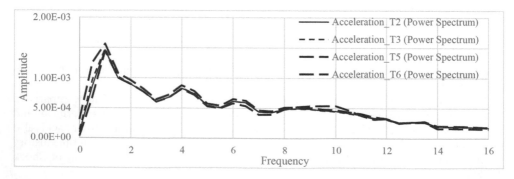

Figure 6.   FFT analysis of acceleration time history of various locations.

74

Table 2. Comparison of peak acceleration for various degrees of pendulum swing.

	Peak acceleration (g)	
Angle of awing	Experiment	FE analysis
10	9.45	10.4
20	26.35	27.08
30	35.39	50
40	–	85

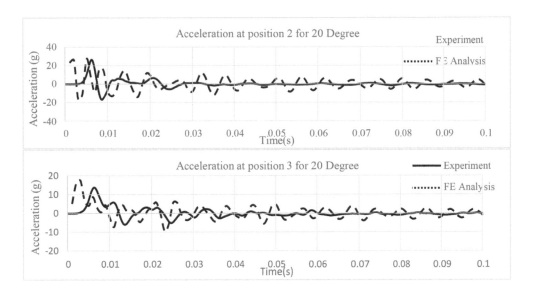

Figure 7. Comparison of experimental and FE analyses results at various locations.

4 CONCLUSION

An experimental evaluation of acceleration at various points on the Shock Table is carried out in order to characterize it. An analytical solution to determine acceleration of various points on the Shock Table is carried out using FE analysis through ANSYS. We derived the following conclusions based on the study:

- Peak acceleration value obtained through experiment and analytical solutions shows good agreement.
- Acceleration time history derived through experimental and analytical solutions shows good agreement.
- Shock Table is capable of producing a force of about 1500 kN.
- Predominant frequency of Shock Table ranges between 0.75–2 Hz.
- Shock Table has the capacity to produce peak acceleration as high as 35 g.

REFERENCES

Chopra, A.K. 3 edition, 2009. *Dynamics of structure*. Pearson Education, Inc.
Ersubasi F. & Korkmaz H., Shaking table tests on strengthening of masonry structures against earthquake hazard, *Natural Hazard and Earth System Sciences*, Volume 10 June 2010.
Purohit, S.P. & Patel, P.V., Assessment of Non-engineered structures through Shock Table Test, *Journal of Engineering and Technology*, Sardar Patel University, Volume 18 December 2005.

Multi-disciplinary Sustainable Engineering: Current and Future Trends – Tekwani, Bhavsar & Modi (Eds)
© 2016 Taylor & Francis Group, London, ISBN 978-1-138-02845-6

Effect of different parameters on the compressive strength of rubberized geopolymer concrete

Salmabanu Luhar & Sandeep Chaudhary
Malaviya National Institute of Technology, Jaipur, India

Urmil Dave
Nirma University, Ahmedabad, Gujarat, India

ABSTRACT: The main objective of this study is to develop rubberized geopolymer concrete by using the optimum content of rubber fiber. Finally, the effect of various parameters, such as ratio of sodium silicate solution to sodium hydroxide solution, ratio of alkaline liquid to fly ash, by mass, curing temperature, concentration of sodium hydroxide solution, in terms of Molar, curing time, addition of superplasticizer, water content of mix, and rest period prior to curing on compressive strength of rubberized geopolymer concrete and control geopolymer concrete, are compared. In total, 24 geopolymer concrete mixes were cast to evaluate the effect of various parameters on the compressive strength of geopolymer concrete. The test results indicate that rubberized geopolymer requires more water as compared to the control geopolymer concrete. Compressive strength increases with increase in the curing temperature, curing time, and concentration of sodium hydroxide solution during rest period and decreases with increase in the superplasticizer dosage.

Keywords: rubberized geopolymer concrete; Na_2SiO_3; NaOH; fly ash; superplasticizer

1 INTRODUCTION

Day by day, the demand of concrete is increasing which is increasing the production of cement. Cement production releases high amount of carbon dioxide gas in to the atmosphere. Around 275 MT cement was produced in India in 2014 which released approx 275 MT CO_2 in to the atmosphere.

In India, around 64% energy is produced by coal based thermal power stations which produce a huge amount of fly ash that is not effectively used. For disposal of this fly ash, several hectares of land are required. To overcome this problem, we need to utilize fly ash in construction.

Cement, the second most consumed product in the world, contributes nearly 7% of the global carbon dioxide. One ton of CO_2 is emitted into the atmosphere while manufacturing one ton OPC due to the calcination of limestone and combustion of fossil fuel [1].

Geopolymer concrete, which has unique properties, such as high early strength, low shrinkage, and sulfate and corrosion resistance, could become a viable alternative to conventional cement and hence substantially reduce CO_2 emission caused by cement and concrete industries.

Nowadays demand of natural sand has increased with increase in construction activity which results in scarcity of natural sand. However, we need to search for other alternative materials of sand that are eco-friendly and inexpensive. On the other side, disposal of waste tires creates health and environmental problems. Recycling of waste tires is difficult because of its highly complex structure. Accumulation of discarded tires in the environment is therefore a serious problem. Utilization of waste rubber tires in concrete, therefore, may be one of the better solutions for the disposal of this waste as this saves natural resources and produces a more efficient material [17].

2 MATERIALS

2.1 *Sodium hydroxide*

Sodium hydroxide (NaOH) in flake form with 98% has been used. Sodium hydroxide solution was prepared by dissolving the flakes into the water.

2.2 *Sodium silicate*

Locally available water glass type sodium silicate was used.

2.3 *Aggregate*

Aggregates of 10 mm and 20 mm in size having 2.59 specific gravity have been used as coarse aggregates. Locally available river sand of 2.56 specific gravity was used as fine aggregate in the mixes.

2.4 *Superplasticizer*

Naphthalene sulfonate based superplasticizer has been used to improve the workability of the fresh geopolymer concrete.

2.5 *Rubber fiber*

Rubber fibers of specific gravity 1.07 were used in this study. Ten percent of rubber fibers were partially replaced as fine aggregates.

3 MIX PROPORTION FOR GEOPOLYMER CONCRETE

	Unit	Mix-1		Mix-2		Mix-3		Mix-4		Mix-5		Mix-6		Mix-7		Mix-8	
		A	B	A	B	A	B	A	B	A	B	A	B	A	B	A	B
Fly ash	Kg/m³	442.31	442.31	425.93	425.93	410.71	410.71	410.71	410.71	410.71	410.71	410.71	410.71	410.71	410.71	410.71	410.71
F.A	Kg/m³	673.75	606.37	673.75	673.75	673.75	673.75	673.75	673.75	673.75	673.75	673.75	673.75	673.75	673.75	673.75	673.75
Rubber fibre	Kg/m³		26.318		26.318		26.318		26.318		26.318		26.318		26.318		26.318
C.A 20 mm	Kg/m³	750.75	750.75	750.75	750.75	750.75	750.75	750.75	750.75	750.75	750.75	750.75	750.75	750.75	750.75	750.75	750.75
10 mm		500.50	500.50	500.50	500.50	500.50	500.50	500.50	500.50	500.50	500.50	500.50	500.50	500.50	500.50	500.50	500.50
Na₂SiO₃	Kg/m³	88.46	88.46	99.38	99.38	109.52	109.52	98.57	98.57	109.52	109.52	117.35	117.35	109.52	109.52	109.52	109.52
NaOH	Kg/m³	44.23	44.23	49.69	49.69	54.76	54.76	65.71	65.71	54.76	54.76	46.94	46.94	54.76	54.76	54.76	54.76
Extra water	Kg/m³	88.46	88.46	85.19	85.19	82.14	82.14	82.14	82.14	82.14	82.14	82.14	82.14	82.14	82.14	82.14	82.14
Super-plasticizer	Kg/m³	8.85	8.85	8.52	8.52	8.21	8.21	8.21	8.21	8.21	8.21	8.21	8.21	8.21	8.21	8.21	8.21
Varying parameter	–	Alkaline/fly ash ratio –0.3		Alkaline/fly ash ratio –0.35		Alkaline/fly ash ratio –0.4		Na₂ sio₃/NaoH ratio –1.5		Na₂ sio₃/NaoH ratio –2.0		Na₂ sio₃/NaoH ratio –2.5		Curing time –24 hr		Curing time –48 hr	

	Unit	Mix-9		Mix-10		Mix-11		Mix-12		Mix-13		Mix-14		Mix-15		Mix-16	
		A	B	A	B	A	B	A	B	A	B	A	B	A	B	A	B
Fly ash	Kg/m³	410.71	410.71	410.71	410.71	410.71	410.71	410.71	410.71	410.71	410.71	410.71	410.71	410.71	410.71	410.71	410.71
F.A	Kg/m³	673.75	606.375	673.75	673.75	673.75	673.75	673.75	673.75	673.75	673.75	673.75	673.75	673.75	673.75	673.75	673.75
Rubber fibre	Kg/m³		26.318		26.318		26.318		26.318		26.318		26.318		26.318		26.318
C.A 20 mm	Kg/m³	750.75	750.75	750.75	750.75	750.75	750.75	750.75	750.75	750.75	750.75	750.75	750.75	750.75	750.75	750.75	750.75
10 mm		500.50	500.50	500.50	500.50	500.50	500.50	500.50	500.50	500.50	500.50	500.50	500.50	500.50	500.50	500.50	500.50
Na₂SiO₃	Kg/m³	109.52	109.52	109.52	109.52	109.52	109.52	109.52	109.52	109.52	109.52	109.52	109.52	109.52	109.52	109.52	109.52
NaOH	Kg/m³	54.76	54.76	54.76	54.76	54.76	54.76	54.76	54.76	54.76	54.76	54.76	54.76	54.76	54.76	54.76	54.76
Extra water	Kg/m³	82.14	82.14	82.14	82.14	82.14	82.14	82.14	82.14	82.14	82.14	82.14	82.14	82.14	82.14	61.61	61.61
Super-plasticizer	Kg/m³	8.21	8.21	12.32	12.32	16.43	16.43	8.21	8.21	8.21	8.21	8.21	8.21	8.21	8.21	8.21	8.21
Varying parameter	–	Curing time –72 hr		Superplasticizer –2.0%		Superplasticizer –3.0%		Superplasticizer –4.0%		Rest period –0 days		Rest period –1 days		Rest period –2 days		Extra water content –15%	

	Unit	Mix-17		Mix-18		Mix-19		Mix-20		Mix-21		Mix-22		Mix-23		Mix-24	
	Kg/m³	A	B	A	B	A	B	A	B	A	B	A	B	A	B	A	B
Fly ash	Kg/m³	410.71	410.71	410.71	410.71	410.71	410.71	410.71	410.71	410.71	410.71	410.71	410.71	410.71	410.71	410.71	410.71
F.A	Kg/m³	673.75	606.375	673.75	673.75	673.75	673.75	673.75	673.75	673.75	673.75	673.75	673.75	673.75	673.75	673.75	673.75
Rubber fibre	Kg/m³	–	26.318	–	26.318	–	26.318	–	26.318	–	26.318	–	26.318	–	26.318	–	26.318
C.A 20 mm	Kg/m³	750.75	750.75	750.75	750.75	750.75	750.75	750.75	750.75	750.75	750.75	750.75	750.75	750.75	750.75	750.75	750.75
10 mm		500.50	500.50	500.50	500.50	500.50	500.50	500.50	500.50	500.50	500.50	500.50	500.50	500.50	500.50	500.50	500.50
Na$_2$SiO$_3$	Kg/m³	109.52	109.52	109.52	109.52	109.52	109.52	109.52	109.52	109.52	109.52	109.52	109.52	109.52	109.52	109.52	109.52
NaOH	Kg/m³	54.76	54.76	54.76	54.76	54.76	54.76	54.76	54.76	54.76	54.76	54.76	54.76	54.76	54.76	54.76	54.76
Extra water	Kg/m³	82.14	82.14	102.68	102.68	82.14	82.14	82.14	82.14	82.14	82.14	82.14	82.14	82.14	82.14	82.14	82.14
Super-plasticizer	Kg/m³	8.21	8.21	8.21	8.21	8.21	8.21	8.21	8.21	8.21	8.21	8.21	8.21	8.21	8.21	8.21	8.21
Varying parameter	–	Extra water content –20%		Extra water content –25%		NaoH concentration –10M		NaoH concentration –12M		NaoH concentration –14M		Curing temperature –60°C		Curing temperature –75°C		Curing temperature –90°C	

4   PRELIMINARY INVESTIGATION

Parameters affecting the performance of geopolymer concrete and rubberized geopolymer concrete are as follows:

1. Alkaline liquid to fly ash ratio, by mass.
2. Na_2Sio_3 to NaOH ratio, by mass.
3. Curing time
4. Addition of superplasticizer
5. Rest period prior to curing
6. Water content of mix
7. Concentration of NaOH solution, in terms of Molar.
8. Curing temperature

5   PARAMETER STUDY

5.1   *Alkaline liquid to fly ash ratio*

The effect of alkaline liquid to fly ash ratio by mass on compressive strength of concrete at 7 days has been observed by comparing results of Mix 1, Mix 2, and Mix 3. From Figure 1 it is observed that the ratio of alkaline liquid to fly ash does not much affect the compressive strength of the control and rubberized geopolymer concrete.

5.2   *Na_2Sio_3 to NaOH ratio, by mass*

The effect of Na_2Sio_3 to NaOH ratio, by mass on compressive strength of concrete has been observed by comparing results of Mix 4, Mix 5, and Mix 6. Figure 2 shows that Na_2Sio_3 to NaOH ratio of 2.0 gives the highest compressive strength as compared to the ratios of 1.5 and 2.5.

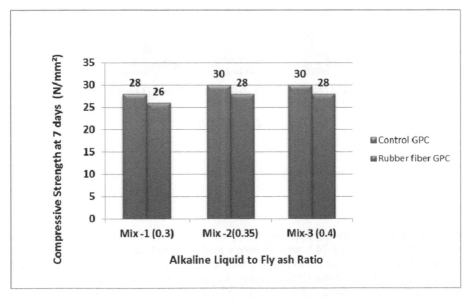

Figure 1.   Effect of alkaline liquid to fly ash ratio on compressive strength.

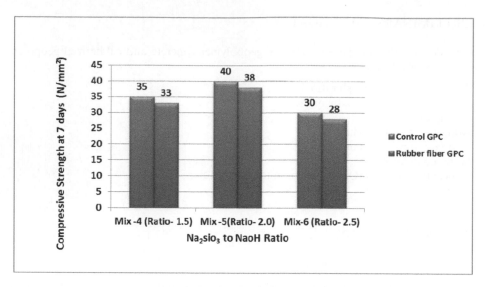

Figure 2. Effect of sodium silicate solution to sodium hydroxide solution ratio on compressive strength.

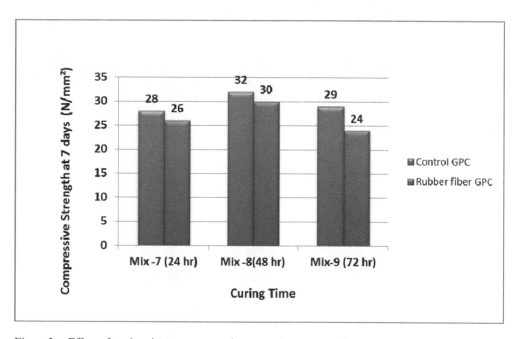

Figure 3. Effect of curing time on compressive strength.

5.3 *Curing time*

To evaluate the effect of curing time on geopolymer concrete, $100 \times 100 \times 100$ mm cubes were cured for curing periods of 24 hours, 48 hours, and 72 hours, respectively. Concrete Mix 7, Mix 8, and Mix 9 were cast for evaluating the curing time on the compressive strength of geopolymer concrete and rubberized concrete. As per the figure, 48 hrs curing time improved the polymerization process which results in higher development of compressive strength of control and rubberized geopolymer concrete.

5.4 Addition of superplasticizer

The effect of superplasticizer on compressive strength of control and rubberized geopolymer concrete has been evaluated by comparing the results of 2%, 3%, and 4% dosage of admixture.

It has been observed that the superplasticizer increases the workability of control and rubberized geopolymer concrete as shown in Figure 4. Concrete Mix 8 with 2% dosage of superplasticizer achieved higher compressive strength as compared to the concrete mixes 11 and 12. Thus, the addition of naphthalene based superplasticizer has been able to increase the workability of geopolymer concrete. However, higher dosage of admixture up to 4% has resulted into reduction of the compressive strength of the geopolymer concrete.

5.5 Rest period prior to curing

"Rest Period" is the time between the completion of casting of geopolymer concrete specimens to the curing of geopolymer concrete specimen.

The effect of rest period of geopolymer concrete was evaluated by Mix 13, 14, and 15. The results indicate that a one-day rest period gives higher compressive strength of geopolymer concrete as compared to that for 0 days and 2 days rest period.

5.6 Water content of Mix

The chemical reaction between the alkaline solution and source materials produces water that is expelled from the binder. However water does not take any part in the chemical reaction of geopolymer [10].

The effect of water content on compressive strength of control and rubberized geopolymer concrete can be observed by comparing results of 15%, 20%, and 25% water content. The water requirement is more for rubberized geopolymer concrete as compared to the control geopolymer concrete.

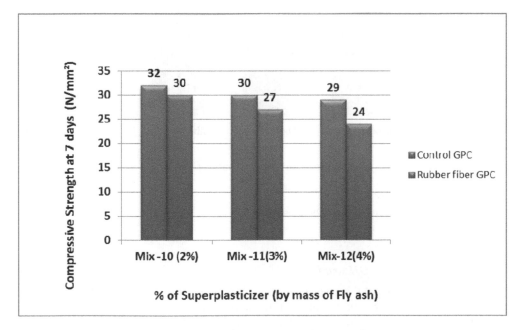

Figure 4.   Effect of addition of superplasticizer on compressive strength.

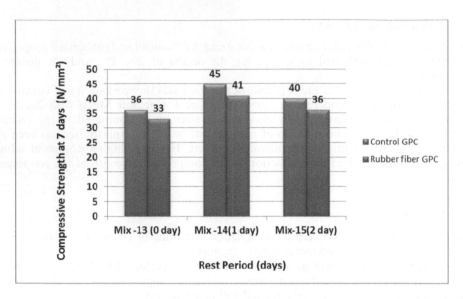

Figure 5   Effect of rest period on compressive strength.

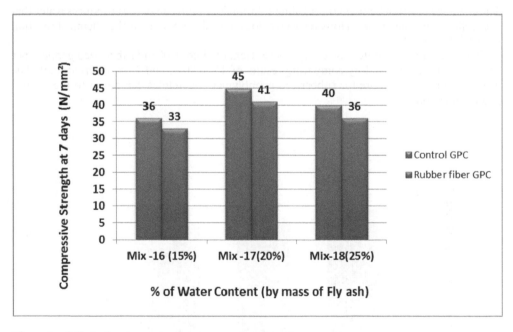

Figure 6.   Effect of water content on compressive strength.

5.7   *Concentration of NaOH solution*

The difference between the variation in the molar concentration of NaOH solution of 10 M, 12 M, and 14 M has been evaluated. It can be concluded that a higher concentration of NaOH solution gives higher compressive strength of control and rubberized geopolymer concrete.

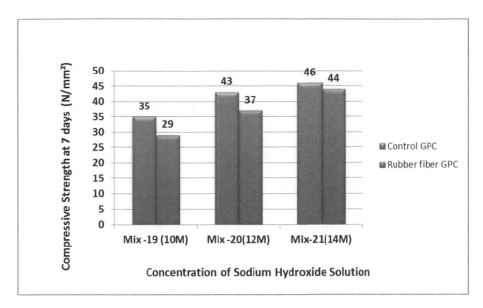

Figure 7.   Effect of concentration of sodium hydroxide solution on compressive strength.

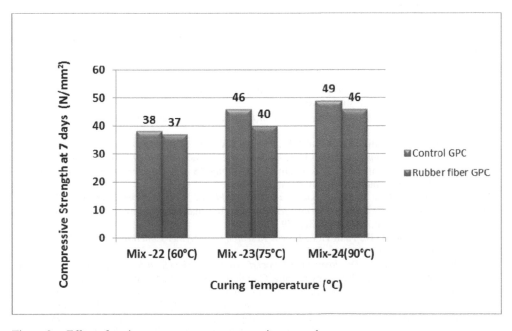

Figure 8.   Effect of curing temperature on compressive strength.

5.8   *Curing temperature*

The effect of curing temperature on compressive strength of geopolymer concrete has been evaluated by comparing results of concrete mixes 22, 23, and 24.

Results indicate that higher curing temperature achieved higher compressive strength for control and rubberized geopolymer concrete. Curing temperatures beyond 75°C caused minor increment in the compressive strength.

6 CONCLUSION

1. The alkaline liquid to fly ash ratio, by mass, does not affect the compressive strength of the control and rubberized geopolymer concrete.
2. The Na_2Sio_3 to NaOH ratio of 2.0 gives higher compressive strength as compared to the other ratios.
3. The increase in NaOH increased the compressive strength of geopolymer concrete.
4. The increase in curing temperature and curing time increased the compressive strength of control and rubberized geopolymer concrete.
5. The compressive strength of rubberized geopolymer concrete increases with increase in the curing time.
6. The compressive strength of geopolymer concrete increases with increased rest periods up to some extent. Results show that a one-day rest period give better results as compared to zero- and two-day rest periods.
7. The workability of the control geopolymer concrete is more as compared to that of rubberized concrete.

Rubberized geopolymer concrete gives similar strength as compared to control fly ash based geopolymer concrete. Hence, practical use of rubberized geopolymer concrete can serve as the solution to two major problems, i.e., disposal of fly ash and waste rubber and carbon dioxide emission into the atmosphere.

REFERENCES

[1] Malhotra V.M., *Introduction: Sustainable development and concrete technology*, ACI Concrete International, 24(7), 2002.
[2] Mehta P.K., *Reducing the Environmental impact of Concrete*, Concrete International, 23(10), 2001, Pp: 61–66.
[3] Davidovits J., *Chemistry of geopolymeric systems, terminology*, Geopolymer 99 international conference, France, 1998, Pp: 3077–3085.
[4] Malhotra V.M., *Making concrete greener with fly ash*, ACI concrete International, 21, May 1999, Pp: 61–66.
[5] Sumajouw D.M., Hardjito D., Wallah S.E., Rangan B.V., *Fly ash-based geopolymer concrete: study of slender reinforced columns*, Advances in geopolymer science and technology, 2006, Pp: 3124–3130.
[6] Rajamane N.P., nataraja M.C., Lakshmanan N., Ambily P.S., *Geopolymer concrete- An ecofriendly concrete*, The Masterbuilder, 11, 2009, Pp: 200–206.
[7] Chakravarthi M., Bhat V., *Utility bonanza from dust -Fly ash*, ENVIS Newsletter, 6(2), 2007.
[8] Davidovits J., *Properties of Geopolymer cements* First international conference on alkaline cements and concretes, Ukrain, 1994, Pp: 131–149.
[9] Chanh N.V., Trung D.B., Tuan D.V., *Recent Research Geopolymer Concrete*, The 3rd ACF international conference, 2008, Pp: 235–241.
[10] Hardjito D., Rangan B.V., *Development and properties of low-calcium fly-ash- based geopolymer concrete, Research Report GC1*, Faculty of engineering, Curtain University, Perth, Australia, 2005.
[11] Wallah S.E., Rangan B.V., *Low-calcium fly ash-based geopolymer concrete: Long term properties, Research Report GC3*, Faculty of engineering, Curtain University, Perth, Australia, 2006.
[12] Devidovits J., *30 years of successes and failures in geopolymer applications. Market trends and potential breakthroughs*, Geolymer 2002 conference, Melbourne, Australia, October 2002.
[13] Balaguru P., *Geopolymer for repair and rehabilitation of reinforced concrete beams*, Geopolymer, 1997.
[14] Lloyd N.A., Rangan B.V., *Geopolymer concrete with fly ash*, SS econdinternational conference on sustainable construction material and technologies, 2010.
[15] Cheema D.S., Lloyd N.A., Rangan B.V., *Durability of geopolymer concrete box culverts-A green Alternative*, Proceedings of the 34*th* Conference on our world in concrete and structures, Singapore, 2009.
[16] Hardjito D., Wallah E., Sumajouw D.M.J., Rangan B.V., *Factors influencing the compressive strength of fly ash-based Geopolymer concrete,* Civil engineering dimension, No. 2, Vol. 6, 2004, Pp: 88–93.
[17] Gupta T., Chaudhary S., *Assessment of mechanical and durability properties of concrete containing waste rubber tire as fine aggregate*, Construction and Building Materials, 2014, Pp: 562–574.

Multi-disciplinary Sustainable Engineering: Current and Future Trends – Tekwani, Bhavsar & Modi (Eds)
© 2016 Taylor & Francis Group, London, ISBN 978-1-138-02845-6

Experimental study on dry precast beam column connections under column removal scenario

Dhaval D. Patel, Digesh D. Joshi & Paresh V. Patel

Department of Civil Engineering, Institute of Technology, Nirma University, Ahmedabad, Gujarat, India

ABSTRACT: Progressive collapse results in substantial loss of human lives and natural resources due to its catastrophic nature. Therefore, it is important to prevent progressive collapse in structures which is also referred as disproportionate collapse. Recently, there is an increasing trend toward construction of buildings using precast concrete. Connections are the most critical elements of any precast structure because in the past, a major collapse of precast structure took place due to connection failure. In this study, the behavior of three different one-third scaled dry precast concrete beam column connections under progressive collapse scenario is examined and their performance is compared with monolithic connection. Precast connections are constructed by adopting different connections detailing at the junction. Each test specimen includes two span beams and three columns with the middle column removed, which actually represents a progressive collapse scenario. Performance of the specimen is evaluated on the basis of ultimate load carrying capacity, maximum deflection, and deflection measured along the span of the beam. From the results, it is observed that ultimate load carrying capacity of dry precast connections, considered for the study, is lower than that of monolithic connections, which necessitates improved detailing of precast connections to enhance their performance during progressive collapse scenario.

1 INTRODUCTION

Progressive collapse of building structures is initiated when one or more vertical load carrying members are seriously damaged or collapse during any of the abnormal events. Once a local failure takes place, the building's gravity load transfers to neighboring members in the structure. If these members are not properly designed to resist and redistribute the additional load, that part of the structure fails. As a result, a substantial part of the structure may collapse causing greater damage to the structure than the initial impact. Thus, it is necessary to provide sufficient redundancy, ductility, and continuity, which helps the structure to find alternate paths for load distribution during undesired failure events and thus reduce progressive collapse.

Nowadays, there is an increasing trend toward construction of buildings using precast concrete. In precast concrete construction, all the components of structures are produced in controlled environments and they are being transported to the site. At the site, such individual component is connected appropriately. This leads to faster construction, reduced formwork and scaffolding, less requirement of skilled labors, massive production with reduced amount of construction waste, better quality. and better surface finishing as compared to normal reinforced concrete construction. Because of such advantages, precast concrete construction is being adopted worldwide including India. In precast concrete construction, connections are the most critical elements of the structure because in the past, major collapse of precast buildings took place due to connection failure. Therefore, it is very important to study the performance precast beam column connections under a progressive collapse scenario.

Many authors have performed experiments to study the behavior of RCC or steel frame under the column removal scenario. However, performance of precast connections under progressive collapse scenario is not explored in detail by researchers. Progressive collapse resistance of precast concrete buildings was examined by Main et al. (2014) through experimental and analytical investigations on full scale test specimens. In their study, precast components were connected using steel link plates that were welded to steel angles embedded in precast beams and steel plates embedded in precast columns. Kang and Tan (2015) have performed experiments to examine the behavior of precast concrete beam-column subassemblages under column removal scenario. The effect of reinforcement detailing and reinforcement ratio on performance of connections were also studied by the author. Nimse et al. (2014 & 2015) investigated the performance of different reduced scale precast beam column connections under progressive collapse scenario. Performance of wet and dry precast connections constructed by adopting different connection detailing under progressive collapse scenario was studied by authors.

Kai & Li (2012) carried out experimental and analytical studies of progressive collapse resistance on four full scale RC beam column assemblies, which were part of an eight-storey building. Behavior of steel and RC beam column assemblies with different seismic design and detailing under a progressive collapse scenario was studied by Sadek et al. (2011) and Lew et al. (2013) through experimental and analytical investigations. Progressive collapse resistance of RC structure under column removal scenario was examined by Yu and Tan (2013). Experimental studies were carried out on reduced scaled specimens, such as beams, and beam-column assemblies prepared with different design and detailing to observe the behavior under progressive collapse scenario by authors.

2 CONNECTIONS DETAILING CONSIDERED FOR THE STUDY

A six-storey symmetrical building with a rectangular plan is considered for the study. The building has four bays in longitudinal direction and three bays in transverse direction with 4-m c/c spacing in each direction. Overall plan dimensions of the building are 16 m × 12 m. A typical floor height of building is 3.2 m with height at the bottom story equal to 3.5 m. Perimeter frames of any building are exposed to the highest risk of occurrence of any undesired event due to ease of accessibility. Therefore, the prototype of test specimens is assumed to be located at the middle of the perimeter frame in a longer direction subjected to column loss at the bottom storey. Dimensional analysis has been carried out to establish sizes of specimen. Each test specimen contains two span beam and three columns with removed middle column as shown in Figure 1. Removed middle column represents the progressive collapse scenario.

Two column stubs of 900 mm height and 135 mm × 135 mm cross section are provided at the ends of specimen for providing sufficient anchorage for the longitudinal reinforcement. The beam is having a cross-section dimension equal to 100 mm × 135 mm. Total length of test specimen is kept as 2700 mm. Design and detailing of all specimens are carried out according to relevant Indian standards. Reinforcement detailing of a specimen with monolithic connection is shown in Figure 1, which is very much similar to detailing of precast test specimens.

In precast dry connection-1 (PC-D-CAO) and precast dry connection-2 (PC-D-CAD), column with RC corbel and beam are cast separately. For precast dry connection-1 (PC-D-CAO), precast

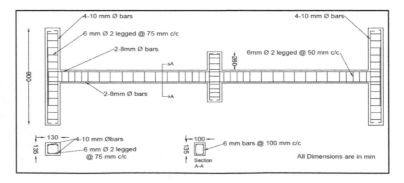

Figure 1.   Reinforcement detailing of test specimen—Monolithic Connection (MC).

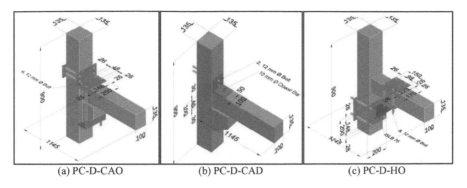

| (a) PC-D-CAO | (b) PC-D-CAD | (c) PC-D-HO |

Figure 2. Typical connection detailing for precast specimens.

elements are connected by means of cleat angles through 12-mm diameter bolts passing from the outer side of precast elements to avoid provision of holes in concrete at the time of casting. For precast dry connection-2 (PC-D-CAD), a dowel bar of 10 mm diameter is inserted in column through concrete corbel at the time of casting. A precast beam is placed on this dowel bar and subsequently it is connected with a column using cleat angles. For precast dry connection-3 (PC-D-HO), a column without a concrete corbel and beam is cast separately and they are connected using ISLB 75 as haunch section. The typical connection details for all the precast specimens are shown in Figure 2. M25 grade of concrete and Fe415 grade of steel is used for casting of all specimens.

3 TEST SETUP AND INSTRUMENTATION

Schematic diagram of the test setup is shown in Figure 3. To simulate the exact condition as in the prototype building, two triangle frames are fabricated to prevent horizontal movement of the end of column. These triangle frames are attached with the existing loading frame that enables them to transfer the load from the column to existing loading frame. The end columns are restrained vertically by providing equal reactive force through the hydraulic jack at bottom of it. The load is applied at the top of the removed middle column with the help of the hydraulic jack of capacity 500 kN till the complete failure of specimen takes place. The response of specimen under column removal scenario is observed in terms of vertical deflection measured along the span of the beam at seven different locations. The arrangement of the instrumentation system to measure vertical deflection consists of two different parts: (i) five dial gauges placed along the span of the beam, i.e., near both end junctions, near both the sides of middle junction and at center of right end beam and (ii) two Linear Variable Differential Transducers (LVDT) at the location of removed middle column and at the center of left end beam.

4 RESULTS AND DISCUSSION

Deflection measured along the span of the beam for monolithic and precast specimens at different loads are presented in Figure 4. These curves indicate the deflected shape of all specimens at different loading. The deflected shape of the beam at various loads gives an indication about the fixity of connection. During the initial phase of loading, deflection at junction is very less and the specimen almost behaves as fixed beam. The change in the curve pattern revealed that beam column connection at both the extreme ends gradually loses fixity as load is increased. It is also observed that the specimen behaves in an almost symmetrical manner with maximum displacement at the location of removed column.

For Monolithic Connections (MC), the minor cracks are observed at a load of 8 kN on the bottom side of right beam near the location of removed middle column is shown in Figure 5. As crack is initiated near the right side of middle junction; the deflection of beam toward

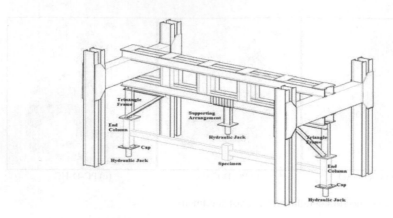

Figure 3.   Schematic diagram of test setup.

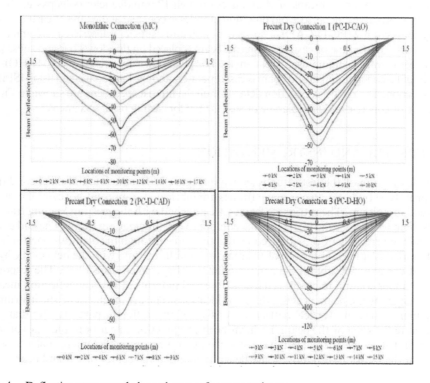

Figure 4.   Deflections measured along the span for test specimens.

that side is slightly more as compared deflection observed at the other side. Tension cracks are also developed on the top surface near both the extreme end beam-column junction. As the load increases, cracks continued to propagate till complete failure of specimen. Specimen failed at load of 17 kN and not capable to resist further load but deflection still continued as reinforcements are contributing in resisting axial tensile forces developed in the beam. After maximum load carrying capacity is reached, dial gauges and LVDT are taken out and deflection is measured with the help of a measuring scale. Crushing of concrete is also observed at the top surface just at the right side of the location of the removed middle column. At deflection of 99.4 mm, two longitudinal reinforcement bars provided at the bottom ruptured just at the right side of the middle junction, which indicates development of catenary action.

For precast dry connection-1 (PC-D-CAO), flexural cracks are initiated at the bottom of the beam near the left beam-column junction at 5 kN load. As the load is increased, flexural

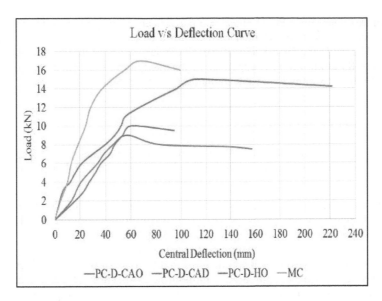

Figure 5.   Load v/s Deflection Curve.

Table 1.   Comparison of maximum load and corresponding deflection.

Specimen	Maximum Load (kN)	% Decrease in load	Deflection corresponding to maximum load (mm)	% Difference	Maximum Deflection (mm)	% Difference
MC	17	–	68.1	–	99.4	–
PC-D-CAO	10	41.18	60.2	−11.60	94.3	−5.13
PC-D-CAD	9	47.06	82.5	21.15	157.5	58.45
PC-D-HO	15	11.76	111.7	64.02	221.7	123.04

cracks are propagated in the upward direction from the bottom of the beam. At the 10-kN load, both ends of the beam failed because of expansion of flexural crack near the extreme beam-column junction. After that point, the deflection of the specimen increases without much increase in the load and maximum deflection of 94.3 mm is observed for the specimen.

For precast dry connection-2 (PC-D-CAD), minor cracks are occurred at a load of 6 kN, at the bottom of the beam near the middle beam-column junction. The specimen is capable to resist maximum load of 9 kN with corresponding deflection at 82.5 mm. As in this specimen, only a single dowel bar is used for connection; it has lowest load carrying capacity. For this specimen, cracks are also observed in column. Though the specimen is not capable to resist further load after 9 kN, the deflection continues, and it deforms up to 157.5 mm.

For precast dry connection-3 (PC-D-HO), minor cracks are observed near the bottom of the right end column at 6 kN load. As load is increased, shear cracks are propagated diagonally in the upward direction from the bottom of the end column. At the 8-kN load, crushing of concrete is observed in both end columns as well as at the top of the beam near the middle beam-column junction. As the load is increased, the beam near the end column is moved away from the column because they are connected by the haunch section and there is no specific connection near the beam-column junction. After a load of 9 kN, bending of steel plate is observed near the end column. The specimen PC-D-HO failed at a load of 15 kN with corresponding deflection at 111.7 mm. Though the specimen is not capable to resist further load after 15 kN, deflection continues, and it deforms up to 221.7 mm.

The comparison of load v/s central deflection for all the four specimens is shown in Figure 5. From the graph, it is observed that central deflection measured at the location of the removed column is more for specimen PC-D-HO as compared to all other specimens. The graph also indicates that precast connections adopted for the study are flexible as compared to the monolithic connection. The load carrying capacity of precast connections needs to be improved by ensuring proper fixity at the connections. Comparison of the maximum value of load and deflection corresponding to the maximum load value for all the specimens is presented in Table 1.

5 CONCLUSION

Based on the experimental study carried out to investigate the behavior of different dry precast beam column connections during progressive collapse condition, following conclusions are drawn:

- From the comparison of maximum load, it has been seen that the load carrying capacity of dry precast connections is less than that of monolithic connection. Load carrying capacity of specimen PC-D-CAO, PC-D-CAD, and PC-D-HO is 41.18%, 47.06%, and 11.76% less respectively, as compared to monolithic connection.
- From the results of deflections, it is concluded that the type of precast connections adopted for the study are flexible as compared to monolithic connection. Deflection corresponding to maximum load for specimen MC is 11.6% more as compared to specimen PC-D-CAO, while it is 21.15% and 64.02% less as compared to specimen PC-D-CAD and PC-D-HO, respectively.
- From the results of current study, it is observed that precast connections adopted for the study has less progressive collapse resistance compared to monolithic connection. However, to use precast connections, improved detailing of connection should be developed to enhance performance during progressive collapse scenario.

ACKNOWLEDGMENT

The authors gratefully acknowledge the financial support through research project number SB/S3/CEE/0028/2013, provided by Science & Engineering Research Board (SERB), Department of Science & Technology (DST), New Delhi, India.

REFERENCES

Kai, Q. & Li, B. 2012. Experimental and Analytical Assessment on RC Interior Beam-Column Sub-assemblages for Progressive Collapse. *Journal of Performance Constructed Facilities* 26: 576–589.

Kang, S.B. & Tan, K.H. 2015. Behaviour of precast concrete beam-column sub-assemblages subject to column removal. *Engineering Structures* 93: 85–96.

Lew, H.S., Main, J., Robert, S., Sadek, F. and Chiarito, V. 2013. Performance of Steel Moment Connections under a Column Removal Scenario. I: Experiments. *Journal of Structural Engineering* 139: 98–107.

Main, J., Bao, Y., Lew, H.S. & Sadek, F. 2014. Robustness of Precast Concrete Frames: Experimental and Computational Studies. *The Structures Congress 2014; Proc. Intern. Conf., Boston, 3–5 April 2014*. Massachusetts: USA.

Nimse, R.B., Joshi, D.D. & Patel, P.V. 2014. Behavior of wet precast beam column connections under progressive collapse scenario: an experimental study. *International Journal of Advanced Structural Engineering* 6: 149–159.

Nimse, R.B., Joshi, D.D. & Patel, P.V. 2015. Experimental Study on Precast Beam Column Connections Constructed Using RC Corbel and Steel Billet under Progressive Collapse Scenario. *The Structures Congress 2015; Proc. Intern. Conf., Oregon, 23–25 April 2015*. Portland: USA.

Sadek, F., Main, J., Lew, H.S., & Bao, Y. 2011. Testing and Analysis of Steel and Concrete Beam-Column Assemblies under a Column Removal Scenario. *Journal of Structural Engineering* 137: 881–892.

Yu, J. & Tan, K. 2013. Experimental and numerical investigation on progressive collapse resistance of reinforced concrete beam column sub-assemblages. *Engineering Structures* 55: 90–106.

Durability studies of fly ash and GGBFS based geopolymer concrete at ambient temperature

S. Thakkar & T. Pandit
Institute of Technology, Nirma University, Ahmedabad, Gujarat, India

ABSTRACT: Concrete usage around the globe is second only to water. An important ingredient in the conventional concrete is the Portland cement. One ton of cement emits approximately one ton of carbon dioxide in the atmosphere. Moreover, cement production is not only highly energy-intensive, but also consumes significant amount of natural resources. Therefore, it is essential to find alternatives to concrete for sustainable development. Geopolymer concrete is one such attempt. In this paper, geopolymer concrete is prepared by using fly ash and slag as source material. Medium strength ambient cured geopolymer concrete was prepared and is subject to durability tests like Chloride resistance, sorptivity and accelerated corrosion test. Durability studies of geopolymer concrete was compared with Ordinary Portland cement concrete specimen of same strength. It is found that geopolymer concrete had better durability properties as compared to OPC based concrete.

1 INTRODUCTION

Concrete usage around the world is second only to water. Cement is one of the ingredient of concrete. It is produced in huge quantity due to easily available raw material. This has lead to extensive usage of Ordinary Portland Cement (OPC) concrete. About 1.5 tons of raw materials is needed in the production of every ton of OPC. At the same time about the amount of carbon dioxide released during the manufacturing of Ordinary Portland Cement (OPC) and due to the calcinations of limestone and combustion of fossil fuel is in the order of one ton for every ton of OPC produced (Malhotra, 2002). Growth of any nation is judged by the power it consumes. In India, power is mainly available by burning of coal, which produces a by-product called fly ash. The abundant availability of fly ash worldwide creates opportunities to utilize this as a substitute for OPC to manufacture concrete. When used as a partial replacement of OPC, in the presence of water and in ambient temperature, fly ash reacts with the calcium hydroxide during the hydration process of OPC to form the calcium silicate hydrate (C-S-H) gel. To produce more environmentally friendly concrete, need is to replace the amount of cement in concrete with materials admixtures such as fly ash.

Geopolymer concrete is an alternative to OPC concrete. It is environmentally friendly concrete developed by inorganic alumina-silicate polymer, which is synthesized from materials of geological origin or by-product materials such as fly ash and slag that is rich in silicon and aluminum (Davidovitis, 1994). By using the fly ash based geopolymer, two issues related to environment i.e. high amount of CO_2 release during production of cement and utilization of fly ash waste is solved. Also other industrial waste like slag can be used to produce a useful material. Drawback of fly ash geopolymer concrete is that it requires heat curing to gain early strength. This can be negated when strength at ambient curing is achieved.

2 EXPERIMENTAL PROGRAM

Fly ash and slag were taken as source material for mix design as ambient curing was attempted. The grade of concrete was kept M25. To achieve this characteristic strength, various percentage

Table 1. Properties of fly ash.

Properties	Unit	Result
Total sulphur as sulphur trioxide (SO$_3$)	%	0.62
Available alkalies as sodium oxide	%	0.74
Silicon dioxide as SiO$_2$	%	60.12
SiO$_2$+Al$_2$O$_3$+Fe$_2$O$_3$	%	91.44
Reactive silica	%	34.26
Magnesium oxide	%	1.90
Total chlorides	%	0.02

Table 2. Mix design of concrete.

Contents	Geopolymer Concrete	Control Concrete	Unit
Cement	–	383.2	kg/m^3
Fine Aggregate	630.5	630.5	kg/m^3
Coarse Aggregate 20 mm	702	733.2	kg/m^3
Coarse Aggregate 10 mm	600	488.8	kg/m^3
Fly Ash and GGBFS	428	–	kg/m^3
Na$_2$SiO$_3$/NaOH	2.5	–	–
Mass of (GGBFS+ fly ash) + alkaline liquid	0.4	–	–
Sodium Silicate	114.2	–	kg/m^3
Sodium Hydroxide	49.0	–	kg/m^3
Extra Water	42.9	–	kg/m^3
Super plasticizer	4.3	–	kg/m^3

combination of source material was tried and it was found that equal amount of slag and fly ash gave the required strength at 28 days at ambient curing for geopolymer concrete (Thakkar, 2014). Fly ash was procured from Ukai power plant in Gujarat. Table 1 shows properties of fly ash. Slag used was procured from JSW steel cement plant. Ground Granulated Blast Furnace Slag (GGBFS) had 37% calcium oxide content while 35% of silica content.

Combination of Sodium hydroxide and sodium silicates were used as alkaline activators. Sodium hydroxide was obtained in flakes form and solution was prepared by dissolving it in water depending upon the molarity of solution required. Sodium silicate was obtained in form of solution. Locally available river sand was used as a fine aggregate and coarse aggregates of two size were used in the concrete i.e. 10 mm and 20 mm. Napthelene Sulphonate based super plasticizer, Rehobuild 1125 from BASF was used to increase workability of geopolymer concrete. The mix design for geopolymer concrete and control concrete is given in Table 2. Control concrete follows mix design guidelines as per IS 10262: 2009.

Fly ash, GGBFS and the aggregate were mixed together in concrete pan mixture. The mixing was allowed to continue for about 3 to 4 minutes. The alkaline solution which was prepared one day before was added with additional water in the mix. Then liquid component and super plasticizer has been added to the dry material and mixing was continued for another 3 to 4 minutes. The fresh concrete was cast into the molds immediately after mixing in three layers (Hardjito, 2005).

3 TEST FOR DURABILITY

3.1 *Chloride resistance test*

The chloride attack is generally in marine structures and penetration of chloride into concrete causes corrosion in reinforcement. Chloride resistance was evaluated on cubes of size 150 mm × 150 mm × 150 mm. The control concrete and geopolymer concrete cubes were immersed in 3% concentration Sodium chloride solution for different exposure period and observation was done for change in mass and change in compressive strength for both type of concrete.

3.2 Sorptivity test

The sorptivity test is a simple and rapid test to determine the tendency of concrete to absorb water by capillary action. In this test 3 cylinders each of 100 mm diameter with 50 mm depth were prepared for both geopolymer and control concrete. They were first over dried and then the sides of cylinders are coated with epoxy coating or N.C. putty so that water absorption from sides does not occur. After that, initial weight of the cylinders was first taken and then cylinders were immersed in 5 to 10 mm water depth and the samples were removed at selected time interval like 1, 2,3,4,5,9,12,15,20,25 minutes. The excess water was wiped away. Then gain in mass per unit area over the density of water was plotted versus the square root of the elapsed time (Adam, 2009).

3.3 Accelerated corrosion test

Reduction of life of structure occurs in concrete due to reinforcement corrosion, which causes spalling and cracking. For coastal and marine structures, resistance of corrosion is an important factor. This test helps in evaluating corrosion resistance performance of control concrete and geopolymer concrete. The cylinder size of 150 mm × 300 mm were casted of control concrete and geopolymer concrete. Cylinders were immersed in tank containing sodium chloride solution of 5% concentration up to 2/3 height after curing it for 28 days. The steel bars and HYSD bars were erected in the cylinder while casting and approximately 150 mm length bar was left exposed from top surface of the cylinder. Steel bar was connected to the positive terminal of power supply and it acted as anodes while the negative terminal was connected to HYSD bar which act as cathode DC power supply. Power supply was of 30 V DC current. The cracks were initiated due to corrosion in the cylinders. After crack initiation, sodium chloride solution had free path to penetrate in the cylinder to steel bar and this led to sudden increase in current. At different time intervals the current was measured.

4 RESULTS AND DISCUSSION

4.1 Chloride resistance test

Chloride resistance property of geopolymer concrete mixes has been studied by exposing both the types of concrete specimens to sodium chloride solution with 3% concentration for 30 and 90 days periods. No mass reduction or change in visual appearance observed. As the immersion period was only of 3 months, no major change in compressive strength observed. Only slight reduction in compressive strength took place for geopolymer concrete which was about 4.33% and for control concrete it was 5.73% as shown in Figure 1.

4.2 Sorptivity test

At different time interval of 1, 2, 3, 4, 5, 9, 12, 16, 20 and 25 minutes properties of sorptivity was found out for control concrete and geopolymer concrete by evaluating change in weight after every interval. Sorptivity curve was found to be less linear for Geopolymer Concrete (GC) compared to that of Control Concrete (CC). The rate of absorption has significant effect on the durability of concrete which was found to be less in geopolymer concrete than the control concrete. Figure 2 shows the graph of rate of absorption v/s time for both control and geopolymer concrete.

4.3 Accelerated corrosion test

Figure 3, shows the graph of time v/s current reading for control concrete and geopolymer concrete. Corrosion resistance was evaluated at regular interval by measurement of current.

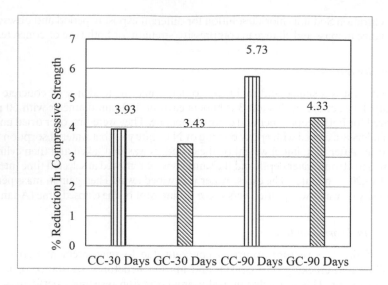

Figure 1.   Reduction in compressive strength for chloride attack.

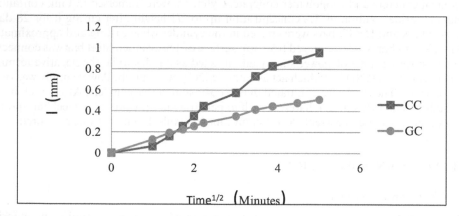

Figure 2.   Sorptivity curve.

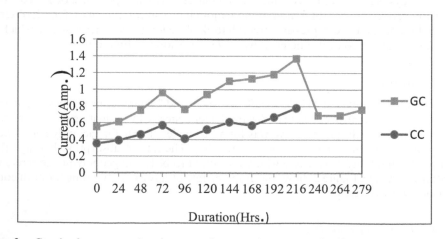

Figure 3.   Graph of current vs. duration curve for control concrete and geopolymer concrete.

It was observed that after 216 Hrs, the first crack initiated in control concrete due to change in current and visual inspection while geopolymer concrete had more resistance to corrosion and first crack was initiated after 279 hrs. Thus, geopolymer concrete had more resistance towards chloride attack as it takes long time for crack formation compared to control concrete. To study the rate and extent of corrosion in the specimen, the readings of half-cell potentiometer were taken for both control concrete and geopolymer concrete cylinders. In the control concrete the initial readings were lower compared to geopolymer concrete.

5 CONCLUSION

The following conclusions can be drawn from the present investigation.

- The test result of chloride attack demonstrate that geopolymer concrete has an excellent resistance to chloride. There are no significant change in mass and the compressive strength after exposure up to three months.
- The curve of Sorptivity for geopolymer concrete was found to be less linear compared to that of control concrete. The absorption rate is less in the geopolymer concrete compared to control concrete which indicates better durability property.
- The crack was initiated in the control concrete at 216 hrs and in the geopolymer concrete 279 hrs. indicating geopolymer concrete has an excellent resistance to chloride attack compared to control concrete

ACKNOWLEDGMENTS

The authors would like to thank the Civil Engineering Department, Nirma University for its constant support throughout the project. Special thanks must be mentioned to Dr. P.V. Patel (Head of Department, Civil Engineering), Dr. Urmil Dave (Professor), and P.N. Raval (laboratory supervisor) for their support throughout the project.

REFERENCES

Adam, A.A, 2009. Strength and durability properties of alkali-activated slag and fly ash based geopolymer concrete. RMIT University, Melbourne Australia.
Davidovits J. 1994, Properties of geopolymer cements, First international conference on alkaline cement and concretes, Ukrain, Pp:131–149
Hardjito D., Rangan B.V. 2005. Development and properties of low-calcium fly ash based geopolymer concrete, Research Report GC1, Faculty of engineering, Curtain University, Perth, Australia.
IS: 10262–2009, Concrete mix proportion guidelines, Bureau of Indian Standards, New Delhi (2009).
Malhotra, V. M. 2002, Introduction: Sustainable Development and Concrete Technology. ACI Concrete International 24(7): 22.
Thakkar Sonal & Bhorwani Darpan. 2014. Ambient Curing of Geopolymer Concrete Using Combination Of Fly Ash and GGBFS, VJTI International Conference, 2014 International Civil Engg Symposium March 14–16.

Multi-disciplinary Sustainable Engineering: Current and Future Trends – Tekwani, Bhavsar & Modi (Eds)
© 2016 Taylor & Francis Group, London, ISBN 978-1-138-02845-6

Durability studies for ambient cured geopolymer concrete using GGBFS as source material

S. Thakkar & R. Shah
Institute of Technology, Nirma University, Ahmedabad, Gujarat, India

ABSTRACT: Concern for environment protection and sustainable development are two major issues focused today worldwide. In the construction industry mainly the production of Portland cement causes the emission of pollutants that results into huge issue of environmental pollution. We can reduce the pollution effect on environment, by increasing the usage of waste industrial by-products in our construction industry. Geopolymer concrete is one such attempt in the area of sustainable development. In present study source material of geopolymer concrete is slag and it is cured at ambient temperature. In this paper a study of durability aspects in form of acid resistance test, sorptivity test and accelerated corrosion test is performed on geopolymer concrete and is compared to ordinary Portland cement concrete of same strength.

1 INTRODUCTION

Concrete is the most commonly used construction material in the world due to its properties such as strength, ease of molding and cheap availability of its ingredients. Customarily, concrete is produced by using Portland cement as the binder. The increasing worldwide production of OPC to meet infrastructure developments indicates that concrete will continue to be a preferred material for construction in the near future. Portland cement production is under critical review due to high amount of carbon dioxide gas it releases into the atmosphere. The cement industry contributes about 5% of total global carbon dioxide emission. The climate change due to global warming has become a major concern (Malhotra, 2002). Also due to tremendous industrial growth lot of waste is generated and its disposal is a big issue. To make a sustainable development, waste from industries if utilized as building material will lead to environment friendly solution and its disposal issue will also be resolved. Geopolymer concrete is one such alternative method where cement is totally replaced. During the geopolymerisation process, silicon and aluminum atoms are combined to form the building blocks that are chemically and structurally comparable to those binding the natural rocks (Davidovits, 1999). It uses any material rich in silica and aluminum ions as its source material and alkaline medium to activate it.

One of the drawbacks of geopolymer concrete is that it requires heat curing to give high early strength. To eliminate this drawback ambient cured geopolymer concrete was developed using slag as source material and sodium hydroxide and sodium silicate combination as alkaline liquid. M 25 grade of concrete was chosen, as this grade is commonly applied in regular structures. Mix design was carried out for geopolymer and Ordinary Portland cement concrete keeping same target mean strength. Durability studies is an important aspect in the development of any kind of new concrete. For the implementation of geopolymer concrete into real life application certain durability test like acid resistance, sorptivity and accelerated corrosion tests were carried out on both geopolymer concrete and control concrete made from OPC and their results were compared.

Table 1. Mix design of geopolymer concrete and control concrete.

Contents	Geopolymer Concrete	Control Concrete	Unit
Density	2400	2400	kg/ m³
Cement	–	383.2	kg/m³
Fine Aggregate	630.5	630.5	kg/m³
Coarse Aggregate 20 mm	702	733.2	kg/m³
Coarse Aggregate 10 mm	468	488.8	kg/m³
Mass of GGBFS	428.6	–	kg/m³
Na_2SiO_3/NaOH	2.5	–	–
Mass of GGBFS + alkaline liquid	0.4	–	–
Sodium Silicate	122.4	–	kg/m³
Sodium Hydroxide	42.9	–	kg/m³
Extra Water	42.9	–	kg/m³
Super plasticizer	8.6	–	kg/m³

2 EXPERIMENTAL PROGRAM

Ground Granulated Blast Furnace Slag (GGBFS) was used as source material with Silica (35%), Calcium Oxide (37%) and Aluminum Oxide (10%). It was procured from JSW cement works. Locally available 10 mm and 20 mm crushed aggregates were used as coarse aggregates and river sand was used as fine aggregate in both the mixes. The alkaline liquid used was a combination of sodium hydroxide and sodium silicate solution. Sodium hydroxide (NaOH) in flakes form with 98% purity was purchased from local supplier. Sodium Silicate was water glass type having Na_2O (16.84%), SiO_2 (35.01%) and water (46.87%). Naphthalene Sulphonate based super plasticizer, supplied by BASF, under the brand name Rheobuild 1125, has been used to improve the workability of the fresh geopolymer concrete.

2.1 Geopolymer concrete mix proportioning

Table 1 shows the mix design of geopolymer concrete for M 25 grade of concrete and control concrete. First GGBFS and the aggregate have been mixed together in concrete pan mixture and dry mixing was done of materials. This mixing was allowed to continue for about 3 to 4 minutes. After the dry mix of aggregate and GGBFS, alkaline solution which was prepared one day before was added with additional water in the mix. Then mixing was continued for another 3 to 4 minutes followed by casting (Hardjito 2005). For control concrete mix design steps given in IS 10262:2009 was followed.

3 DURABILITY TESTS PROCUDURE

3.1 Acid resistance test

Acid resistance property of geopolymer concrete mixes has been studied by exposing the concrete specimens in sulphuric acid for 30, 60 and 90 days periods. 9 cubes of each of control concrete and geopolymer concrete were cast and immersed in 5% concentration of sulphuric acid after 28 days of curing and tested after 30, 60, 90 days to compare results. Acid Resistance of concrete was to be evaluated by measuring the change in mass and change in compressive strength after the exposure of sulphuric acid solution (Wallah, 2006).

3.2 Sorptivity test

Three cylinders each of control concrete and geopolymer concrete were cast. Size of specimen was kept as 100 mm diameter and 50 mm height. Measurement for change in mass was done. The sides of the specimen were sealed in order to achieve unidirectional flow. Locally available wax and resin with 50:50 proportions was used as sealant. Weights of the specimen after sealing were taken as initial weight. At selected times i.e. after 1, 2, 3, 4, 5, 9, 12, 16, 20 and 25 minutes, the sample was removed from the water, the stopwatch stopped, excess water blotted off with a damp paper towel or cloth and the sample was weighed again. It was then replaced in water and stopwatch was started again. The gain in mass per unit area over the density of water was plotted versus the square root of the elapsed time. The slope of the line of best of these points (ignoring the origin) is reported as the sorptivity (Adam, 2009).

3.3 Accelerated corrosion test

Corrosion of reinforcement causes cracking and spalling of concrete and results into reduction of life of structure. Corrosion resistance is an important factor for the marine and coastal structures. Three cylinders of 150 mm in diameter and 300 mm in height were taken as control concrete and geopolymer concrete. Each were cast with stainless steel and HYSD bar embedded in it. The specimens were submerged in Sodium chloride solution of 5% concentration and upto 2/3 height at after curing for 28 days. Then the exposed steel bars were connected to the positive terminal of a constant 30 volt DC power supply, to make the steel bars act as anodes. This high voltage was used to accelerate the corrosion rate and shorten the test period. The negative terminal of the DC power source was connected to a HYSD, to make the bar act as cathode. When crack was initiated in the specimen by stresses caused by buildup of corrosion products, the electrolyte solution has a free path to the steel. This results in a sudden increase in current. So, in order to determine the time at which crack occurs, the current was recorded at different time intervals. Corrosion resistance of specimens was evaluated by change in current, visual inspection, change in UPV and half-cell potential meter readings before and after test.

4 RESULTS AND DISCUSSION

4.1 Acid resistance test

Visual appearance: It can be seen that the specimens exposed to sulphuric acid undergoes erosion on the concrete surface. The damage observed in control concrete was significantly higher than the geopolymer concrete for the same exposure period. Figure 1 shows both the cubes at 60 days exposure to sulphuric acid exposure

Change in Mass: Table 2 shows change in mass of control concrete and geopolymer concrete after various exposure period. It can be seen that loss of mass is higher in control concrete than geopolymer concrete for same exposure period.

Change in compressive strength: This was determined by testing the concrete specimens after 30, 60 and 90 and 120 days of exposure to sulphuric acid solution. Specimens were allowed to dry at room temperature and then tested in saturated surface dry condition. Figure 2, shows comparison of reduction in compressive strength for control and geopolymer concrete.

Figure 1.   Visual appearance of geopolymer concrete and control concrete.

Table 2. Change in mass of control concrete and geopolymer concrete after expousre to acid.

Days	Mass before exposure (kg)		Mass after exposure (kg)		% loss of mass	
	Control Concrete	Geopolymer Concrete	Control Concrete	Geopolymer Concrete	Control Concrete	Geopolymer Concrete
30	8.41	8.61	7.97	8.52		
	8.38	8.72	7.91	8.64	5.7	1.15
	8.45	8.68	7.93	8.55		
60	8.51	8.42	7.85	8.30		
	8.59	8.36	7.88	8.21	8.40	1.59
	8.63	8.38	7.82	8.25		
90	8.45	8.43	7.65	7.89		
	8.52	8.31	7.62	7.78	14.05	3.13
	8.54	8.36	7.69	7.73		
120	8.61	8.52	6.90	8.20		
	8.59	8.63	7.10	8.24	19.09	4.09
	8.57	8.49	6.87	8.15		

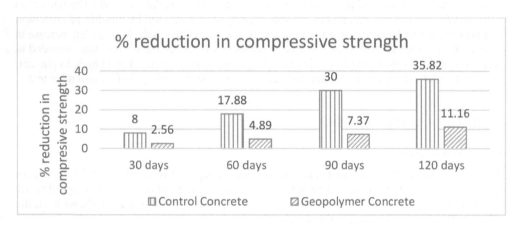

Figure 2. Percentage reduction in compressive strength after exposure to acid.

4.2 Sorptivity test

The Sorptivity property of geopolymer concrete and control concrete for time interval of 1,2,3,4,5, 9,12,15,20,25 minutes and gain in weight of each specimen has been plotted in Figure 3. It was found out the rate of absorption of water in geopolymer concrete was less as compared to control concrete.

4.3 Accelerated corrosion test

The corrosion resistance of concrete has been evaluated by measuring the current readings of the specimens at regular intervals and by visual inspection. The current was stopped at the time when a crack was seen in the concrete. The initiation of crack was seen in control concrete at 216 hours while in geopolymer concrete it was seen at 266 hours. This shows that geopolymer concrete has more resistance to chloride attack as compared to control concrete. Figure 4 shows graph of current and time for both control and geopolymer concrete.

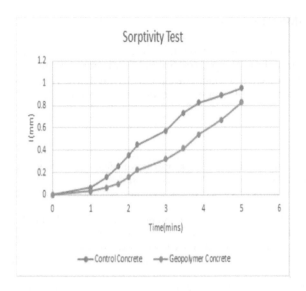

Figure 3. Sorptivity of concrete.

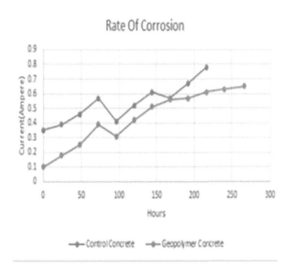

Figure 4. Accelerated corrosion test.

5 CONCLUSION

The following conclusions can be drawn from the present investigation.

- In case of acid resistance, ordinary concrete shows high loss in compressive strength and mass loss compared to geopolymer concrete.
- Sorptivity of geopolymer concrete was found to be less as compared to control concrete. This means that rate of absorption for geopolymer concrete is less as compared to control concrete.
- Rate of corrosion of geopolymer concrete was found to be much less as compared to control concrete. The crack initiated in control concrete was observed at 216 hours and geopolymer concrete was much more at 266 hour.

ACKNOWLEDGMENTS

The authors would like to thank the Civil Engineering Department, Nirma University for the constant support throughout the project. Special thanks must be mentioned to Dr. P.V. Patel (Head of Department, Civil Engineering), Dr. Urmil Dave (Professor), and P.N. Raval (laboratory supervisor) for their support throughout the project.

REFERENCES

Adam, A.A. 2009. Strength and durability properties of alkali-activated slag and fly ash based geopolymer concrete. RMIT University, Melbourne Australia.

Davidovits, J. 1999. Chemistry of Geopolymeric Systems, Terminology. Geo-polymer '99 International Conference, France.

Hardjito D., Rangan B.V. 2005. Development and properties of low-calcium fly ash based geopolymer concrete, Research Report GC1, Faculty of engineering, Curtain University, Perth, Australia.

IS: 10262–2009, Concrete mix proportion guidelines, Bureau of Indian Standards, New Delhi.

Malhotra, V.M. 2002. Introduction: Sustainable Development and Concrete Technology. ACI Concrete International 24(7): 22.

Wallah, S.E, Rangan B,V. 2006. Low-calcium fly ash-based geo-polymer concrete: Long term properties, Research Report GC2, Faculty of engineering, Curtain University, Perth, Australia.

Multi-disciplinary Sustainable Engineering: Current and Future Trends – Tekwani, Bhavsar & Modi (Eds)
© 2016 Taylor & Francis Group, London, ISBN 978-1-138-02845-6

Study of mode shapes of multistoried buildings with rooftop water tanks considering with and without infill wall panel stiffness

Sangram K. Nirmale

Urban Planning, Department of Architecture and Planning, Visvesvaraya National Institute of Technology, Nagpur, India

ABSTRACT: Rooftop water tank is mostly located at a higher elevation. The height and location of these structures often provide a desirable performance. There are various types of structural configurations which gives reduced response under earthquake loading. The Seismic Modal Contribution Factor (SMCF) and eigenvalues like period and frequency are the significant parameters which are used in a professional structural design software for analysis and design of multistoried buildings. In this paper, eight different configuration models of a G+9 building are analyzed for seismic loading by considering the infill wall panel stiffness and rooftop water tank at different locations. In this study first ten mode shapes are evaluated with the help of SMCF and the period of the structure. This research is carried out to study the seismic behavior of the structure by considering the infill wall panel and different positions of rooftop RCC water tanks. Conclusions are made for the best placement of the rooftop water tanks.

1 INTRODUCTION

The analysis of a multistoried building subjected to dynamic loading needs the understanding of its natural frequency and corresponding mode shapes (Rubinstein, 1965). Mode shapes are very useful for a designer because they represent the various shapes of buildings that will vibrate in free motion and these same shapes tend to dominate the motion during the earthquake. Understanding of mode shape will help for better design. The significance of mode shape is denoted by the seismic modal contribution factor and eigenvalues of each mode. Ideally, structures shall seismically behave at a torsional mode for shorter periods, and the first three mode shapes shall contribute at least 80% of mass participation factor (Mode shape, 1999).

Water tanks are an integral part of multistoried buildings, but they severely affect the building during an earthquake (Durgesh, 2003). From literature review, it can be said that, the study for the reducing response of a structure is carried out on elevated water tanks, not on rooftop water tanks. Therefore, in this study, a multistoried building along with a rooftop RCC water tank is considered for realistic simulation. Eight models of a G + 9 building are analyzed by the response spectrum method. In total, 72 m^3 capacity is assumed for a rooftop water tank. Eight simulations are done for different placement and configuration of the structure. The main objective of the research is to study the effects of water tank positions and infill wall stiffness on mode shape under seismic loading. All these models are checked with respect to the above stated ideal condition and the best model of a G + 9 buildings with rooftop water tank is suggested.

2 METHODOLOGY

This research is carried out in two phases. The main difference in these two phases is that the first phase considers the infill wall stiffness, whereas the second phase does not consider any stiffness effect from infill wall panel. The details of the two phases are as follows:

2.1 *Phase 1*

Phase 1 involves modeling and analysis in STAAD Pro software. Figure 1 shows a plan for all eight models. All joints at the same floor level are constrained by the floor diaphragm. This causes all joints to move together along the same plane level. At the below plinth level, the structure is rigidly fixed. Table 1 shows assumed data for modeling and analysis purpose.

2.1.1 *Phase 1 models description*

Out of eight models, the first four models are considered in this phase, which are shown in Figure 2. The stiffness modeling is done using the equivalent diagonal strut method. The New Zealand Code specifies the width of the diagonal strut equal to one-quarter of the diagonal length of the panel (NZS:4230, 2004). Therefore, the width of the external wall strut is assumed to be 1.2 m and thickness to be 0.23 m, and for internal wall strut, width is assumed to be 1.2 m and thickness 0.15 m. Table 2 illustrates phase 1 model details. All eight models are G + 9 storied building structures.

Figure 1. Building plan for all models.

Table 1. Assumed preliminary data required for analysis of structure.

Parameters	Details	Parameters	Details
Size of building	24.46 m x 16.23 m	Slab thickness	140 mm
Grade of concrete	M 20	External wall thickness	230 mm
Floor to floor height	3 m	Internal wall thickness	150 mm
Plinth level above foundation	2 m	Live load on floor	3 KN/m^2
Seismic Zone	IV	Floor finish Load	1 KN/m^2
Importance Factor	1	Type of Soil	Medium Soil
Response combination method	CSM	Density of Concrete	25 KN/m^2
Young's Modulus of infill wall	10500 MPa	Poisson's Ratio of infill	0.239586
Frame Type	RC moment resisting	Density of infill wall	20 KN/m^2

Table 2. Phase 1 and Phase 2 model description.

Model No.	Description
Model 1	No rooftop water tank on this structure.
Model 2	Four 18 m^3 capacity RCC water tanks are placed on the roof at four corners.
Model 3	One 72 m^3 capacity RCC water tank is placed at the center of the plan of structure.
Model 4	One 72 m^3 capacity RCC water tank is placed on the roof at the corner of the structure.

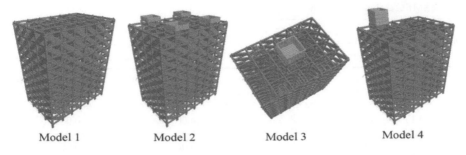

Model 1   Model 2   Model 3   Model 4

Figure 2.   3D view of phase 1 models.

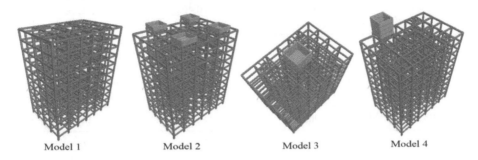

Model 1   Model 2   Model 3   Model 4

Figure 3.   3D view of phase 1 models.

2.1.2 *Selection of structural member size and loads assumed on the structure*

The following criteria are used to decide structural member size for phase 1 models. The criteria are as per Indian standards (IS 1893, 2002):

1. Storey drifts within the 0.004 x storey Height
2. Soft storey shall not be detected during analysis.
3. The ratio of design base shear and base shear calculated using the fundamental period is 1.
4. For analysis of buildings, a number of modes are considered for minimum 90% of SMCF

The various load combinations used in the analysis are as per IS 1893 (IS 1893: 2002).

2.1.3 *Phase 2*

Phase 2 involves the same models as phase 1 but without considering the infill wall stiffness. The details of phase 2 models are the same as given in Table 2. Figure 3 shows the 3D models of phase 2.

3   RESULTS AND DISCUSSIONS

The results obtained from analysis of all models are tabulated below:

3.1 *Observations*

1. SMCF is more along the direction of the smaller plan dimension of the building.
2. Soft storey at the seventh floor and ninth floor is detected for Model 2 of phase 2.
3. For the phase 1 models, natural period is less as compared with phase 2 models.
4. First four modes of the Model 1, Model 2, and Model 4 of phase 1 along Z axis contribute more than 95% modal mass.

Table 3. SMFC along X direction.

Mode Shape	Model 1		Model 2		Model 3		Model 4	
	Phase 1	Phase 2	Phase 1	Phase 2	Phase 1	Phase 2	Phase 1	Phase 2
1	0.16	0.00	0.14	0.00	0.12	0.00	0.12	0.08
2	71.50	0.00	71.42	0.00	70.31	0.01	67.76	5.66
3	4.95	76.51	4.58	76.56	3.40	76.48	7.79	70.81
4	0.17	0.00	0.18	0.00	0.00	0.00	0.00	0.01
5	14.40	0.00	14.75	0.00	6.15	0.00	16.01	0.69
6	2.72	10.72	2.73	10.72	0.06	10.60	0.87	10.00
7	0.55	0.00	0.54	0.00	0.01	0.00	0.45	0.00
8	0.53	0.00	0.41	0.00	10.37	0.00	0.88	0.23
9	0.34	3.82	0.40	3.86	3.58	3.65	0.20	3.61
10	0.47	0.00	0.49	0.00	0.46	0.00	1.79	0.00
Total	95.79	91.05	95.64	91.14	94.46	90.74	95.87	91.09

Table 4. SMFC along Z direction.

Mode Shape	Model 1		Model 2		Model 3		Model 4	
	Phase 1	Phase 2	Phase 1	Phase 2	Phase 1	Phase 2	Phase 1	Phase 2
1	76.58	77.80	75.90	77.80	73.33	77.76	76.40	73.40
2	0.00	0.00	0.00	0.03	0.00	0.01	0.00	4.33
3	2.42	0.00	2.41	0.00	2.07	0.00	0.86	0.09
4	17.00	10.38	17.48	10.42	6.91	10.31	17.84	9.89
5	0.36	0.00	0.38	0.00	0.00	0.00	0.06	0.54
6	0.64	0.00	0.68	0.00	13.94	0.00	1.21	0.01
7	0.34	3.46	0.36	3.50	0.01	3.33	0.47	3.33
8	0.53	0.00	0.81	0.00	0.33	0.00	0.94	0.18
9	0.81	0.00	0.69	0.00	0.46	0.00	0.14	0.01
10	0.35	1.61	0.25	1.62	0.36	1.31	0.02	1.52
Total	99.03	93.25	98.96	93.37	97.41	92.72	97.94	93.30

Table 5. First ten natural time periods of mode shapes.

Mode Shape	Model 1		Model 2		Model 3		Model 4	
	Phase 1	Phase 2	Phase 1	Phase 2	Phase 1	Phase 2	Phase 1	Phase 2
1	0.674	2.439	0.700	2.517	0.719	2.534	0.711	2.572
2	0.542	2.191	0.563	2.25	0.575	2.196	0.582	2.276
3	0.423	1.957	0.433	2.02	0.425	2.035	0.432	2.033
4	0.239	0.818	0.247	0.842	0.356	0.853	0.256	0.859
5	0.196	0.733	0.202	0.751	0.279	0.735	0.209	0.759
6	0.17	0.652	0.175	0.671	0.232	0.681	0.180	0.676
7	0.145	0.483	0.149	0.496	0.225	0.514	0.162	0.506
8	0.117	0.433	0.120	0.443	0.191	0.435	0.144	0.448
9	0.113	0.381	0.115	0.391	0.177	0.406	0.127	0.396
10	0.102	0.338	0.104	0.346	0.145	0.381	0.119	0.354

5. First four modes of the Model 1, Model 2, and Model 4 of phase 1 along X axis contribute more than 90% modal mass.
6. First nine modes of all phase 2 models along X axis contribute more than 90% modal mass.
7. First seven modes of all phase 2 models along Z axis contribute more than 90% modal mass.

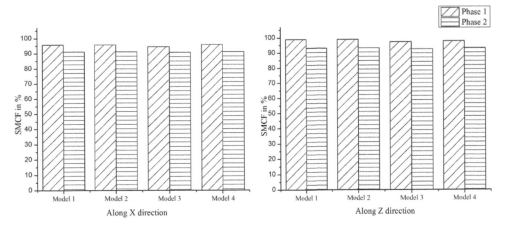

Figure 4. Total SMCF along X and Z directions for phase 1 and phase 2.

4 CONCLUSION

In this study, an attempt is made for simulation of realistic models by considering rooftop RCC water tanks and effects of infill wall on multistoried structures. The comparative study has been done on eigenvalue and modal mass participation factor. From this study, following conclusions can be drawn:

1. If the floor weight at an elevated water tank level is less than the floor weight of just below the floor, then there will be no adverse effect on the building, while earthquake and no soft storey will be detected at the floor level. Therefore, an elevated water tank can be designed such that the weight of the water tank will be less as compared to the top floor's seismic weight. As the capacity of the water tank is based on the number of residential units and functional requirement, the extra water requirement should be made at ground level.
2. Infill wall panel plays an important role in increasing SMCF for less number of modes and decreases the natural time period. This will result in reducing base shear.
3. Maximum response reduction is obtained when the rooftop water tank is constructed at the center of the building plan; further reduction is obtained when stiffness of infill wall panel is considered during analysis.

REFERENCES

Durgesh C.R. 2003. Performance of elevated tanks in Mw7:7 Bhuj earthquake of January 26th, 2001. *Proceedings of the Indian Academy of Sciences, Earth and Planetary Sciences* 112 (3): 421–429.
IS 1893 (Part 1): 2002. 2002. Criteria for earthquake resistant design of structures. New Delhi: Bureau of Indian Standards.
Mode shape. 1999. [Online]. Available: http://www.engr.sjsu.edu/mcmullin/courses/ce265/tutorials/mode.htm.
NZS 4230:2004. 2004. Design of Reinforced Concrete. Wellington: Standards New Zealand
Rubinstein, M.F. 1965. An engineering approach to computing the natural modes and frequencies of a tall building. *Proceedings of the third world conference on earthquake engineering New Zealand 1965*.2: 516–532.

Figure ... Hydrographs and calibration by phase I and phase II.

CONCLUSION

REFERENCES

Multi-disciplinary Sustainable Engineering: Current and Future Trends – Tekwani, Bhavsar & Modi (Eds)
© 2016 Taylor & Francis Group, London, ISBN 978-1-138-02845-6

Vibration control of structures using piezoelectric material

Pawan Pandey & Sharadkumar P. Purohit
Department of Civil Engineering, Institute of Technology, Nirma University, Ahmedabad, Gujarat, India

ABSTRACT: Structural elements and systems used in spacecraft and antenna structures are flexible and light in weight. Vibration in flexible structures may produce secondary effects in it and affect the performance. Therefore, it is important to control structural response of flexible structures. This can be achieved by either Passive control or Active control. It has been found that active control of flexible structures may lead to instability, if there is an error in feedback control. In such situations passive control of flexible structures coated with thin viscoelastic material offers promise. Apart, smart materials like Piezoelectric Ceramic Powder, Magneto-rheological Fluids etc. are potential alternatives. In present study cantilever structure made up of aluminum material is considered. Bare specimen is coated with Piezoelectric Ceramic Powder. Acceleration is captured using Data Acquisition System and LabVIEW through Free vibration test. Results shows piezoelectric powder coating increases damping into the system.

1 INTRODUCTION

Mechanical Vibrations have harmful effects on structures. Vibration of a structure can be controlled by either active method or passive method of controlling response. Active method includes a sensor, a processor and an actuator. Passive method is simple and it don't use a processor. Examples of passive method are Base Isolation, Tune Mass Damper, Viscoelastic damper, Frictional Damper, Viscous Damper etc.

Many researchers are working to access the potential of new generation smart material for the response control of structures. Ferroelectrically soft piezoelectric ceramic has the potential to control the vibrations passively but it can be used for thin flexible structures. Thin layers of highly sensitive ferroelectrically soft piezoelectric ceramic and piezo ceramic powder coating shows vibration damping effects (Munjal, B.S et al.).

In the present study Piezoelectric Ceramic SP-4 and SP-5 A are surface coated on aluminum cantilever beam in the form of thin layers with the help of Araldite Epoxy Hardener and Resin (HV 953 & AW106). Bare (un-coated) and coated specimens are tested under time domain for calculating vibration damping. Log decrement method is used to calculate damping coefficient.

2 PIEZOELECTRIC MATERIAL AND PROPERTIES

Piezoelectricity originated from the Greek word "piezo" which means pressure electricity. It is the property of certain crystalline substances that on the application of mechanical stress they generate electrical charges. Piezoelectric Ceramic SP-4 and SP-5 A are considered for the present study has following properties.

3 EXPERIMENTAL WORK

Dynamic properties of coated and un-coated aluminum cantilever beam are determined. SP-4 and SP-5 A are procured from Sparkler Ceramics Pvt. Ltd. Cantilever structural system

Table 1. Properties of piezoelectric powder SP-4 and SP-5 A.

Property	SP-4	SP-5 A
Particle Size	0.9 micron	0.9 micron
Density	7600 kg/m^3	7650 kg/m^3

Table 2. Physical and mechanical properties of aluminum cantilever beam.

Length	600 mm
Width	25 mm
Thickness	3 mm
Modulus of Elasticity	60000 MPa

Figure 1. Bare aluminum cantilever beam.

Table 3. Natural frequency of bare aluminum cantilever beam.

Method	Frequency (Hz)
Experimental	5.70
Numerical (Ansys)	5.73
Beam Theory	5.76

is used for determination of natural frequency and damping ratio. Damping ratio is determined by using free vibration.

3.1 *Bare aluminum cantilever beam*

Un-coated aluminum cantilever beam is tested for mechanical and dynamic properties. Physical, Mechanical and Dynamics properties obtained for bare aluminum cantilever beam are described in the present section. Table 2 shows physical and mechanical properties of aluminum cantilever beam.

Natural frequency of aluminum cantilever beam is calculated by two different methods, Analytical approach and Experimental approach. Beam Theory is used for calculating frequency analytically. Experimentally frequency of cantilever beam is obtained by capturing acceleration response of beam using Accelerometer and Data Acquisition System. Natural frequency determined using different approaches are described in Table 3. Results shows natural frequency is determined from different methods shows good agreement. There is not much difference in natural frequencies. Figure 1 shows bare aluminum cantilever beam used as host material on which coating of SP-4 and SP-5 A piezoelectric powder is applied and tested.

Damping ratio of aluminum cantilever beam is calculated using free vibration test. Displacement and velocity are provided as initial conditions to the structure and response of the structure is recorded using an Accelerometer which is attached at the top of aluminum cantilever beam. LabVIEW software is used for analyzing the recorded signal. Figure 2 shows response of aluminum cantilever beam under free vibration in LabVIEW.

Damping ratio is determined from the free vibration plot of aluminum cantilever beam, some portion of waveform is extracted and peaks are detected from the extracted waveform using LabVIEW software. Logarithmic decrement method is used to determine damping ratio. Apart, a log curve is developed compiling peak acceleration through MATLAB Curve Fitting Tool. An Average damping value obtained for bare aluminum specimen is 1.52%.

3.2 *Piezoelectric ceramic coated aluminum cantilever beam*

Piezoelectric Ceramic coated aluminum cantilever beam is tested for mechanical and dynamic properties. SP-4 and SP-5 A Piezoelectric Ceramic are used for surface coating. Physical and Dynamic properties determined and described in Table 4. Figure 3 and Figure 4 shows SP-4 and SP-5 A coated aluminum cantilever beam respectively.

Damping ratio of coated aluminum cantilever beam is calculated similar to bare aluminum cantilever beam. Figure 4 (a), (b) and (c) shows response of aluminum cantilever beam under

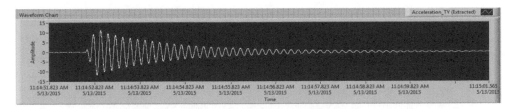

Figure 2.   Response of aluminum cantilever beam under free vibration.

Table 4.   Physical and dynamic properties of coated aluminum cantilever beam.

Property	SP-4	SP-5 A	SP-4+SP-5 A
Thickness of Coating	340 microns	170 microns	260 microns
Frequency (Hz)	5.64	5.81	5.70

Figure 3.   SP-4 Coated aluminum cantilever beam.

Figure 4.   SP-5 A Coated aluminum cantilever beam.

free vibration in LabVIEW. It is clearly visible on comparing Figure 1 and Figure 4 (a), (b) and (c) the damping increases for coated cantilever aluminum beam as compared to bare aluminum cantilever beam with very less mass penalty.

Damping ratio of coated aluminum cantilever beam is determined from the free vibration plot using logarithmic decrement method. Table 5 shows damping ratio of bare and coated aluminum cantilever beam. SP-4 coated beam shows significant increase in damping whereas SP-5 A alone and mixture of SP-4 and SP-5 A coating shows marginal increase in damping with very less mass penalty.

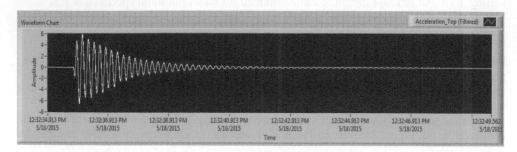

Figure 4 (a).   Free vibration response of SP-4 coated aluminum cantilever beam.

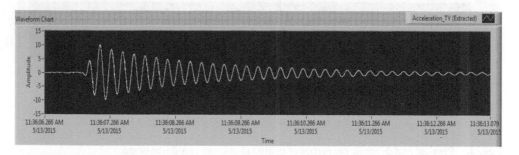

Figure 4 (b).   Free vibration response of SP-5 A coated aluminum cantilever beam.

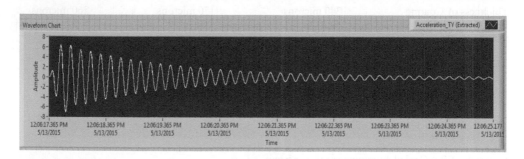

Figure 4 (c).   Free vibration response of SP-4 and SP-5 A coated aluminum cantilever beam.

Table 5.   Average damping of SP-4, SP-5 A and SP-4 + SP-5 A coated aluminum cantilever beam.

Coating	SP-5 A	SP-4	SP-4 & SP-5 A
Damping	1.622%	2.046%	1.642%
Increase in Damping	6.44%	34.27%	7.756%

4 CONCLUSION

Structural response of flexible structure can be controlled passively by many ways. Piezoelectric powder coating on structures helps in increasing damping of the structure. A bare and coated aluminum beams are considered for the present study. A free vibration test is carried out to determine damping ratio.

The study carried out indicates that

- Coating with smart materials on light flexible structures shows significant damping effects.
- As compared to bare specimen, specimen coated with piezoelectric powder SP-4, SP-5 A and mix of SP-4 and SP-5 A shows increment in damping.
- Maximum damping is achieved for aluminum specimen coated with SP-4 piezoelectric powder.

REFERENCES

Banerjee, S., Du, W., Wang, L. & Cook-Chennault, K.A. Fabrication of dome-shaped PZT- epoxy actuator using modified solvent and spin coating technique. *Journal of Electroceramics*. Publisher Springer US, Volume 31, Page 148–158.
Chopra, A.K. 3 edition, 2009. *Dynamics of structure*. Pearson Education, Inc.
Jordan, T.L. & Ounaies, Z. 2009. Piezoelectric Ceramics Characterization.
Munjal, B.S., Trivedi, H.V. and Sharma, P.V.B.A.S., Passive Damping Characterization of Parabolic Composite Reflectors with Hybrid PZT- Coated Layers. *Journal of Intelligent Materials Systems and Structure*. November 2008 19: Pages 1281–1294.
Panchal, D, Purohit, Sharad, Dynamic Response Control of a Building Model using Bracings, *Journal of Procedia Engineering, Chemical, Civil and Mechanical Engineering Tracks of the 3rd Nirma University International Conference on Engineering (NUiCONE-2012)*, Volume 51, 2013, Pages 266–273.

Experimental evaluation of dynamic properties of controlled building

Rushi J. Patel & Sharadkumar P. Purohit

Department of Civil Engineering, Institute of Technology, Nirma University, Ahmedabad, Gujarat, India

ABSTRACT: Nowadays, taller, more flexible, and lighter structure are built due to space constraints. These structures are more sensitive to dynamic loadings like wind, earthquake, etc. Therefore, it is necessary to control the responses of such structures to protect them against hazards. This paper includes the experimental determination of dynamic properties of uncontrolled as well as controlled building. Controlled building includes Tuned Mass Damper (TMD) and Base Isolation System. Two types of Buildings, one represented as Single Degree of Freedom (SDOF) and other as Multi Degree of Freedom (MDOF) system are considered. Dynamic properties like natural frequency and damping ratio are extracted for both types of buildings. It is found that both TMD as well as base isolation system shows increment in damping ratio for SDOF and MDOF systems and thus shows their efficacy.

1 INTRODUCTION

Most structures that are constructed in seismically active areas are subjected to an earthquake at least once during their design life. These structures, when subjected to earthquake, respond dynamically which results into additional stresses in structural members. Thus, understanding the behavior of structures under dynamic loading is quite important.

The response of the structures can be controlled by various means like passive control, active control, and hybrid control systems. Many researchers have found that passive control technique is quite effective and it controls the structural response of the building. Passive control technique includes base isolation, bracings, viscous damper, Tuned Mass Damper (TMD), etc.

To determine the responses of structures with or without control techniques, generally scaled down models of structures are used which are excited using the Shake Table that produces motions similar to earthquake.

2 SINGLE DEGREE OF FREEDOM (SDOF) BUILDING MODEL

A building representing the SDOF system is shown in Figure 1. Dynamic properties of the SDOF system is determined by the free and forced vibration test. Table 1 shows natural frequency of the SDOF system determined analytically by considering shear building and experimentally by determining mass and lateral stiffness of the model.

It is evident from Table 1 that analytical and experimental results show good agreement for lateral stiffness and natural frequency.

Inherent damping ratio of the SDOF building model is extracted by applying logarithmic decrement method (Chopra 2009) to free vibration response. Free vibration response is captured at the top of the building model through piezo-based accelerometer, data acquisition system, and LabVIEW software. Part of free vibration waveform is extracted and peak of each cycle is extracted using the peak detector of LabVIEW. Figure 2 shows free vibration waveform of the SDOF building model.

Figure 1. SDOF building model.

Table 1. Stiffness and natural frequency of SDOF model.

Method	Stiffness (N/m)	Frequency (Hz)	Circular frequency (rad/s)
Analytical	2567.54	6.075	38.17
Experimental	2572	6.07	38.13

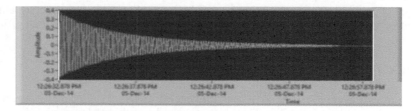

Figure 2. Response of SDOF model under free vibration.

Table 2. Damping ratio of SDOF model with logarithmic decrement method.

Amplitude (g)	Damping ratio							
1.1321	0.0103	0.0106	0.0108	0.0110	0.0112	0.0114	0.0115	0.0117
1.0610	0.0108	0.0110	0.0112	0.0114	0.0116	0.0117	0.0119	
0.9911	0.0112	0.0114	0.0116	0.0117	0.0119	0.0121		
0.9238	0.0116	0.0118	0.0119	0.0121	0.0122			
0.8586	0.0119	0.0121	0.0122	0.0124				
0.7965	0.0122	0.0123	0.0125					
0.7378	0.0125	0.0127						
0.6823	0.0129							
0.6292								

Table 2 summarizes damping ratio that is obtained for SDOF building model by considering subsequent and distance peak of waveforms. Average damping ratio is calculated to be 1.15% of critical damping ratio.

Damping ratio is also evaluated through forced vibration test, where the SDOF building model is harmonically excited by various frequencies. A frequency response plot is determined by capturing acceleration responses through accelerometer and data acquisition system (Panchal 2012). Figure 3 shows the frequency response curve for the SDOF building model. Using half power bandwidth method, the damping ratio extracted is 1.7% of critical damping ratio.

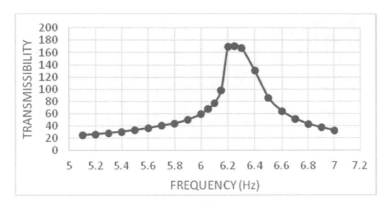

Figure 3.   Frequency response curve of SDOF building model.

Figure 4.   MDOF building model.

Table 3.   Natural frequency of MDOF building model.

Method	Frequency (Hz)	Circular frequency (rad/s)
Analytical	2.9667	18.63

The damping ratio value obtained through free vibration test is observed to be higher as compared with forced vibration test. This may be due to difficulty in capturing the peak of the frequency response curve of forced vibration test.

3   MULTI DEGREE OF FREEDOM (MDOF) BUILDING MODEL

A building represented as MDOF system is as shown in Figure 4. Dynamic properties of MDOF system are also determined by free and forced vibration tests. Table 3 shows the natural frequency of the MDOF system determined analytically by solving the Eigen value problem using MATLAB.

Inherent damping ratio of the MDOF building model is extracted by applying the logarithmic method to free vibration response. Free vibration response is captured at each storey of the building model through a piezo-based accelerometer and data acquisition system and LabVIEW software. Part of free vibration waveform is extracted and peak of each cycle is

119

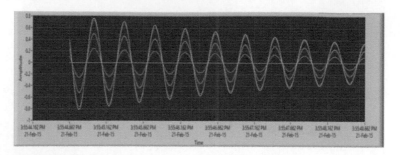

Figure 5.   Free vibration response of MDOF system.

Table 4.   Damping ratio of the MDOF model with logarithmic decrement method.

Amplitude (g)	Damping ratio						
0.2551	0.0144	0.0130	0.0132	0.0134	0.0139	0.0144	0.0147
0.2330	0.0116	0.0126	0.0131	0.0138	0.0143	0.0147	
0.2166	0.0135	0.0139	0.0145	0.0150	0.0154		
0.1989	0.0142	0.0150	0.0155	0.0158			
0.1820	0.0158	0.0162	0.0164				
0.1647	0.0166	0.0166					
0.1485	0.0167						

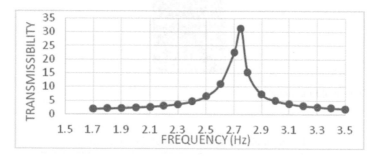

Figure 6.   Frequency response curve for MDOF building model.

extracted using a peak detector of LabVIEW. Figure 5 shows free vibration waveform of the MDOF building model. Part of free vibration waveform is extracted and peak of each cycle is extracted using peak detector of LabVIEW. Figure 5 shows free vibration waveform of SDOF Building model.

Table 4 summarizes damping ratio obtained for MDOF building model by considering subsequent and distance peak of waveform. Average damping ratio is calculated to be as 1.54% of critical damping ratio.

Damping ratio is also evaluated through forced vibration test, where in MDOF building model is harmonically excited by various frequencies. A frequency response plot is determined by capturing acceleration responses through the accelerometer and data acquisition system. Figure 6 shows frequency response curve for SDOF building model. Using half power bandwidth method, the damping ratio extracted is 1.45% of critical damping ratio.

4   CONTROLLED RESPONSE WITH TMD

TMD consists of spring, mass, and damper system which is attached to primary mass such that the response of the mass damper is out of phase with the primary mass. Three important

Table 5. Design parameters of TMD for SDOF and MDOF models (Bakre & Jangid 2007).

Design parameters	SDOF model	MDOF model
Stiffness	121 N/m	75 N/m
Mass	90 g	280 g
Mass ratio	4.5%	3%

Table 6. Damping ratio of SDOF and MDOF model with TMD.

	SDOF with TMD	MDOF with TMD
Damping ratio	2.26%	1.8%

Table 7. Time period of SDOF and MDOF with and without base isolation.

	Time period without base isolation (s)	Time period with base isolation (s)
SDOF system	0.167	0.45
MDOF system	0.36	0.769

parameters considered for the design of TMD are mass ratio, frequency ratio, and damping ratio. Referring to previous works by many researchers on TMD, the design parameters for SDOF and MDOF system as shown in Table 5 are calculated.

The above parameters are considered such that TMD for the SDOF system is tuned for the frequency of 6.1 Hz, i.e., natural frequency of SDOF model, and TMD for MDOF system is tuned for 3 Hz, i.e., natural frequency of MDOF model. Models are fixed on the shake table and are subjected to harmonic base excitation with various frequencies. Response is captured at each storey level and at TMD. Damping ratio is evaluated using the half power bandwidth method applied on frequency response. The damping ratio of the SDOF and MDOF models with and without TMD is shown in Table 6.

It is evident from Table 6 that damping ratio has increased for buildings with TMD as compared to uncontrolled (bare) building.

5 CONTROLLED RESPONSE WITH BASE ISOLATION

Base isolation is a technique in which the structure is isolated from the ground with isolators so that energy is not transferred from the ground to the structure. The design of base isolation is carried out such that the time period of isolated buildings shifts to 2.5–3 times the time period of the building without isolation (Sanghani 2004).

An experimental study is carried out to evaluate the reduction in the response of SDOF and MDOF models with base isolation. Models are subjected to free and forced vibration tests and responses are captured using an accelerometer at each floor level. The damping ratio for SDOF and MDOF models is evaluated using the logarithmic decrement method and the half power bandwidth method is tabulated in Table 7.

Table 7 shows the time period of MDOF and SDOF systems with and without base isolation. It is evident from Table 7 that the natural time period of SDOF system has shifted 2.7 times whereas, the natural time period of MDOF system has shifted 2.1 times.

6 CONCLUSION

A building represented using SDOF system and MDOF systems is considered for the present study. Dynamic properties of SDOF and MDOF systems like natural frequency and damping ratio are extracted using free and forced vibration tests.

The following conclusions are derived on the basis of the study:

- Analytical and Experimental results of lateral stiffness show good agreement.
- Damping ratios obtained for SDOF and MDOF building models are 1.15% and 1.45%, respectively.
- Moderate increment in damping ratio is achieved for SDOF and MDOF building models with TMD.

REFERENCES

Bakre S.V. & Jangid R.S, 2007. *Optimum parameters of tuned mass damper for damped main system.* Structural control and Health monitoring 14.
Chopra, A.K. 3 edition, 2009. *Dynamics of structure.* Pearson Education, Inc.
Panchal, D 2012. *Experimental evaluation of dynamic properties of building system.* Master's thesis, Nirma University.
Sanghani, D.B 2004. *Design of base isolation systems for an R.C.C. building. Master's thesis*, Nirma University.

Infrastructure project planning and management

Multi-disciplinary Sustainable Engineering: Current and Future Trends – Tekwani, Bhavsar & Modi (Eds)
© 2016 Taylor & Francis Group, London, ISBN 978-1-138-02845-6

Flood inundation mapping in the GIS environment: A case study of the west zone, Surat city

Chandresh G. Patel
Department of Civil Engineering, Ganpat University, Gujarat, India

Pradip J. Gundaliya
Department of Civil Engineering, LDCE, Ahmedabad, Gujarat, India

ABSTRACT: This paper attempts to apply modern techniques such as Geographical Information System (GIS) and Remote Sensing (RS) for the assessment of flood hazard in the west zone of Surat city situated on the right bank of the river Tapi in the state of Gujarat, India. A Digital Elevation Model (DEM), prepared in the GIS environment, was used as the basic input for preparing the flood inundation map of the west zone, which was created from the contour data. Using the geometric data of the river Tapi, the discharge carrying capacity of the river sections was found to have a maximum discharge of 25,768.09 cumecs, which was released in the river that caused the occurrence of flood events in the study area during August, 2006; consequently, most of the area was under water at a depth more than 3 m. Flood inundation maps of the west zone, comparatively low elevation in Surat city, were developed based on the flood return period of 25 years, 32 years, and the worst flood events. The validation of DEM was done with the GPS survey of the zone, and remedial measures were suggested to minimize the flood effect.

1 INTRODUCTION

Flooding is one of the serious natural hazards in the world (Seyedeh 2008), and accounts for 40% of deaths caused by natural disasters, with most flood events occurring in developing tropical regions of India. About 4,00,000 km² of land is prone to flooding, which is about 12% of the total geographical area of 32,80,000 km². About 1,20,000 km² of the flood-prone land has been provided with some reasonable protection against floods by providing flood embankments and drainage channels (Apte 2009). Advances in Geographical Information System (GIS) and Remote-Sensing (RS) technologies, and a new satellite platform developed by recently launched sensors have widened the application of satellite data. These technologies can be applied to validate flood inundation models (Samarsinghel 2010). GIS not only facilitates model development and analysis, but also is critical in helping local residents and municipal engineers to understand the existing flooding extent, as well as its potential changes, as a result of the proposed improvements (Baumann 2011).

2 STUDY AREA

Surat city, shown in Figure 1, is situated at the latitude of 21°06′ to 21°15′ N and the longitude of 72°45′ to 72°54′ E on the bank of the river Tapi, with the coastline of the Arabian Sea on its west coast. Surat is divided into seven zones. The elevation of the west zone varies from 4 m to 12.5 m above mean sea level, and the zone lies at a low level compared with other six zones of the city. Population growth increased by 74% in the west zone from 2001 to 2011, and it has an area of 51.27 km². During the yearly monsoon period, the water in the river near Surat city rises above the danger level and enters the peripheral area of the west zone. As a result, few slum areas situated in this zone, along the river bank, are adversely affected and

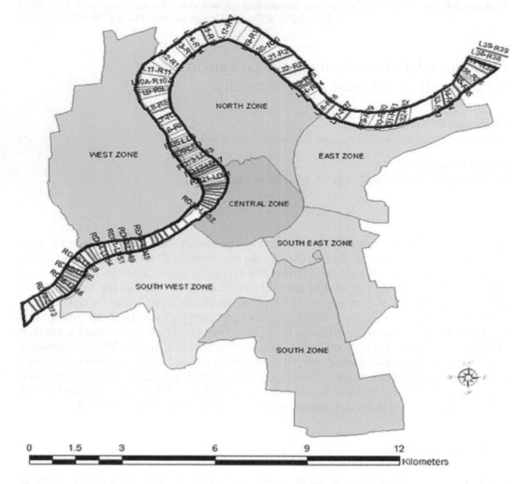

Figure 1. Zone-wise map of Surat city with the river Tapi and river cross-sections.

the people are temporarily shifted elsewhere (Saxena 2011). The topography in the Lower Tapi Basin (LTB) comprises a narrow valley and a gently sloping ground (CWC 2001). In the worst flood event of the year 2006, more than 400 people died and approximately property worth 4.5 billion US$ was damaged. The flood occurs typically during the monsoon period, mainly due to heavy rains and discharge released from the Ukai dam. The low discharge carrying capacity of the Tapi river sections also results in flood.

3 DATA COLLECTION

Reliable data are necessary to realize the objective. The study was based on both primary and secondary data. In addition to this, a frequent field observation was done together with the GPS survey of the west zone to generate primary information. Data collected for the research included: satellite images of the study area obtained from National Remote Sensing Centre (NRSC), Hyderabad; contour map of the study area collected from the Survey of India (SOI) Department; river cross-section details and the city map collected from Surat Municipal Corporation and Irrigation Circle; and hydrological data collected from Central Water Commission (CWC)/State Water Data Centre (SWDC), Gujarat.

4 METHODOLOGY

Reduction in the discharge carrying capacity is one of the major factors for the occurrence of flood in this region. Immediately after the flood in 2006, irrigation department measured the cross-sections of the river Tapi including the bed and bank levels at various points from the center line across the river and at an average interval of 150 m to 200 m along the length of the river from the Ukai dam to the Arabian Sea. Using the geometric data, the safe discharge carrying capacity of all 309 river sections were calculated by the continuity equation. Historical data show that the flood had occurred about 20 times since the year 1895 in Surat. In the last two decades, the major flood event occurred in 1994, 1998, and 2004, with the worst flood being in the year 2006. Using Gumbel's probability method, the return period of occurrence of the flood at a 5-year interval with relative discharge was calculated. It was found that for the return period of 32 years, a discharge capacity of 25768.09 cumecs was released in the river Tapi in August 2006. The section-wise discharge carrying capacity superimposed with the return period of 15 years, 20 years, 25 years, 32 years, and 35 years, as shown in Figure 2.

Digital Elevation Model (DEM) is a computerized representation of the Earth's terrain. GIS and RS provide a broad range of tools for determining the areas affected by floods or for forecasting the areas that are likely to be flooded due to high river water levels. DEMs are increasingly used for visual and mathematical analysis of topography, landscapes, and landforms, as well as for modeling surface processes (Millaresis 2000, Tucker 2001). The accuracy of a DEM is determined by the data type and the actual sampling technique of the surface when creating the DEM. A DEM offers the most common way of showing topographic information and even enables the modeling of flow across topography, a controlling factor in distributed models of landform processes (Dietrich 1993, Desmet 1995, Wang 2006). In this paper, a DEM of the west zone was prepared using ArcGIS software. A contour map of scale 1:50,000 of the west zone was scanned at 400 dpi resolution to convert it into an image. The scanned image was geo-referenced and overlaid on the satellite image of the study area. Necessary operations such as buffering were carried out for adjustment. Contours were digi-

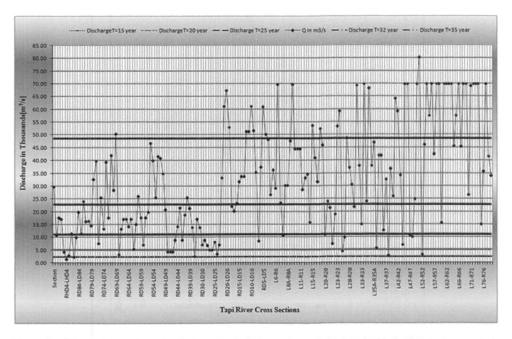

Figure 2.   River cross-section wise discharge carrying capacity superimposed with the flood return period.

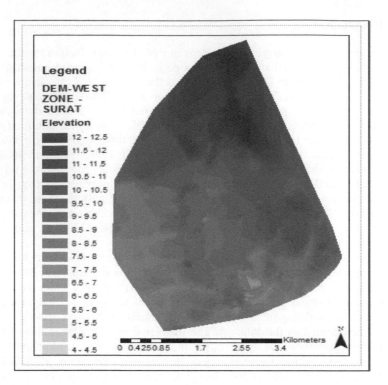

Figure 3.   DEM of the west zone.

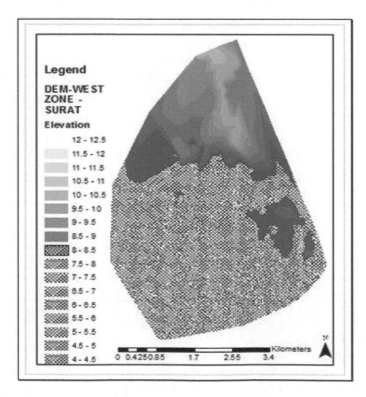

Figure 4.   Flood inundation map of the west zone for the flood return period of 25 years.

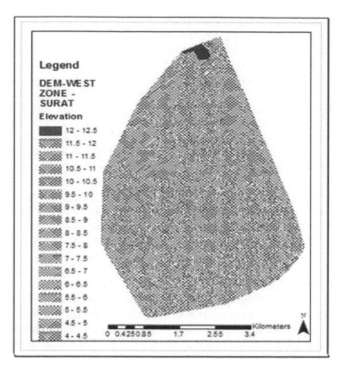

Figure 5.   Flood inundation map of the west zone for the flood return period of 32 years.

tized by the line tool, and errors such as dangling, overshoot, undershoot, and silver polygon were removed for developing an accurate DEM. Spatial and temporal data were inter-linked through ArcGIS and customized database management system. A 3D analyst tool was used to convert digitized contours into the Triangulated Irregular Network (TIN). Interpolation was done to convert vector data (TIN) into raster data (DEM), as shown in Figure 3.

In order to prepare a flood inundation map of the west zone for different discharges related to the return period, the surface analysis sub-tool of the 3D analyst was used. For finding out the probable submergence area, the feature file of the west zone was used for further operation in Arc Toolbox to create different shape files such as fill, flow direction, flow accumulation, and watershed.

To find out the submergence area of the west zone, raster data were converted into polygon to create the "calculated area" shape file. Finally, query functions F AREA and GRID-CODE were used to calculate the area of the closed polygon. The flood inundation map at the return period of 25 years and 32 years is shown in Figures 4 and 5.

5   RESULTS AND DISCUSSION

The length of the river Tapi is 724 km, of which the last lap of 214 km is in the state of Gujarat and meets the Arabian Sea near the city of Surat. Insufficient river and rain gauge stations, gradually increasing siltation, extensive encroachment, and low height and uncompleted embankment near and in the upstream of Surat city are the issues associated with the river Tapi. After the worst flood event of the year 2006, the river was measured and about 309 cross-sections were taken. The safe carrying capacity of each cross-section was calculated and found that the least discharge carrying capacity was 4531 cumecs near Surat city and the highest discharge carrying capacity was more than 70,000 cumecs near the mouth of the Arabian Sea. The west zone is developed on the right side of the river Tapi, looking from the upstream, and river sections close to the zone are named as L9 A-R9 A to

LD49- RD49, as shown in Figure 6. The Nehru bridge connects the west zone to Surat city, near which the river Tapi is gauged. During August 2006, the discharge observed near the Nehru bridge (river cross-section LD26-RD26) was 25,768.09 cumecs and the water surface level in the river was found to be about 12.5 m according to the river monitoring government agency. The safe discharge carrying capacity of the river cross-section LD26-RD26 was found to be very low compared with 25,768.09 cumecs discharge, released in the year 2006. It was found that the right-hand bank height of the section was just 8.55 m, which clearly indicated that the water overtopped from this section and other river sections had low embankment height. The average bank height of the river sections close to the west zone was about 9.3 m. Moreover, the bed width of most river sections was also narrow. To identify poor bank height sections, it is essential to calculate the water surface level at 25,768.09 cumecs discharge.

A graphical representation of river sections versus bed level, right-hand bank height, and calculated water surface is shown in Figure 7. It is clearly indicates that the river cross-section RD-18 to RD-21 have a sufficient bank height to carry discharge and the cross-section RD-8 to RD-15 also have enough geometry to pass the discharge safely. The RD-7 section has low height and thus water can easily spill from there even at a discharge of less than 25,768.09 cumecs. The modeling phase involved the combination of spatial, hydrologic, and hydraulic data to build a flood model for the catchment area. The basic data input for the model was a DEM prepared from the topographic map. After the implementation of operations in the ArcGIS, the flood inundation map generated data for flood return periods including worst flood events. In order to find out the probable submergence depth in different areas, a special statistical tool in ARC Toolbox was used to obtain the area enclosed between specific contour intervals.

The results obtained from ArcMap clearly indicate that the major area of the west zone has a reduced level of 6 m to 12.5 m. If 48,421 cumecs discharge is equal to the flood return period of 35 years, the entire zone will be submerged because the relative water surface level in the river may rise up to 16.5 m. The reduced level (1 m interval) and the probable submergence area of the west zone are listed in Table 1.

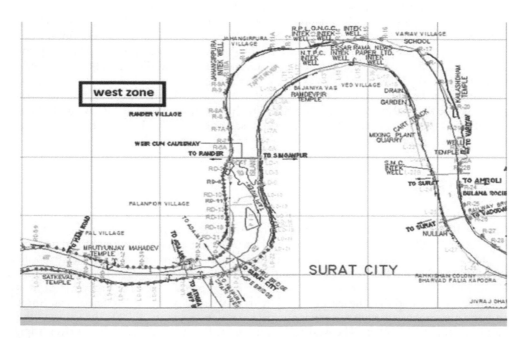

Figure 6.   River Tapi with river cross-sections in AutoCAD format.

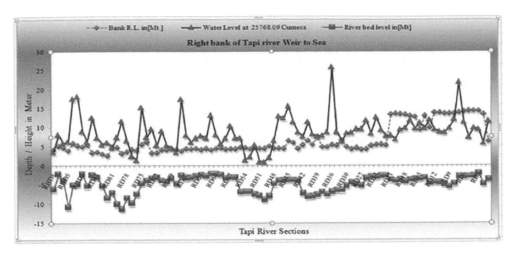

Figure 7. Graph of river sections versus bed level, right hand bank height, and calculated water surface of the river Tapi.

Table 1. Reduced level (M) and probable submergence area of the west zone at the discharge of 25,768.09 Cumecs.

RL (M)	Submergence area (M²)	RL (M)	Submergence area (M²)	RL (M)	Submergence area (M²)
4–5	14,439.1177	7–8	64,31,672.414	10–11	20,14,084.62
5–6	60,800.7645	8–9	51,32,725.985	11–12	23,56,755.009
6–7	29,70,535.12	9–10	40,82,536.673	12–12.5	7,91,509.3267

6 CONCLUSION

The research study demonstrated an integration of GIS technology with hydrological data to generate a high risk of flood inundation areas for disaster risk management of the west zone. The GIS helped in data updating, visualization of different scenarios, and risk mapping. A GPS survey was conducted to check the reliability of the DEM generated from the contour data of the west zone. The model predicted flood inundation at the return period of 25 years, 32 years as well as the worst flood event in the history, and the extreme future flood event at the return period of 35 years. Due to an increase in flooding frequency, the population residing near the river banks and valuable infrastructures such as roads and bridges are found to be at a high risk of flood inundation. Hydraulic modeling using the GIS technique proved to be useful in simulating flood water depth levels and inundation areas for various return periods in the low-lying area of Surat city. The research clearly indicates that at the 35-year return period, the maximum increase in flood water elevation predicted in the river Tapi is over 16.5 m, which is more than the highest elevation point of the west zone. To minimize the effect of flood, the embankment height of river sections close to the west zone should be increased on a priority basis to improve the discharge carrying capacity of the river Tapi.

REFERENCES

Apte. 2009. Urban Floods in Context of India. *Indo-US workshop on Urban Flood Disaster Management.* Hyderabad, 7–9 Jan. 2009.

Baumann, C. & Halaseh, A. 2011. Utilizing Interfacing Tools for GIS, HEC-GeoHMS, HEC-GeoRAS, and ArcHydro. *World Environmental and Water Resources Congress 2011*: pp. 1953–1962.

Central Water Commission. 2000–2001. Water Year Book 2000–2001 Tapi Basin. Hydrological Observation Circle, Gandhinagar, Gujarat, India.

Desmet, P.J.J. & Govers, G. 1995. GIS-based Simulation of Erosion and Deposition Patterns in an Agricultural Landscape: A Comparison of Model Results with Soil Map Information. *Catena*, Vol. 25, 389–401.

Dietrich, W.E., Wilson, C.J., Montgomery, D.R. & McKean, J. 1993. Analysis of Erosion Thresholds, Channel Networks, and Landscape Morphology using a Digital Terrain Model. *Journal of Geology*, Vol. 101, pp. 259–278.

Millaresis, G.C. & Argialas, D.P. 2000. Extraction and Delineation of Alluvial Fans from Digital Elevation Models and Landsat Thematic Mapper Images. *Photogrammetric Engineering and Remote Sensing* Vol. 66, 1093–1101.

Samarsinghe S.M.J.S. et al. 2010. Application of Remote Sensing and GIS for flood Risk Analysis. *Remote Sensing & Spatial Information Science*, pp. 110–115.

Saxena Vijay K. & Verrghese, T. Observation on urban ecology of Surat and bubonic plague transmission in the city via Google search

Seyedeh, S.S., Thamer, A.M., Mahmud, A.R.B., Majid, K.K. & Amir, S. 2008. Integrated Modelling for Flood Hazard Mapping Using Watershed Modelling System. *American Journal of Engineering and Applied Sciences* 1 (2): 149–156.

Tucker, G.E., Catani, F., Rinaldo, A. & Bras, R.L. 2001. Statistical Analysis of Drainage Density from Digital Terrain Data. *Geomorphology*. Vol. 36, pp. 187–202.

Wang, L. & Liu, H. 2006. An Efficient Method for Identifying and Filling Surface Depressions in Digital Elevation Models for Hydrologic Analysis and Modelling. *International Journal of Geographical Information Science*. Volume 20, Issue 2, pp. 193–213.

Multi-disciplinary Sustainable Engineering: Current and Future Trends – Tekwani, Bhavsar & Modi (Eds)
© 2016 Taylor & Francis Group, London, ISBN 978-1-138-02845-6

Variation in population density using spatio-temporal techniques—a case study of Shirur Tahsil, western Maharashtra

A.M. Zende
Annasaheb Dange College of Engineering and Technology, Ashta, Sangli, Maharashtra, India

R. Jadhav
Chandmal Tarachand Bora College, Shirur, Pune, Maharashtra, India

ABSTRACT: Population growth, density, sex ratio, literacy, etc., determine the qualitative and quantitative characteristics of population. In the present paper, an attempt has been made to analyze the variations in the population density of Shirur Tahsil, Pune District, western Maharashtra. To study the variations and patterns of density, secondary data has been obtained from the town and village directory of Pune District for the three different periods of census years, namely 1991, 2001, and 2011. ArcGIS 9.3 software was applied to prepare the base map and thematic maps. It is found out from this research that the density of Shirur Tahsil and all the circles was continuously increasing from 1991 to 2011. The highest density was observed in Shirur and Talegaon Dhamdhere circle. Shirur Municipal Council, Ranjangaon, Shikrapur, Koregaon Bhima, and Sanaswadi are densely populated villages due to the favorable physical and socio-economic factors. The villages and circles that are located near the industrial zone and are well connected through road transportation are characterized by a high density of population, whereas, the remote villages and circles showed the lowest density in study area. Circles are classified into low, moderate, and high density patterns. Shirur and Talegaon Dhamdhere circles had highest density whereas Takali Haji and Nhawara circles had the lowest density in 1991 and 2001.

Keywords: population; density; density pattern; industrial zone

1 INTRODUCTION

Man being a powerful geographical factor on the surface not only determines the economic pattern of resource utilization but also is a very dynamic and important resource for society (Mahajan, 2012). The distribution, growth, and density of population are the characteristics of the population. The term "density of population" was used by Henry in 1837, while preparing railway maps (Barkade, 2011). The concept of density of population is the most rarely used tool in the analysis of the diversity of man's distribution in space (Clarke, 1972). It is one of the important indices of the concentration of population; nature of balance between population of the region and its natural resources; and the magnitude of social, cultural, and economic development of a region. Population density is a simple concept of relating population size to the land area with a view to assessing crudely the pressure of population upon the resources of the area (Chandana, 2000). However, the density of population is expressed in different ways to understand the relationship between population and resources (Narke, 2010). It is stated in terms of persons per unit area. The population of any area or region is the outcome of its physical and socio-economic environment (Mahesha D. et al. 2012).

The density of population increased in all the States and Union territories from 1951 to 2011. As per census data of 2011, Delhi ranked first (11,320 persons/km²) in Union territories, whereas, among the states, Bihar stood first (1106 persons /km²), followed by West

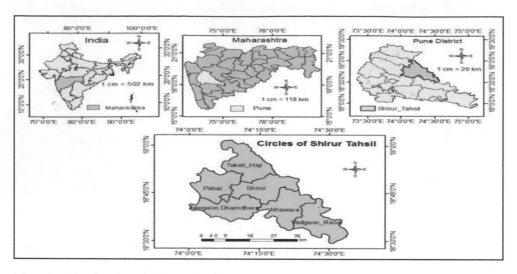

Figure 1.   Map location of Shirur Tahsil.

Bengal (1028 persons /km^2) and Kerala (860 persons /km^2) in the third place. The density of Maharashtra and Pune District was 365 and 603 persons/km^2, respectively. Though the study area had comparatively less density than Pune District and Maharashtra, the density of Shirur Tahsil increased from 159 persons/km^2 in 1991 to 257 persons/km^2 in 2011. These variations at the tahsil level and circle-wise level were analyzed for the period 1991–2011. High increase in population density is a great concern as it puts immense pressure on our natural resources. Differences in climatic conditions, availability of resources, agriculture, industrialization, educational facilities, etc., brings variation in population density.

Objective: To assess spatiotemporal variations and pattern of population density of the study area.

2   STUDY AREA

The Shirur Tahsil of Pune District is selected for the present research work. It occupies the eastern place of western Maharashtra and Pune district. It lies entirely in the basin of Ghod-nadi and Bhima and extends between 18°50″ N to19°02″ N latitude and 74°01″ E to 74°57″ E longitudes, comprising an area of 1514.30 km^2. The study area is included in SOI Topographic Index Numbers 47J/1, 47J/2, 47J/5, 47J/6, 47J/10, and 47J/11on 1:50,000. This tahsil is confined by Ahmadnagar District to east and north-east, Khed Tahsil to west, Ambegaon tahsil to north-west, and Haveli tahsil to South. Shirur tahsil is divided into six circles, namely Pabal, Shirur, Takali-Haji, Talegaon Dhamdhere, Nhavara, and Vadgaon-Rasai and covers 117 villages. Shirur is the administrative headquarter of Shirur Tahsil and it is only 67 km from Pune, 55 km from Ahemadnagar, and is well-connected by Major State Highway (MSH) 5. The total length of this SH5 is 45 km which connects the villages Koregaon Bhima, Sanaswadi, Shikrapur, Kondhapuri, Ranjangaon, Khandale, Pimpri Dumala, Kardilwadi, Ganegaon, Karegaon, Saradwadi, and Shirur. State Highway 55, 103, 117, 118, 128, and 129 are well distributed in the study area and so the Shikrapur, Ranjangaon, Shirur, Nhawara, Malthan, Kawathe, Kanhur, and Pabal villages are well connected by MSH and SH.

3   DATA AND METHODOLOGY

To study the characteristics of population, secondary data has been obtained from the town and village directory of Pune district for the three different periods of census years for 1991,

134

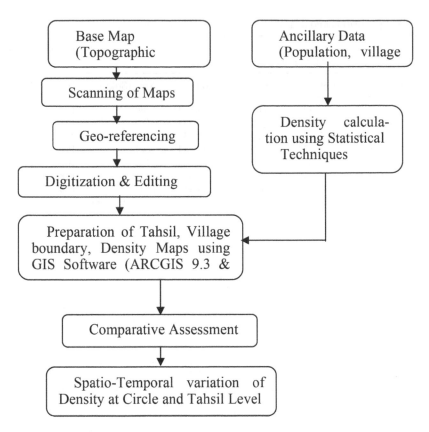

Figure 2.   Flow chart of methodology.

2001, and 2011. A comparative analysis has been done to assess the spatiotemporal variations in density at circle and tahsil levels. Simple statistical techniques and formulae are referred to calculate and represent the density. The village-wise data is compiled to compute the circle-wise data of Shirur tahsil. ArcGIS 9.3 and ILWIS 3.3 softwares were applied to prepare the base map and density maps. The detailed methodology for this work is shown in Figure 2.

4   RESULT AND DISCUSSION

Population distribution is the geographical arrangement of the population within the physical space of the state boundaries. The major factors that determine the pattern of population distribution are: (1) geographical factors, such as climate, terrain, soils, and natural resources; (2) economic, social, and political factors, such as the type of economic activity and the form of social organization; and (3) demographic factors, such as the different growth rates that exist between areas as a result of differences in births, deaths, and migration rates (Mule, 2011).

4.1   *Population density*

The culture of the study area, the age, the economic functions, size of population, and the physical setting all determines the spatial patterns of the population densities (Borude, 2013). Historical and cultural factors of Shirur Tahsil, geomorphology of the area, old flood plain of Ghod, Kukadi and Bhima rivers, dry and hot climate, well-connectivity with other major

Table 1. Total population and density of Shirur Tahsil (From 1991 to 2011).

Circles of Shirur Tahsil	Total Population				Population Density per km^2		
	1991	2001	2011	Area (Sqkm)	1991	2001	2011
Takali Haji	32,895	41,222	49,593	266	124	155	186
Pabal	36,048	41,566	44,834	230	157	180	195
Shirur	52,117	70,246	1,00,836	253	206	278	399
Talegaon Dhamdhere	48,023	69,560	95,944	247	194	282	388
Nhawara	25,682	35,744	38,589	253	102	141	153
Vadgaon Rasai	44,640	52,252	55,618	251	178	208	221
Shirur Tahsil	2,38,609	3,10,590	3,85,414	1500	159	207	257

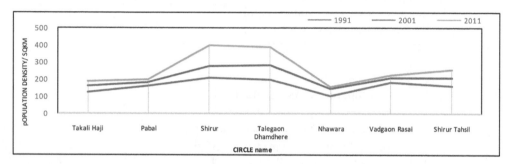

Figure 3. Population density of Shirur Tahsil (1991–2011).

cities, economic, and agricultural development, etc., support the spatial patterns of population density in Shirur tahsil. The ever increasing population is creating continuous pressure on the available resources. Table 1 shows the distribution of total population and the variations in overall density among the circles of Shirur Tahsil. The density of Shirur Tahsil has been increasing since 1991. It was 159 persons /km^2 in 1991 which then reached 257 persons/ km^2 in 2011. During the span of two decades, the density of Shirur Tahsil increased by 62%. The circle-wise variation in population density during 1991 to 2011 is shown in Table 1. The circle with the least population density was Nhawara Circle. In 1991, Shirur Circle was at the top with a density of 206 persons per km^2 and Nhawara Circle was at the bottom with 102 persons /km^2 of density in 1991.

As per the census of 2001, among the Shirur Tahsil circles, Talegaon Dhamdhere Circle had the highest density of 282/km^2, and it was followed by Shirur Circle (278 persons/km^2) in 2001. In 2011, the highest density was observed in Shirur Circle (399 persons/km^2). It indicated the highest growth in density among all the circles and it was 110.40% growth during the span of 20 years. Shirur circle includes 23 villages and Shirur Municipal Council urban centre. The density of Shirur Municipal Council was 2922 persons/km^2 in 1991 and it increased to 5890 persons/km^2 in 2011. Maximum density was recorded in the villages of Ranjangaon, Saradwadi, Karegaon, Dhoksangavi, Kardelwadi, and Khandale. Educational, health facilities, Major State Highway 5, Ranjangaon MIDC, and cultural and religious tourist centers led to the rise in the density in this circle. At the same time, Karanjawane, Pimpri Dumala, Ganegaon Khalsa, Bhambarde, and Babulsarkh villages of Shirur Circles recorded the least density due to the absence of transportation facilities.

Talegaon Dhamdhere circle has 18 villages with 250 km^2 area. Sanaswadi, Koregaon Bhima, and Shikrapur are the suburban places of this Circle and the industrial areas are

Table 2. Circle-wise density pattern of Shirur Tahsil.

Density per km^2	Density Pattern	1991	2001	2011
100–150	Low Density	Takali Haji, Nhawara	Nhawara	Nil
151–250	Moderate Density	Shirur Tahsil, Talegaon Dhamdhere, Vadgoan Rasai Pabal	Takali Haji, Pabal, Vadgaon Rasai, Shirur Tahsil	Takali Haji, Pabal, Vadgaon Rasai, Nhawara
>250	High Density	–	Shirur, Talegaon Dhamdhere	Shirur, Talegaon Dhamdhere, Shirur Tahsil

well connected by Major State Highway 5. Nhawara circle had the lowest density but ranked second in growth of density during the study period. Out of 15 villages of Nhawara Circle, Nhawara, Uralgaon, Ambale, AlegaonPaga, RanjangaonSandas, Karade, and Nimone villages had the highest density due to fertile soil, Chaskaman irrigation facilities, sugar industry, road network, and seasonal immigration of sugarcane cutters.

Takali Haji circle stood second with the lowest density (186 persons/km^2). This is the northern part of the Shirur Tahsil and is located between Kukadi and Ghod rivers. Vadner Khurd village ranked first in density while Jambut and Ichakewadi villages also recorded high density. Lift irrigation of Ghod river and fertile soil are the main reasons of the agriculture-based economy of this circle. Vadgaon Rasai depicted the lowest growth in population density during the first decade, whereas population density of Shirur Circle decreased by 4.72% in the second decade.

4.2 Spatial pattern of density

Density is taken into consideration while grouping the density pattern. Circle-wise density is classified into low, moderate, and high literacy categories. Density pattern depicts the concentration of population and the causes of it.

4.2.1 Low density pattern
The circles having 100–150 persons /km^2 are included in low density pattern. In 1991, Takali Haji and Nhawara circles had low densities because of less transportation and educational facilities. Nhawara circle had the lowest density in 2001, whereas there was no area under this category in 2011.

4.2.2 Moderate density pattern
The circles which recorded the density between 151 and 250 persons /km^{2}were included in the moderate density category. Table No. 2 specifies that among the circles, Talegaon Dhamdhere, Vadgaon Rasai, and Pabal circles showed moderate density in 1991. In 2001, Takali Haji, Vadgaon Rasai, and Pabal Circle recorded moderate density. In 2011, Nhawara and Pabal circles were added in this class along with the Takali Haji and Vadgaon Rasai circles.

4.2.3 High density pattern
The circles which have more than 251 persons /km^2 were mentioned in the high density category. In 1991, there was not even one circle in this category, but in 2001 and 2011 Shirur and Talegaon Dhamdhere circles were included in this category (Figure 4). Kondhapuri, Ranjangaon Ganapati, Saradwadi, Koregaon Bhima, Sanaswadi, Shikrapur, and Talegaon Dhamdhere villages are well connected by SH 60. The availability of management and professional educational institutions, Ranjangaon, Sanaswadi MIDC's, more job opportunities, road network attracted population from the surrounding areas.

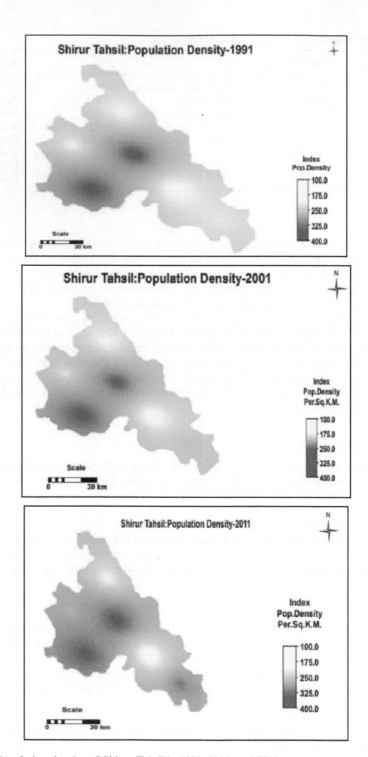

Figure 4.   Population density of Shirur Tahsil in 1991, 2001, and 2011.

5 CONCLUSION

1. The density of Shirur Tahsil and all the circles continuously increased from 1991 to 2011. The highest density was observed in Shirur and Talegaon Dhamdhere circles due to the older flood plains of Ghod-Bhima, fertile soil, transportation network, Chaskaman canal irrigation, Shikrapur and Sanaswadi industrial, zone etc.
2. The villages having a good transportation network indicated high density of population, whereas, the remote villages and circles showed the lowest density in study area. Low, moderate, and high density patterns are observed in the study area.
3. Shirur and Talegaon Dhamdhere circles had the highest density whereas Takali Haji, Nhawara circles had the lowest density in 1991 and 2001. There were no high density circles in 1991 and no low density circles in 2011.

REFERENCES

Barakade, A.J. 2011. Changing pattern of population density in Satara district of Maharashtra, *Review of Research*, Vol 1, Issue 2: 1–4.
Borude, S.A. and Gaikwad, S.D. 2014. Application of Spatial variation Urban Density Model: A study of Ahmednagar City, Maharashtra, India, *Scholarly Research Journal For Interdisciplinary Studies*, Vol 2, Issue 14: 2081–2090.
Clarke, J.I. 1972. Population Geography, Pergamon Press Oxford: 29.
Mahajan, Y. 2012. A study of population growth and agriculture intensification of Jalgoan District, Sustainable rural development with inclusive approach, *Insight publications*, Nashik: 45–49.
Mahesha, D. and Shivalingappa, B.N. 2012. An Appraisal of Population Characteristics in Union Territories of India, *International Journal of Research in Management*, Vol. 3, Issue. 2: 25–30.
Mule, B.M. and Barkade, A. J. 2011. Growth of Population Change in Maharashtra (India), *Geoscience Research*, Vol. 2, Issue 2, pp. 70–75.
Narke, S.Y. 2010. Population density patterns in Ahmednagar District, Maharashtra, Bhugolshastra Parishad: 1–4.
Vikasvedh 1988. Chandmal Tarachand Bora College, Shirur, Dist Pune.

Multi-disciplinary Sustainable Engineering: Current and Future Trends – Tekwani, Bhavsar & Modi (Eds)
© 2016 Taylor & Francis Group, London, ISBN 978-1-138-02845-6

Structural evaluation of flexible pavement using Falling Weight Deflectometer

Ujjval J. Solanki
R.K. University, Rajkot, Gujarat, India
Department of Civil Engineering, Darshan Institute of Engineering and Technology, Rajkot, Gujarat, India

P.J. Gundaliya
Department of Civil Engineering, L.D. College of Engineering, Ahmedabad, Gujarat, India

M.D. Barasara
Department of Civil Engineering, Darshan Institute of Engineering and Technology, Rajkot, Gujarat, India

ABSTRACT: The structural evaluation of flexible pavement using FWD is carried out in Punjab from Mansa to Talwandi Sabo section having a length of 24.970 Km. The moduli values are back calculated as per IRC:115-2014 guidelines.

To identify the pavement condition directly from the deflection observe the differentiation of all the seven deflection observation is carried out. The differentiation of deflection value shows good correlation with pavement condition. The observed moduli value after fifth order differentiation of deflection values for Bituminous and Granular layers are in the range of −10 to −130 micron and the 6th order differentiation value range −60 to −90 micron shows the good condition of both the layers in terms of its stiffness. The same is presented in the study. It is observed that the old thin pavement layer shows high moduli of bituminous layer, as well as granular layer due to time being consolidation of pavement layers.

Keywords: falling weight deflectometer; moduli; structural evaluation

1 INTRODUCTION

Pavement evaluation is carried out to determine the existing condition of pavements in terms of its surface and structural adequacy. The data obtained from such studies are used for deciding the type of maintenance operations required, prioritization of maintenance works and for establishing a pavement maintenance management system. Evaluation of pavement surface condition can be studied with reference to the riding quality in terms of surface roughness, undulations cracking, potholes and other surface distresses. Such surveys are termed as Functional Evaluation Methods. Diagnosing the existing structural condition of individual layers of pavements as well as examining the overall pavement strength is termed as Structural Evaluation Of The Pavements.

Structural evaluation of in-service pavements is generally carried out to

– Assess the structural strength
– Estimate the remaining life of pavements
– Determine the thickness of overlay required
– Establish a pavement management system based on the performance of the road.

2 SCOPE AND OBJECTIVES OF THE STUDY

– To study the effect of pavement distresses on deflections.
– To study the moduli of the pavement with reference to pavement layer condition.
– To study the pavement condition directly from deflection observation.

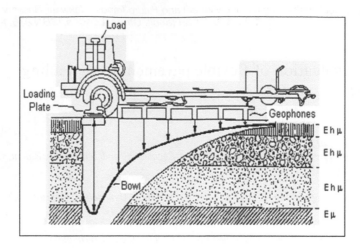

Figure 1. Deflection bowl in pavement layers due to FWD loading.

3 WORKING PRINCIPLES OF FALLING WEIGHT DEFLECTOMETER

Falling Weight Deflectometer (FWD) works on the principle of applying the impulse load to the pavement and measuring the shape of the deflection bowl. Since more than one surface deflection is measured, more information is available about the pavement. Using surface deflection obtained at suitably selected locations, interpretation can be made of different layers of the pavement. The schematic diagrams is showed in Figure 1. Surface deflections are normally measured at radial distances measured from the center of the load plate of 0, 300,600,900,1200 and 1500 mm.

4 STUDY AREA INTRODUCTION

The study area is located in the State Punjab, District: Sangrur, Mansa & Bathinda. The work contract Type—Output & Performance Based Road Contract-(OPRC) is first of its kind in India and supported by the World Bank. The study area as showed in Figure 2 is located in all the three districts. The study stretch is ODR-9, the section is S4, length is 24.97 Km Mansa to Talwadi sabo and treatment is improvement in riding quality.

5 PAVEMENT EVALUATION SURVEY AND DATA COLLECTION–(IRC:115-2014 GUIDELINES)

5.1 *Historical data of existing pavement layers*

To assess the existing sub-grade material properties, large trial pits of size 1.0 m × 1.0 m × 1.0 m were dug at every 5.0 km interval in staggered manner at the interface of carriageway edge and earthen shoulder for section S4. Cross section details are showed in Table 1.

The subgrade soil sample is collected and tested in the laboratory for the physical properties. Result is tabulated in Table 2. It is found that subgrade soil is low plasticity silty soil and sandy soil. Plasticity index is below 5 and major area has non plastic soil. Four days soaked CBR tested as per IS:2720 (part: 16) The observed CBR of sub grade for the section S-4 varies from 12.13% and 18.80%. As per the cross section details simultaneous layer of BT,WBM,BT,GSB and bricks layer were found. For the analysis, three layers have been divided as top BT first layer, and WBM + BT + GSB + Bricks are considered together as second layer and the subgrade is considered as the third layer for infinite depth.

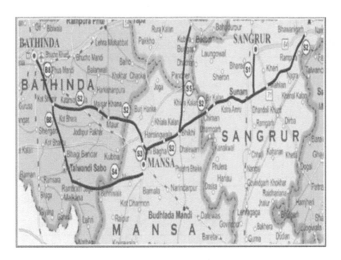

Figure 2.   Site location at Punjab, district Bathinda, Mansa and Sangrur—S4 Road.

Table 1.   Existing cross section detail of pavement.

Sr. No	Chainage-Km	Direction	BT (mm)	WBM (mm)	BT (mm)	GSB (mm)	Bricks (mm)	Sub grade (mm)	Total Thickness (mm)
1	5	LHS	45	80	–	–	70	300	495
2	10	RHS	45	140	30	120	–	300	635
3	15	LHS	30	55	30	85	70	300	570
4	19.7	RHS	25	60	30	70	70	300	555
5	24.15	LHS	30	70	30	90	70	300	590

BT-Bituminous layer, WBM-Water bound macadam, GSB-Granular sub base.

Table 2.   Subgrade soil physical properties.

| | | | | | | | | Grain size analysis | | | |
Sr. No	Chainage-Km	Direction	MDD gm/cm³	OMC (%)	Liquid limit (%)	Plasticity index	FSI (%)	Gravel (%)	Sand (%)	Silt and clay (%)	I.S. Soil Classification
1	5	LHS	1.94	11.7	23.8	2.4	9.38	1.08	44.3	54.6	ML
2	10	RHS	1.83	8.9	23.8	NP	3.12	2.1	47.9	50.0	SM
3	15	LHS	1.87	11.0	21.7	NP	3.33	0.7	51.3	48.0	SM
4	19.7	RHS	1.99	10.2	21.8	4.3	8.82	0.1	35.9	64.0	ML
5	24.15	LHS	1.94	10.0	23.9	NP	Nil	0.9	56.6	42.5	SM

5.2   *Pavement condition survey criteria (As per IRC:SP-19-2001)*

Pavement condition survey has been carried out as per IRC: SP: 19-2001 at an interval of 100 m. It revealed the extent of potholes, cracking, patching, ravelling which have been grouped into 5 categories based on the percentage area of distress and the yardstick adopted is given in Table 3.

Table 3. Adopted yardstick for pavement condition.

Sr. No	Condition	Pot holes %	Cracking (%)	Patching (%)	Raveling (%)	Rut depth
1	Excellent	Nil	≤ 5	Nil	≤ 1.0	10 mm
2	Good	≤ 0.05	>5 ≤ 10	≤ 0.5	>1.0 ≤ 2.0	10 mm
3	Fair	>0.05 ≤ 0.10	>10 ≤ 2.0	>0.5 ≤ 2.0	> 2.0 ≤ 5.0	10–20 mm
4	Poor	>0.10 ≤ 0.50	>20 ≤ 3.0	> 2 ≤ 6.0	>5.0 ≤ 10.0	10–20 mm
5	Very Poor	> 0.50	> 30	6.0	> 10.0	>20 mm

Table 4. Observed pavement condition.

		Pot holes		Cracking		Patching (%)		Raveling (%)		Rut depth
Sr. No	Condition	Km	(%)	Km	(%)	Km	(%)	Km	(%)	Km %
1	Excellent	17.3	69.2	10.1	40.4	21.9	87.6	25	100	10 mm
2	Good	0.0	0.0	5.1	20.4	1.3	5.2	0.0	0.0	10 mm
3	Fair	0.1	0.4	3.9	15.6	0.1	0.4	0.0	0.0	15 mm
4	Poor	0.1	0.4	2.3	9.2	0.7	2.8	0.0	0.0	15 mm
5	Very Poor	7.5	30.0	3.6	14.4	1.0	4.0	0.0	0.0	>20 mm

5.3 Pavement condition survey observation (As per IRC:SP-19-2001)

As shown in Table 4, about 70% length of the project road is in good to excellent condition and remaining 30% length is in fair to very poor condition.

5.4 Deflection measurement (IRC:115-2014 guidelines) and its differentiation

The deflection observation at 90 kN is normalized at 40 kN standard loads and the deflection profile is available as showed in Figure 4. The deflection observation is differentiated i.e $\Delta^1 d_1 = D_1 - D_2$ is first order, $\Delta^2 d_1 = \Delta^1 d_1 - \Delta^1 d_2$ is second order $\Delta^3 d_1 = \Delta^2 d_1 - \Delta^2 d_2$ is third order, $\Delta^4 d_1 = \Delta^3 d_1 - \Delta^3 d_2$ is fourth order, $\Delta^5 d_1 = \Delta^4 d_1 - \Delta^4 d_2$ is fifth order and $\Delta^6 d_1 = \Delta^5 d_1 - \Delta^5 d_2$ is sixth order differentiation.

As per FWD deflection observation, following three homogeneous sub-section observed as per deflection value shown in Table 5.

5.5 Determination of layer moduli

In order to be able to model the pavement into a linear multi-layered pavement model, layer thicknesses and material types have to be known. The crust thickness for S-4 (ODR-6/9) road is shown in Table 2. The moduli determination is carried out using KGP-BACK. The KGP-BACK is a specific version of BACKGA program, which was developed for the research scheme R-81 (2003) of the Ministry of RoadTransport and Highways, is recommended in these guidelines for back calculation. KGPBACK is a Genetic Algorithm based model for back calculation of layer moduli. Using KGP-BACK by input the range of moduli of existing old pavement, standard load, passions ratio of three layer back calculated moduli is determined. The back calculation moduli of all the three layer has been found using KGP-BACK.

The deflection observation and back calculated moduli is tabulated.The inputs are standard wheel load, passion's ratio of three layer, old pavement layer thickness and deflection observation, obtained through geophone and probable range of moduli of three layer. This is a very important parameter and it requires skill and experience of pavement layer properties.

144

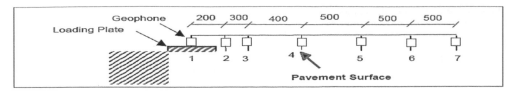

Figure 3.   Arrangement of geophone—spacing between two geophone.

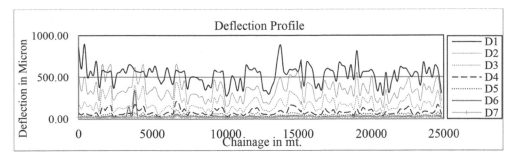

Figure 4.   Deflection profile of study stretch.

Table 5.   Summary of observed three homogenios subsection average normalized deflection.

Sr. No	Sub section From-to (Km)	Distance from load center (mm)						
		0	200	500	900	1400	1900	2400
		Deflection value in µm/m						
1	0–6.2	695	406	169	73	53	32	26
2	6.2–13.4	511	303	148	76	41	28	25
3	13.4–24.8	628	374	180	84	46	31	28

6   CONCLUSION

FWD deflection observation shows that the old thin pavement does not contribute any strength but shows higher moduli value of all the three layer in the range of 1000 to 1700 MPa for bituminous layer, 200 to 600 MPa for granular layer and 20 to 200 MPa for subgrade layer. These show that the consolidated layer of pavement provides structural strength.

For the prediction of pavement condition the deflection value also matches with the pavement structural strength performance and so, one can directly predict the pavement condition based on the deflection value.

The pavement moduli is back calculated which is time consuming and needs expertise for the input value of existing pavement moduli. The effort made here to predict pavement performance directly from the deflection value is carried out by differentiation of deflection value. The differentiation is carried out up to fifth order. The same is represented in Table 6. D_1 to D_7 first raw is actual deflection value and $\Delta^1 d_1$ is difference of D_1 and D_2 like wise up to sixth order differentiation is carried out in between observed seven deflection values and the following relation is found in between moduli value of bituminous layer and granular layer to differentiation of deflection for fifth and sixth order.

The differentiation of fifth order −12 to −130 range shows moduli 1541 to 1796 MPa. It shows the good condition of bituminous layer and 345 to 567 MPa is also a good condition of granular layer.

145

Table 6. Inference of differentiation with moduli value.

Back calculated moduli		Differentiation		
		5th Order		6th Order
BT	GB	$\Delta^5 d_1$	$\Delta^5 d_2$	$\Delta^6 d_1$
E1	E2			
1796.5	375.6	−128.00	−68.44	−59.56
1768.3	567.2	−86.22	−12.44	−73.78
1719	405	−56.89	−24.44	−81.33
1541	345	−41.78	−48.44	−90.22

The sixth order differentiation value −59 to −90 shows good condition of both the layers as its moduli value is considerable. So the pavement performance is directly available with the help of differentiation no need of condition survey it saves the time and energy.

REFERENCES

IRC:115-2014 "Guidelines for Structural Evaluation and Strengtheneing of Flexible pavement using Falling weight deflectometer".
IRC:37-2012 "Tentative guidelines for the design of Flexible pavements".
IRC:81-1997 "Guidelines for Strengthening of Flexible pavement using Bankelman Beam Deflection technique".
Pandey B.B. and Reddy, K.S. Research report on Structural Evaluation of pavements in Eastern India using Falling weight deflectometer Road research scheme (R-81), IIT Kharagpur 2003.
Srinivasa Kumar, R. (2011) "Text book of Pavement Evaluation and Maintenance management Highway" pp 299–357.

Site suitability analysis of solid waste dumping for Jaipur city using remote sensing and GIS

Ruchi Tah
Amity University, Jaipur, Rajasthan, India

Gautam Dadhich & Parul R. Patel
Department of Civil Engineering, Institute of Technology, Nirma University, Ahmedabad, Gujarat, India

ABSTRACT: The management of solid waste has become an acute problem due to enhanced anthropogenic activities and rapid urbanization in megacities of India. The government has paid attention to handle this problem in a safe and hygienic manner. Jaipur, a city of Rajasthan state, is also facing the same problem due to rapid growth of urbanization and industrialization along the periphery of the city. The current land filling site is located at Langadiawas, which is not in a condition to accommodate more waste in the near future. Many problems are arising, such as leachate percolation, odor, pollution, etc., in this area. This study deals with the identification of suitable sites for waste disposal around Jaipur by combining geomatics and Multi Criteria Decision Method (MCDM) tools. For the suitability analysis of sites for solid waste dumping, various factors have been taken into consideration, such as geology, geomorphology, and socioeconomic factors. This paper presents results obtained from geo spatial analysis in the form of suitability map for solid waste disposal site from higher suitability to least suitability around Jaipur city. The outcome of this study showed the efficacy of GIS and MCDM in decision-making for solid waste management.

1 INTRODUCTION

Solid waste management issues for local municipal authorities are a major environmental agenda that is growing due to a high population level, rapid economic growth, unplanned development, and rise in community living standard. Disease transmission, fire hazards, odor, noise, air and water pollution, aesthetic view, and economic losses are some of the problems connected with inappropriate management of domestic solid waste (Nwambuonwo and Mughele, 2012). The management of huge waste in terms of collection, storage, segregation, transportation, and disposal using conventional methods has become very difficult. The proper and scientific solid waste disposal requires the selection of a proper site, which is dependent upon several factors. These factors and governmental regulations make selection of new landfill sites the most complicated tasks for local municipal authorities. Satellite remote sensing images can provide information about the wasteland and other connected features, which help in the site selection for waste disposal. Coupled with GIS, it can offer a chance to assimilate in situ parameters with demographic and other data relevant to site selection (Yogeshwar Singh et al., 2012). The use of a GIS-based framework can be incredibly useful in locating the most appropriate site for a number of operations to ensure the quality of the location selected in a cost effective and timely manner (Javaheri et al., 2006). The major objective of this study is selection of best suitable site for solid waste disposal using GIS and MCDM techniques. For identifying potential waste dump sites, various non-exclusionary and exclusionary criteria (soil type, groundwater level, road network, drainage, and slope) are analyzed. MCDM technique is used to assign weights to various parameters. Maps are overlaid and buffering technique is used to carry out suitability evaluation.

2 STUDY AREA

Jaipur is located in Rajasthan state of India, having a total area of 200.4 km². The available site for dumping the solid waste has been used since the past few decades and within few more years it will not able to take more solid waste. It is already filled up to highest extent. Due to increase in the amount of solid waste, new dumping sites are required for Jaipur city. So, attempts have been made to carry out suitability analysis to find a new suitable site for waste disposal around Jaipur city.

3 METHODOLOGY

The main objective of this study is to identify suitable sites for solid waste dumping around Jaipur city. The methodology is catered into two parts: preparation of land use/cover maps and site selection criteria using spatial MCDM. The detailed flowchart of methodology adopted for this study is shown in Figure 1.

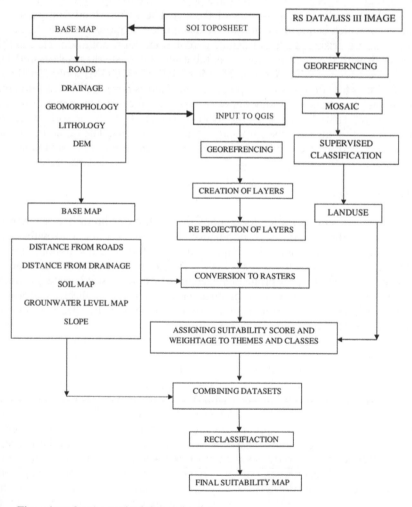

Figure 1. Flow chart for the methodology adopted.

Table 1.   Criteria used for selecting potential sites for dumping.

Suitability	Class		
Criteria	Highly suitable	Moderately suitable	Less suitable
Ground water level	50–70 m	35–40 m	20–30 m
Soil type	Clayey soil	Brown soil	Sandy soil
Slope	3°–15° & 15°–25°	25°–40°	0°–3° & 40°–87°
Distance from major roads in meter	500–2000	More then 2000	0–500
Distance from drainage in meter	300–400	More then 400	0–200
Land Use	Waste land	–	Agriculture

Three toposheets having a scale of 1:5000 are acquired and digitalized for this work. After geo referencing, all the toposheets are arranged in a mosaic in QGIS. A buffer of 18 km is created from the center of the city to extract the study area. The various thematic layers like Roads, Drainage, Ground Water Level, Soil Type, and Slope are created. Different researchers have used different criteria for site suitability for different geographical locations considering local rules and regulations, climatic conditions, and demographic limitations of the study area. To decide the criteria for this research work, solid waste management and handling rules 2000, Central Pollution Control Board (CPCB), and literature of Central Public Health and Environmental Engineering Organization (CPHEEO) are referred. Moreover, many research papers have been referred and found that, parameters like water permeability, ground water table level and distance from water bodies, distance from residential areas, distance from roads, drainage, soil characteristics, land use have been used by different researchers (Abu-Qadis, 2006; Al-Jarrah and H. Manu et al., 2004; Mahini and M. Gholamafard, 2006; Sadek et al. 2006; Stinnette, 1996). In this study, various parameters used are shown in Table 1 with their respective classes.

4   RESULTS AND ANALYSIS

The principal criteria used for the spatial analysis for four different cases are shown in Table 2 with weightage.

4.1   *Road network*

Solid waste dumping sites should not be very close or too far from the road. The road map of the study area is shown in Figure 2. Euclidean distance method is used to create a buffer for roads. A buffer of 0–500 m, 500–2000 m, and more than 2000 m is created. A scale of one to three is used to find out a suitable site for the waste disposal. Five hundred meters is too close in that it will create unhygienic conditions; hence, the score given is one and more than 2000 m is too far away; hence, the score given is two. The highly suitable site is between 500 and 2000 m, which is given a score of three.

4.2   *Drainage map*

From an environmental point of view, dumping sites should be there should be fairly away from the city. A drainage map of the study area is shown in Figure 5. Buffers of 0–200 m, 300–400 m, and more than 400 m have been created. Out of which, a buffer of more than 400 m has the highest score, whereas distances less than 200 m is found unsuitable for dumping waste.

Table 2. Criteria used for selecting potential site for dumping.

Theme	Weightage %			
	Case I	Case II	Case III	Case IV
Soil type	40	30	20	20
Ground water level	20	30	40	20
Distance from roads	15	15	10	20
Distance from drainage	10	15	10	20
Slope	10	05	10	10
Land cover	05	05	10	10

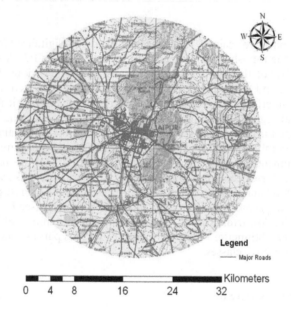

Figure 2. Road network map of the study area.

4.3 Soil type map

The infiltration rate is an important criterion for site selection. The soil map of the study area is shown in Figure 4. As clayey soil is less permeable, it is given the highest score whereas sandy is given the least score because of high permeability.

4.4 Groundwater table map

The ground water table is a substantial factor in assessing the contamination risk. Ground water table map of the study area is shown in Figure 5. The ground water level is observed between 20 and 70 m in the study area. The highest score of 3 has been given to water levels between 50 and 70 m, and 1 is given to water levels between 20 and 30 m.

4.5 Slope map

As per the selection criteria, the site should not have steep slopes. In the monsoon, solid waste will drain with the rain and accumulate below. Hence, higher weightage is given to flat surfaces compared to steep slopes. Slope map of the study area has been prepared using Digital Elevation Model.

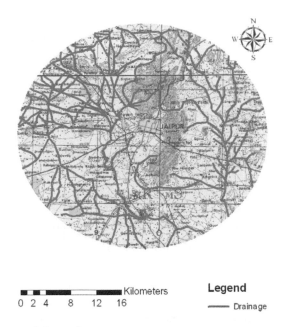

Figure 3.   Drainage map of the study area.

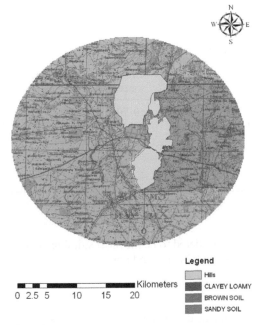

Figure 4.   Soil type map of study area.

4.6   *Land cover*

Two **LISS III** satellite images have been used for generation of the land cover map. The supervised classification in **ERDAS** Imagine is performed using the maximum likelihood algorithm of the classified data. As per the criteria, fallow land/waste land or area covered

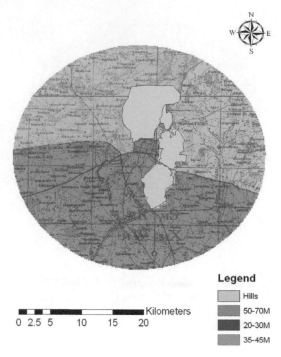

Legend

Hills

50-70M

20-30M

35-45M

Kilometers
0 2.5 5   10   15   20

Figure 5.   Ground water table map.

Table 3.   Suitability score given for each criteria.

Road distance meter	Distance of drainage	Type of soil	Ground water level	Slope in degree	Type	Suitability score
500–2000	>400	Clayey loamy	50–70 m (Deep)	3–15	Fallow Land	3
More than 2000	300–400	Brown soil	30–45 m (Medium)	15–25	Bushes and Shrubs	2
0–500	0–200	Sandy soil	20–30 m (shallow)	25–40	Inactive Agriculture Land	1

with bushes and shrubs are considered to be the most suitable area of disposal of solid waste. Table 3 shows weightage assigned for the land cover.

4.7   *Final suitability map*

Maps of distance from drainage, distance from major roads, soil type, ground water level, and slope are given in the figures, respectively. After creating layers, all the layers were projected and were converted to raster files. Based on rank and weightage separate datasets are created. For analysis, all the raster datasets for different layers having different scores were over layered, and using weighted overlay method, the final result output map is generated as shown in Figure 6. It shows four different cases having different weightages for different criteria. Results obtained from site suitability analysis prove that highly suitable areas can adequately accommodate the waste to be generated for the next five decades.

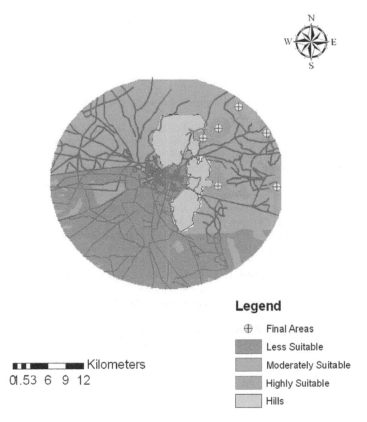

Figure 6.   Final suitability map.

5   CONCLUSION

In this study, an attempt has been made to investigate the most suitable site for the disposal of solid waste in Jaipur city using GIS and MCDM. The study systemized the entire procedure of analysis by reclassifying the criteria, such as soil type, groundwater level, distance from road network, distance from drainage, and slope, giving weightage for each criterion such as 1 for unsuitable and 3 for suitable site. It gives the most suitable site for waste disposal. Based on the four different cases, highly suitable areas are located in the Northeast and East direction of Jaipur city within 15 km from the center of Jaipur City. This area is beyond the hilly terrain and well connected by road. As the roads are passing through hilly areas, the impact on the environment during transportation is yet to be explored. The selected sites comprise clayey soil and ground water level is between 50 and 70 m. Most of the selected sites have fallen to fallow land of the outer periphery of Jaipur city. The distance of these potential sites from the road is between 500 m and 800 m. These selected potential sites for waste dumping are at distance of 300–400 m from drainage. Thus, from this study, a conclusion can be drawn that open source Q-GIS software is an equally effective tool for selecting solid waste dumping sites considering various selection criteria for waste dumping site.

ACKNOWLEDGMENT

We would like to thank all, who have directly or indirectly helped us in carrying out this study. We are also extremely indebted to Director, IT and Management, Nirma University for providing thte necessary infrastructure and resources to accomplish our research work.

We are grateful to Shri O.P. Goyal, SMEC India Pvt. Ltd, Jaipur for providing their support to carry out this work.

REFERENCES

Al-Jarrah, O. & Abu-Qdais, H. 2006. Municipal solid waste landfill sitting using intelligent system. Waste Manage. 26, 299–306.

Javaheri, H. et al., 2006. Site selection of municipal solid waste landfills using analytical hierarchy process method in a geographical information technology environment in giroft 1., 3(3), pp. 177–184.

Mahini S.A. & Gholamafard M. (2006). Siting MSW Landfills with a Weighted Linear combination Methodology in a GIS Environment. Int. J. Environ. Sci. Technol. 3(4):435–445.

Manu, A., Twumasi, Y.A., Coleman, T.L., Maiga, I.A. & Klaphake, K. (2004). Database Development for urban planning using photogrammetry and GIS techniques: the case of Niamey, Niger. Paper Submitted to Proceedings of the 5th African Association of Remote Sensing of the Environment (AARSE) Conference. 18–21 October, Nairobi, Kenya.

Nwambuonwo, O. Jude & Mughele, E.S. (2012). Using Geographic Information System to Select Suitable Landfill Sites for Megacities (Case Study of Lagos, Nigeria). Computing, Information Systems & Development Informatics Vol. 3 No. 4, September, 2012.

Sadek S., Fadel M.E. & Freiha F. (2006) Compliance factors within a GIS-based framework for landfill siting. International Journal of Environmental Studies 63: 71–86.

Stinnette D.S., "10 Steps to Successful Facility Siting," 1996. Waste age, Internet Available: http://wasteage.com.

Yogeshwar Singh, Dr. Chauhan, M.S.& Dr. Katiyar, S.K. (2012), SWM of Kolar Municipality Using Remote Sensing and GIS Techniques, International Journal of Advance Technology & Engineering Research, 2(1), pp 30–33.

Chemical process development and design

Multi-disciplinary Sustainable Engineering: Current and Future Trends – Tekwani, Bhavsar & Modi (Eds)
© 2016 Taylor & Francis Group, London, ISBN 978-1-138-02845-6

Mathematical modeling and simulation of gypsum drum dryer

Bruhad S. Naik
GIDC Degree Engineering College, Abrama, Gujarat, India

A.M. Lakdawala
Institute of Technology, Nirma University, Ahmedabad, Gujarat, India

ABSTRACT: The study aims to solve the problems faced by JK Lakshmi Cement, Kalol. The plant is using wet chemical gypsum as one of the raw materials. The material, being sticky in nature, causes numerous problems in transportation and also overflows the hoppers. The suggested solution of this problem is drying with Inclined Drum Dryer (IDD). The design of IDD is done by generating mathematical model by making heat and mass balance. Two models are generated, One for sizing and one for rating. Both models are converted in to computer codes. Plant requirements are considered as reference and with trial and error method final design is suggested.

Keywords: mathematical modeling; drying of gypsum; simulation with coding

1 INTRODUCTION

Inclined Rotary Drum Dryer (IDD) is one of the most widely used drying techniques in industry. Many authors have suggested mathematical model for thermal design of IDD. Most of them however are specific to the product. The earliest approach of such a design is more dependent on actual data generated from working dryers. Myklested [1] is first to develop an expression for final moisture contains of product. Kamke et al. [2][3] has developed a computer code for retention time (Part I) and mass transfer (Part II) for cascading rotary dryer. The problem with this mathematical model is: they are more specific to a dryer and not general. Iguaz et al. [4] proposed a mathematical model for vegetable wholesale by-product by dividing the drum into 10 control volume and applying the heat and mass balance to each control volume. A similar approach of division of control volume is used for analysis of an industrial ammonium nitrate plant by Abbasfard et al. [5]. The study aims to compare data gathered from actual plant with data generated from mathematical modeling. The retention time equation of Foust et al. when used in mathematical model the generated data is closer to actual data than any other model. Fernandes et al. [6] has presented a study of comparison between the industrial and modeling data. Retention time equation of Friedman and Marshall is used for mathematical modeling.

2 DRYER DESCRIPTIONS

A schematic diagram of concurrent dryer assembly is shown in the Fig. 1. The Feeding of material can be done with the help of hopper and belt conveyor assembly. The exhaust air is passing through the cyclone collector and bag collector for collection of residue. The air at 30°C and having moisture contain of 0.001 kg/kg dry air is heated up to 350°C with final moisture contain of 0.001 kg/kg dry air in furnace. The gypsum is entering from inlet chute in the drum with moisture contain of 30 kg/kg dry gypsum and temperature of 30°C. The drum is equipped with straight radial flights for lifting of material.

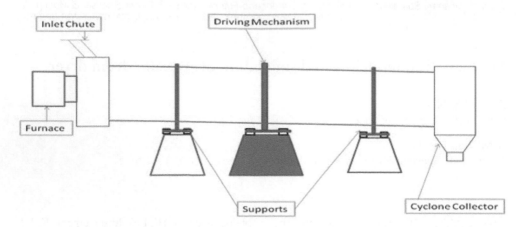

Figure 1. Concurrent Industrial Drum Dryer.

3 MATHEMATICAL MODELING

The thermal design of dryer is done by generating two mathematical models. First model, the sizing model is for sizing of dryer. The sizing model provides the 1st approximation of geometry of drum. Second model, the rating model is used for optimizing the dryer by using data generated from first model.

3.1 Sizing model

3.1.1 Mass balance

The Gypsum having $x_i\%$ moisture is fed with rate of Co kg/hr. Hence the total dry solid in the feed is $\dot{m}_m = C_o(1-x_i)$ and water rate in the feed $\dot{m}_w = Cox_i$. The moisture is reduced to $x_o\%$ at the end of drying. Hence the final water rate in the product is $\dot{m} = \left(\frac{x_o \dot{m}_m}{(1-x_o)}\right)$. The total amount of water evaporated is $\dot{m}_e = \dot{m}_w - \dot{m}$.

3.1.2 Heat balance

The mass flow rate of air required to evaporate the required amount of moisture from product is found by the heat balance as:

Total heat Required = (heat required to raise the Material temp to safe exittemp)
+ (Heat supplied to residual moisture)
+ (Sensible heat supplied to the evoprated moisture)
+ (Latentheat supplied to evopartaed water)

$$\dot{Q}_1 = (\dot{m}_m c_m (T_{m,o} - T_{m,i})) + \left(\dot{m}c_m\left(\frac{T_{m,o}+T_{m,i}}{2}\right)\right) + \left(\dot{m}_e c_m\left(\frac{T_{m,o}+T_{m,i}}{2}\right)\right) + (\dot{m}_e h_{fg})$$

The heat is lost from the surface of the drum that is nearly equal to 10% of the total heat required. Hence, $\dot{Q}_{loss} = 0.1\dot{Q}_1$. Mass of air required to transfer heat then will be

$$\dot{m}_a = \left(\frac{\dot{Q}_1 + \dot{Q}_{loss}}{c_a(T_{a,o} - T_a)}\right)$$

Diameter of the dryer can be found out by following equation

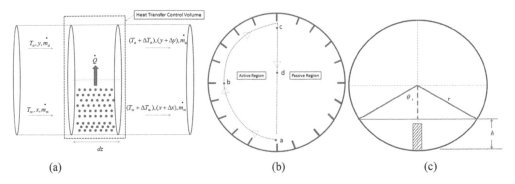

Figure 2.   a) CV for calculating heat and mass transfer (b) Number of flights taking part in drying (c) Geometry of a single flight.

$$D = 2\sqrt{(V_{total} / \pi v_c)}$$

Where, V_{total} is total quantity of gas inside the drum $V_{total} = V_{air} + V_{water}$ and v_c is conveying velocity of air i.e

$$v_c = \frac{6000 \, d_p^{\,0.4}}{(s+1)} \frac{1.2}{\rho_{air}}$$

3.1.3 *Flight design*

With reference to Figs. 2 (b and c), the height of flight $h = R(1 - \cos \theta_i)$. Where θ_i is calculated from $\theta_i = 0.321 \, H_0^{\,0.3544}$. Here, H_o is material holdup selected. Similarly, based on the calculated h the length of flight "l" is also selected. The redial angle covered by a single flight can be calculated from geometry i.e. $\theta_n = tan^{-1}\left(\frac{l \tan \theta_r}{R}\right)$ hence total number of flight is $N_f = \left(\frac{360}{\theta_n}\right)$.

3.1.4 *Length of drum*

Length of the drum depends on LMTD and total heat transfer rate i.e.

$$L = \left(\frac{\dot{Q}_{total}}{U_a A(\Delta T)_m}\right)$$

$$U_a = K\left(\frac{h_c A_m N_p}{A}\right) \qquad \text{where } h_c = \left(k_g \frac{2 + (0.6 Re^{0.5} Pr^{0.33})}{d_p}\right)$$

3.2 *Rating model*

The rating of drum is modeled by dividing the drum in number of control volumes of length d_z and N_p number of practical in each control volume. Fig. 3 (a) shows the control volume for gypsum and Fig. 3 (b) shows control volume for air.

Assuming that the initial values of temperature of gypsum and air are $T_{m,i}$, $T_{a,i}$ respectively and moisture contain of 0.3 (w/w) and 0.1 (w/w) respectively, the heat and mass balance is applied at each and every CV in such a way, that the output of one CV is input to the consecutive control volume. The analysis is done with the following equations

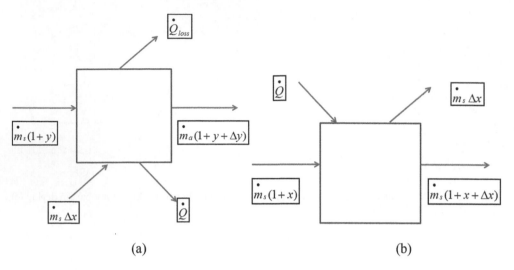

Figure 3.   (a) Gypsum control volume (b) Air control volume.

$$\Delta T_a = \left(-\frac{\dot{m}_s(c_{p,s}+xc_{p,l})+\dot{m}_a\Delta y(h_{fg}+c_{p,v}(T_a-T_s)+\dot{Q}_{loss}}{\dot{m}_a(c_{p,a}+yc_{p,v})} \right)$$

$$\Delta T_a = \frac{\left(hA_p\left\{ T_a-T_s+\left(\frac{A_1}{2}\right)\right\}\right)-\left(h_{fg}A_p\left(\frac{P_{sat}}{R_1T_a}-\frac{Py}{(y+0.62)R_2T_a}\right)\right)}{\left(\left(\frac{\dot{m}_p v}{fdz}(c_{p,s}+xc_{p,l})\right)-\left(\frac{hA_p(A_2-1)}{2}\right)+\left(\frac{h_{fg}k_mA_p}{2R_2T_s}\left(\frac{dP_{sat}}{dT}\right)\right)\right)}$$

$$\Delta x = \left(-\frac{k_mA_pfdz}{\dot{m}_p u_{eff}} \right)\left(\left(\frac{P_{sat}}{R_1T_s}\right)-\left(\frac{yP}{(y+0.62)R_2T_a}\right)\right)$$

$$\Delta y = \left(-\frac{\dot{m}_s}{\dot{m}_a} \right)\Delta x$$

3.2.1   *Retention time*

The retention time of dryer is found out with the help of matchett and Sheikh [7] equation as follows

$$\tau = \left(\frac{L\left(N_{rpm}t_{fall}+\delta\right)}{N_{rpm}t_{fall}\left(U_1+\frac{aD\delta\tan\alpha}{t_{fall}}\right)} \right)$$

Table 1. Sizing model results.

D (m)	L (m)	l (m)	N_f	x_o (w/w)	y_o (w/w)	$T_{a,o}$ (°C)	m_a (kg/hr)
1.88	15.694	0.49	19.31	0.10	0.124	82	9711

Table 2. Rating model results and final selected parameters.

D_p(m)	D (m)	L (m)	l (m)	N_f	N_{rpm}	x_o (w/w)	y_o (w/w)	$T_{a,o}$ (°C)	$T_{m,o}$ (°C)
0.001	1	8	0.51	19	3	0.10	0.16	77.96	63.87

4 RESULTS AND DISCUSSION

The rating and sizing models are converted into codes with C++ program. The sizing model code is used for calculation of geometrical parameters and the calculated geometrical parameters are rated with rating model code. The sizing code results are tabulated in Table 1. The results show that L/D ratio is 8.3 and is within the permissible limit of 5 to 10 and final moisture contains of air and gypsum are within required limits. The final air and gypsum temperature is in good fit with selected range. Hence, the result of sizing model is accepted.

The result of sizing model is reselected with nearly L/D ratio of 8 making length of drum as 8 m and diameter of drum as 1 m. The rating model code is then used for rating of drum dryer. The rating model code gives the final selected parameter. The selection procedure is done in this order. Material screening is performed for average particle diameters and 60% of material is around 0.001 m hence the average particle diameter is selected to be 0.001 m. The code is then used for selection of flight length and number of flights with average particle diameter as 0.001 m. Similarly, the code is used for selection of inlet air temperature and RPM of drum each time the previously finalized parameter is made constant. The result of rating code is tabulated with Table 2.

5. CONCLUSION

The sizing and rating analysis of inclined drum dryer for gypsum drying is carried out. It is shown that the sizing and rating model is robust enough to calculate required dimensions of inclined drum dryer.

REFERENCES

[1] Myklestad, O. (1964). "Heat and Mass Transfer in Rotary Dryers", Chemical Engineering Process Symposium Series 59 (41), 129–137.
[2] Kamke, F.A. & Wilson, J.B. (1986). "Computer Simulation of Rotary Dryer Part-I: Retention Time". AIChE Journal Vol.32 No.2, 263–268.
[3] Kamke, F.A. & Wilson, J.B. (1986). "Computer Simulation of Rotary Dryer Part-II: Heat Transfer". AIChE Journal Vol.32 No.2, 269–275.
[4] Iguaz, A., Esnoz, A., Martiez, G, Lopez, A. & Virseda, P. (2003). "Mathematical Modeling and Simulation for the Drying Process of Vegetable wholesale by-Product in a Rotary Dryer". Journal of Food Engineering 59, 151–160.
[5] Abbasfard, H., Rafsanjani, H.H., Ghader, S. & Ghanbari, M. (2013). "Mathematical Modeling and Simulation of an Industrial Rotary Dryer: A case study of Amonium Nitrate Plant", Powder Technology 239, 499–505.
[6] Fernandes, N.J., Ataide, C.H. & Barrozo, M.A.S. (2009). "Modeling and Experimental Study of Hydrodynamic and Drying Characteristics of an Industrial Rotary Dryer". Brazilian Journal of Chemical Engineering, Vol.26, No.2.
[7] Matchett, A.J. & Sheikh, M.S. (1990). "An Improved Model of Particle Motion in Cascading Rotary Dryers", Trans. Instn. Chern. Engrs. 68. Part A, March, 139–148.s.

Multi-disciplinary Sustainable Engineering: Current and Future Trends – Tekwani, Bhavsar & Modi (Eds)
© 2016 Taylor & Francis Group, London, ISBN 978-1-138-02845-6

Development of a software for design of cyclone separator: CySep

Ronak Patel, Arpit Parikh & Sanjay Patel
Department of Chemical Engineering, Nirma University, Ahmedabad, India

ABSTRACT: Cyclone separator is a widely used industrial equipment for separation of solid and gas mixture. Process design of the cyclone affects the separation efficiency to greater extent. Design of cyclone based on Stairmand's cyclone is an iterative procedure. Software developed on user friendly platform can reduce human effort and increase efficiency. In current study, user friendly software 'CySep' is developed to calculate separation efficiency of cyclone separator. Moreover, results obtained from the 'CySep' is compared and found very accurate with already published literature.

1 INTRODUCTION

Cyclone separator is widely used as a gas-solid separator in process industries. Due to stringent environment regulations, it is now mandatory to separate dust particles from exhaust gases. It is also used in industries for the separation of catalyst particles laden with process gas. No moving parts and easy construction makes cyclone separator an economically viable option as a good gas-solid separator (Elsayed c al. 2010, Funk et al. 2014, Safikhani et al. 2011). Cyclone separator separates gas-solid using centrifugal force. Gas-solid mixture enters into the cyclone tangentially, which creates swirling motion, due to which heavy solid particles are directed towards wall of the cyclone and moves downwards (Cortes et al. 2007). As gas moves further downward more and more solid particles are separated. In cyclone separator, there are two vortices namely inner vortex and outer vortex. Gas-solid mixture, first directed in outer vortex and moves downwards after reaching to the bottom of cyclone gas take inner vortex path as it provides least resistance for gas and relatively pure gas is collected from the top of cyclone separator. Solid particles are collected from the bottom of the cyclone separator which provides least resistance path for solids.

Process design of cyclone separator is relatively easy, but iterative procedure of designing makes it quite cumbersome and less accurate. Design of cyclone is dependent on various factors like, particle size distribution of solids, desired efficiency, as well as pressure drop. Smaller size particles are difficult to separate compared to larger size particles and presence of smaller particles reduces efficiency of cyclone separator. It is recommended to use cyclone above particle size of more than 5 µm as efficiency is better with larger size particles (Elsayed et al. 2010, Sinnot et al. 2005). In order to achieve higher efficiency in less number of cyclones, one can not neglect pressure drop aspect. Actual pressure drop within the cyclone, must be less than or equal to maximum allowable pressure drop set by process constraints. By developing user friendly software, we can reduce human efforts and enhance accuracy. For this purpose, Java platform has been selected for programming language. Designer can use this software for designing and simulation. The design of software is based on Stairmand's standard cyclones. In present study, we have limited our focus to efficiency calculation.

2 METHODOLOGY

Designing and simulation is based on Stairmand's standard cyclone. Stairmand has developed effciciency curves for (i) High efficiency and (ii) High throughput cyclones. Design

Table 1. Design conditions for Stairmand's standard high efficiency and high throughput cyclones (Sinnot et al. 2005).

Sr. No.	Design conditions	High efficiency cyclone	High throughput cyclone
1	Test fluid	Air	Air
2	Density difference between gas and solids	2000 kg/m^3	2000 kg/m^3
3	Temperature	20°C	20°C
4	Pressure	1 atm	1 atm
5	Volumetric flowrate of gas	223 m^3/h	669 m^3/h
6	Viscosity of test fluid	0.018 mNs/m^2	0.018 mNs/m^2

Table 2. Dimensions of standard Starimand's cyclones (Sinnot et al. 2005).

Sr. No.	Dimension	High efficiency cyclone	High throughput cyclone
1	Inlet height	0.5 D$_c$	0.75 D$_c$
2	Inlet width	0.2 D$_c$	0.375 D$_c$
3	Vortex finder diameter	0.5 D$_c$	0.75 D$_c$
4	Total cyclone height	4 D$_c$	4 D$_c$
5	Cylinder height	1.5 D$_c$	1.5 D$_c$
6	Vortex finder length	0.5 D$_c$	0.875 D$_c$
7	Cone tip diameter	0.375 D$_c$	0.375 D$_c$

conditions for standards cyclones are tabulated in Table 1, while dimensions of standard cyclones in terms of diameter of cyclone (D$_c$) are tabulated in Table 2. Diameter for both standard cyclones (High efficiency and High throughput) is 203 mm. Actual design of cyclone is based on these standard cyclones. Scaling Factor (SF) determines deviation from standard cyclone. If SF is near to one, then design is near to standard design. Higher SF values indicate larger deviation from standard (Sinnot et al. 2005).

In present study, we have taken different data points for efficiency curve manually from elsewhere (Sinnot et al. 2005) and using Excel as a computation tool, curve fitting is done.

3 DEVELOPMENT OF SOFTWARE

Graphic User Interface (GUI) is developed on the Java. There are two modes in CySep (i) Design and (ii) Simulate. In designing of cyclone, designer has to design equipment based on feed flowrate, particle size distribution, density of solids, density of gas and desired separation efficiency. So, these parameters will become inputs for designing and the final design will give dimension of cyclone and number of cyclones in parallel required to achieve desired efficiency. In simulation of cyclone, designer will get idea whether given number of cyclones would be able to separate solids from gas at desired efficiency at given process conditions or not. For illustration purpose, snapshot of GUI is shown in Figure 1. On left hand side of GUI for design mode, designer has to input variables like volumetric flow rate, density of solid particles, density of gas, required efficiency and particle size distribution. In simulate mode, designer has to input variables like volumetric flow rate, density of solid particles, density of gas, number of cyclone in parallel and particle size distribution. In design mode, 'CySep' will calculate the number of cyclones in parallel required to achieve desired efficiency, achieved efficiency and diameter of cyclone in design mode.

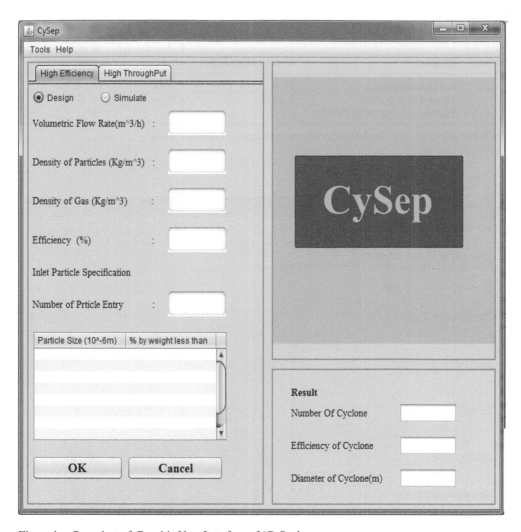

Figure 1.　Snapshot of Graphic User Interface of 'CySep'.

Table 3.　Particle size distribution of solids.

Sr. No.	Particle size, μm	Percentage by weight less than
1	50	90
2	40	75
3	30	65
4	20	55
5	10	30
6	5	10
7	2	4

While in simulate mode, 'CySep' will calculate achieved efficiency and diameter of cyclone for entered number of cyclones.

For comparison purpose, 'CySep' is compared with the solved example elsewhere (Sinnot et al. 2005).

Table 4. Comparative study.

Sr. No.	Parameter	Results from (Sinnot et al. 2005)	CySep Design mode	CySep Simulate mode
1	Achieved efficiency	88.7%	82.649%	88.427%
2	Diameter of cyclone	0.42 m	0.86 m	0.43 m
3	Number of cyclone in parallel	4	1	4*

*Number of cyclones in parallel has to be entered by user in simulate mode

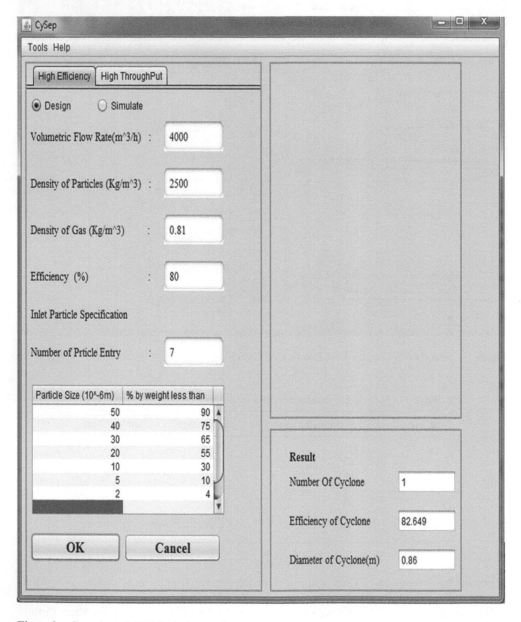

Figure 2. Snapshot of GUI in design mode.

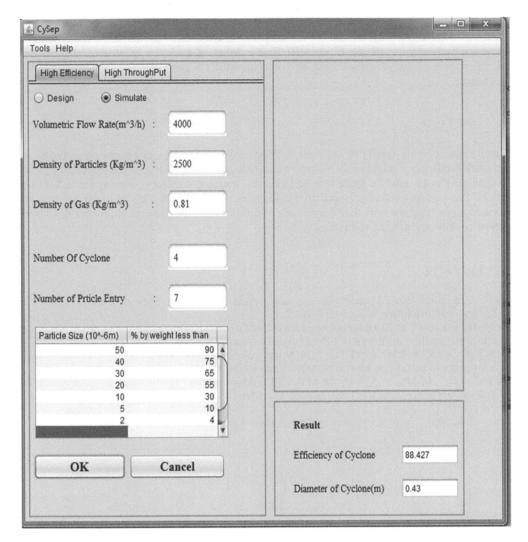

Figure 3.   Snapshot of GUI in simulate mode.

3.1   *Problem statement*

Design a cyclone separator for separation of solid particles form gas stream. Particle size distribution in the inlet gas stream is tabulated in Table 3. Density of solid particles is 2500 kg/m³. Gas is nitrogen at 150°C. Operating pressure within the cyclone is 1 atm. Volumetric flowrate of gas is 4000 m³/h and 80 percent recovery of solids is required. Comparative study of solved example from elsewhere (Sinnot et al. 2005) and 'CySep' is tabulated in Table 4. Snapshots of results obtained through 'CySep' in design and simulate mode are shown in Figures 2 and 3 respectively.

In design mode, 'CySep' will start iteration with minimum number of cyclones i.e. one. It will initiate loop with the calculation of achieved efficiency with minimum number of cyclones. If desired efficiency has been achieved, then it will break the loop. Otherwise it will increase the number of cyclone by one and continue the iteration till the actual efficiency is equal to or more than the desired one. In the given problem, desired separation efficiency is 80 percent. As per the results obtained in design mode, using a single cyclone having diameter of 0.86 m separation efficiency of 82.649 percent can be achieved.

In simulate mode, 'CySep' will calculate achieved efficiency for given number of cyclones. In simulate mode, software will not go for iteration. It will directly predict achieved efficiency for given number of cyclones. If we enter four number of cyclones in parallel, then as per the software achieved efficiency is 88.427 percent, which is quite close to already published literature.

4 CONCLUSION

Results obtained on developed software 'CySep' are found to be quite accurate in terms of separation efficiency calculation. Software can reduce human efforts to considerable extent, as designing is an iterative procedure and can also increase accuracy to many folds. Limitation of this software is that pressure drop calculations and other models have not been incorporated. Continuous efforts have been made to modify the software and obtain results which can be readily accepted by industries.

REFERENCES

Cortes C. & Gil A. 2007, Modeling the gas and particle flow inside cyclone separators. Progress in Energy and Combustion Science 33: 409–452.
Elsayed K. & Lacor C. 2010, Optimization of the cyclone separator geometry for minimum pressure drop using mathematical models and CFD simulations. Chemical Engineering Science 65: 6048–6058.
Funk P.A., Holt G.A. & Whitelock D.P. 2014, Novel cyclone empirical pressure drop and emissions with heterogeneous particulate. Journal of Aerosol Science 74: 26–35.
Safikhani H., Hajiloo A. & Ranjbar M.A. 2011, Modeling and multi-objective optimization of cyclone separators using CFD and genetic algorithms. Computers and Chemical Engineering 35: 1064–1071.
Sinnot R.K. 2005, Chemical Engineering Design. Oxford: Elsevier Butterworth-Heinemann.

Technologies for green environment

All time operating photo voltaic and thermal hybrid solar cooker

S.B. Joshi & A.R. Jani
Department of Physics, G. H. Patel College of Engineering and Technology, Sardar Patel University, Gujarat, India

ABSTRACT: A small-scale photovoltaic and thermal hybrid casserole-type solar cooker has been developed and tested for various food recipes. The solar cooker was tested for cooking rice, pulses, grains, cereals, curry, vegetables, dumplings, pancake etc. The sensory evaluation test, which is performed on the attributes of texture, flavor, appearance, aroma, and mouth feel, scored higher or equal, when compared with the conventionally cooked food. The cooking performance showed that the cooker is ideal for boiling, roasting, and baking purposes. At a time, seven dishes can be prepared. Four to five meals can be cooked on a clear sunny day. This unique cooker is lightweight (1.6 kg), portable, user-friendly, and affordable.

Keywords: solar cooker; photovoltaic effect; thermal effect; solar tracker

1 INTRODUCTION

Conventional solar cookers are not easily acceptable as the time required for cooking is more. Working efficiency of a cooker reduces if any shadow falls on it; therefore, it requires to be located in a shadowless area. Due to rising land prices and population, high-rise residential buildings have become common and affordable. The high-rise buildings do not have sufficient open space, open to the sky, for the use of solar cookers. This has given rise to new designs of various solar cookers that can be used for indoor cooking. There is still a dire need to develop a suitable type of solar cooker for providing relief to the women (Kuhnke, 1988). Hybrid photovoltaic and thermal technology provides solar conversion into electricity and heat. This technology is useful for the future trends in the development of solar greenhouse, water desalination, solar heating, photovoltaic-thermal solar heat pump, solar still, air-conditioning system, and solar power co-generation (Tyagi et al., 2012). Economic and energy efficient feasible (Photovoltaic/Thermal) PV/T systems are desired to be developed for optimizing their structural/geometrical configurations (Xingxing et al., 2012).

An all-time operating photovoltaic and thermal hybrid solar cooker has been developed. The cooker can be preferably kept in a balcony near the kitchen where the available sun rays will provide thermal energy along with the photovoltaic energy that can be received through the PV module. Therefore, Photovoltaic and Thermal hybrid solar cooker makes it possibly a 24-hour usable cooker. The cooking time during the daytime can be reduced a lot compared with the conventional solar cooker. When there is no cooking, the solar PV module would be charging the battery, which can be used for cooking as well as lighting purposes. Scientifically, the present work has a novel and innovative approach towards the designing of an all-time working cooker, which can be used for drying as well as water purification. A dual axis auto tracking system was connected to the PV module to capture the maximum amount of solar radiation. The dual axis auto tracking system, which is installed at the Department of Physics, Sardar Patel University, Vallabh Vidyanagar, can generate

1.5 kilowatt per hour of electrical power using 20 solar modules each with 75 watt capacity. This dual axis tracker works on the principle of sunflower motion. Every day the tracker starts automatically at 5:45 am and shuts down at 6:45 pm. This dual axis sun tracker can increase the maximum power generation capacity by 45%.Of the 20 above mentioned solar PV modules, one was used to supply power to the DC heater of the developed hybrid solar cooker and the performance of the solar cooker was tested in tracking and non-tracking mode of PV module (Joshi & Jani, 2013b).

A very small scale, light weight (1.6 kg), casserole type hybrid solar cooker was developed and tested for its maximum utilization. Different recipes were cooked in the casserole type cooker for testing its actual cooking performance (Joshi & Jani, 2013a). The cooker was tested using a fixed power supply of 30 watt (Joshi & Jani, 2013c).The use of ionic liquids for storing heat during the daytime and releasing the same at nighttime was studied. Ionic liquids increase the heat retaining time of the solar cooker (Joshi & Jani, 2014).The casserole type solar cooker with some modifications was also utilized for drying applications. Bitter gourd chips and potato chips were dried in this solar cooker cum dryer (Joshi et al., 2014).

In this article, the authors report with suitable methodology for cooking various types of dishes in the casserole type solar cooker (Fig. 1) with supporting figures, schematic diagram, photographs, and scientific results.

2 METHODOLOGY

2.1 *Testing with solar PV module*

Figs. 1 and 2 show a schematic diagram of the casserole type (1.6 kg) hybrid solar cooker and the photograph of the solar cooker with test facility respectively. A 75 watt solar PV module was mounted on the terrace of the department and the cooker, battery, and test facility were kept in the laboratory. The PV module was connected to the battery backup using a charge controller. Two multimeters were connected for the measurement of voltage and current to calculate the power input to the heater. The heater was fixed at the bottom of the solar cooker. The pot with the water was kept on the heater. The testing was done in the laboratory, so the cooker was closed with its lid. Pt-100 thermocouples connected with temperature indicator used to measure the temperature. The stainless steel pot was filled with 100 ml water, kept in the cooker, and the cooker was closed with the upper lid. Heater temperature and water temperature were measured at an interval of 10 minutes. The battery was connected for 40 minutes with the cooker, and then it was disconnected and the temperature profile of the cooker was studied. Within 40 minutes, the boiling temperature essential for cooking food was obtained.

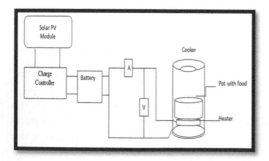

Figure 1.   The schematic line diagram of test facility.

Figure 2.   Solar cooker with test facility.

3 RESULTS AND DISCUSSION

The variation of temperature for simultaneous cooking of rice and pulses is graphically represented in Fig. 3. A pot containing 50 g pulses and 150 ml water was kept on the heater directly, and another pot containing 50 g rice and 150 ml water was kept on it to check the possibility of cooking of two dishes simultaneously. After 90 minutes, the battery was disconnected, and then the cooker was allowed to get cooled. Both pulses and rice were cooked well simultaneously. Fig. 4 shows the rice and pulses cooked in the solar cooker.

It is very well known that cooking of hard nuts requires a lot of time and energy. The hard nuts cooking test was also conducted for the confirmation of better performance of the solar cooker. The battery was connected to the cooker in which 100 g hard nuts with 150 ml water were kept for cooking. Initially at 10:20 am, hard nuts temperature was 20.8°C. By connecting battery to the cooker, the temperature started to rise and the hard nuts were ready by 12:00 pm. The battery was disconnected at 12:00 pm. The cooker took only 80 minutes for cooking hard nuts. Similar cooking tests were conducted for hotdog too. Figs. 5 and 6 show the temperature vs. time plot for hard nuts and the photograph of hard nuts cooked in the solar cooker, respectively.

Fig. 7 shows the plot of time vs. solar radiation for the month of January 2013. The radiation data were measured with a pyranometer tilted at the same angle of the solar panel i.e., 45° at the Department of Physics. In Fig. 8, it is shown that seven recipes (Gujarati thali) can be cooked in a solar cooker simultaneously.

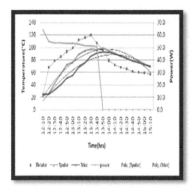

Figure 3. Temperature vs. time plot for rice and pulses cooking test.

Figure 4. Rice and pulses cooked in solar cooker.

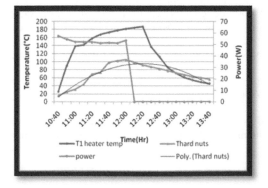

Figure 5. Temperature vs. time plot for hard nuts cooking test.

Figure 6. Hard nuts.

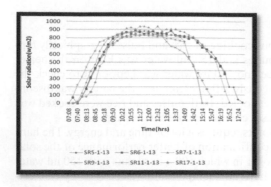

Figure 7. Time vs. solar radiation plot.   Figure 8. Gujarati thali prepared in solar cooker.

4 CONCLUSION

The experimental observations related to the temperature profile indicate that the developed solar cooker can be easily used for cooking soft food materials like rice as well as hard nuts. Cooking needs only 50 watt power which can be easily generated by the solar panel of 75 watt and can be stored in the battery, so that this power can also be utilized for lights and fans in the house in case of power failure, or even it can provide power to night lamps for the whole night. The battery operated cooker does not need any preheating and tracking like the conventional cooker and can make it more user-friendly. There were two types of energy utilized for cooking: 1) thermal energy; and 2) electrical energy generated by PV module. As a result of this, although the food quantity was triple fold, the cooking time was only half compared with the cooking done inside the laboratory.

REFERENCES

Joshi, S.B. & Jani, A.R. (2013a). A new design of solar cooker for maximum utilization. Center for ionics University of Malaya (CIUM), Malaysia, (pp. 287–292). Kuala Lumpur.

Joshi, S.B. & Jani, A.R. (2013b). Certain analysis of a solar cooker with dual axis sun tracker. (p.5). Ahmedabad, India: ieee explore.

Joshi, S.B. & Jani, A.R. (2013c). Photovoltaic and thermal hybridized solar cooker. ISRN, Renewable Energy, 2013(746189), 5. doi:http://dx.doi.org/10.1155/2013/746189

Joshi, S.B. & Jani, A.R. (2014). Development of heat storage system for solar cooker. International Journal of Latest Technology in Engineering, Management & Applied Sciences, 3(4), 247–251.

Joshi, S.B., Thakkar, H.R. & Jani, A.R. (2014). A novel design approach of small scale conical solar dryer. International Journal of Latest Technology in Engineering, Management & Applied Sciences, 258–261.

Kuhnke, K. (1988). Solar cookers developing countries: a worldwide study. Solar Energy Technology, 2678–2682.

Tyagi,V.V., Kaushik, S.C. & Tyagi S.K., Advancement in solar photovoltaic Thermal (PV/T) hybrid collector technology, Renewable and Sustainable Energy Reviewers, Vol. 16, (3), April, 2012, pp. 1383–1398

Xingxing, Z., Xudong, Z., Stefan, S., Jihuan, X. & Xiaotong, Y. (2012). Review of R&D progress and practical application of the solar photovoltaic/thermal (PV/T) technologies. Renewable and Sustainable Energy Reviews, 16(1), 599–617.

Multi-disciplinary Sustainable Engineering: Current and Future Trends – Tekwani, Bhavsar & Modi (Eds)
© 2016 Taylor & Francis Group, London, ISBN 978-1-138-02845-6

Experimental investigation and analysis on a concentrating solar collector using linear Fresnel lens

N.A. Dheela
Department of Mechanical Engineering, Marwadi Education Foundation Group of Institutions, Rajkot, Gujarat, India

B.A. Shah & S.V. Jain
Department of Mechanical Engineering, Institute of Technology, Nirma University, Ahmedabad, Gujarat, India

ABSTRACT: Solar energy is used as an alternative source of energy for several domestic and industrial applications. Solar energy can be used in industries for quick and efficient water heating purpose or to generate low-pressure steam by concentrating more solar energy on a smaller area. This can be done by using concentrating solar thermal collectors. Concentrating collector technologies offer a favorable technique for the enormous use of solar energy. It gives a higher efficiency compared to the non-concentrating collector at high temperature due to its higher concentration ratio and smaller amount of heat losses. Solar energy concentrating technology with linear Fresnel lens is one of the effective ways to utilize solar energy for water heating. The objective of the present work is to investigate and analyze the performance of concentrating linear Fresnel lens solar water collector. In the present work, polymethyl methacrylate sheet was used as Fresnel lens. A heat loss calculation was carried out at various mass flow rates, with and without the Fresnel lens. To study the effects of Fresnel lens on the performance of evacuated tube collector, experiments were performed in the wide range of mass flow rates. The system efficiency was increased by 10–17% due to the presence of Fresnel lens at different flow rates.

1 INTRODUCTION

There are two types of solar thermal collectors, the concentrating type and the non-concentrating type. The concentrating type collector usually contains parabolic mirrors to focus the incident solar radiation on the absorber surface. Flat plate collectors and evacuated tube collectors are the two most commonly used non-concentrating type of collectors. During the recent decade, linear Fresnel lens collector system (LFC) is initially examined and established. The first Fresnel lens developed and invented by French physicist and engineer Augustin-Jean Fresnel was used for lighthouses in 1822. The first Fresnel lens collector was patent and prototyped by Frencia in Italy in 1964. Major investigation of Fresnel lens includes air heater, Fresnel lens for hydrogen generation, photo-bio reactor, solar powered refrigeration, photochemical reaction, linear Fresnel reflector technology and metal surface modification [2].

The linear Fresnel lens is a photosensitive piece, flat on one side and with a groove on the other side, which helps concentrate sun rays by means of a series of parallel grooves, away from the centre line in the form of a linear strip. Compared with conventional lens, linear

Fresnel lens has advantages like lightweight, small volume, less cost, and mass production high concentration radio. Subsequently most Fresnel lens designers of concentrated solar energy application chose PMMA (polymethyl methacrylate) for their lens for high optical quality [3]. In this paper a flat LFC with axis tracking mode and U-tube absorber tube is developed and tested.

2 CONFIGURATION OF FRESNEL LENS SOLAR COLLECTOR

A concentrating liner Fresnel lens solar collector consists of an evacuated U-type absorber tube, linear Fresnel lens, water tank, pump, thermocouple, etc. Fig. 1 shows the experimental setup consisting of two linear Fresnel lens arranged in a series. Each Fresnel lens is 1800 mm long and 250 mm wide. The experimental setup consists of a U-type evacuated absorber tube having a vacuum in annular space between two concentric transparent glasses. The tube has a selective coating on the outer surface of an inner tube to increase the absorption of the incoming solar radiation, and a copper U-tube with an aluminum fin to enhance the heat transfer. The outer diameter of the evacuated glass tube is 58 mm, the inner diameter is 47 mm, the absorber length is 1.8 m, and the outer diameter of the copper U-tube is 8 mm.

2.1 *Design of the Fresnel lens*

A material called PMMA, popularly known as an acrylic glass, is to be used in the lens as it is much lighter. Fresnel lens is designed based on the minimum thickness in such a way that it can neither be bent nor distorted. It has minimal absorptivity and ensures maximum radiation passes [4]. The design of Fresnel lens is done in Auto-CAD as shown in Fig. 2 in which conventional lens of radius 200 mm is divided into equal parts, and then machining is carried out on the lens. CNC machining is used to generate longitudinal groove on conventional lens.

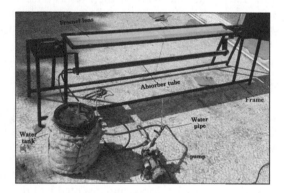

Figure 1. Experimental setup.

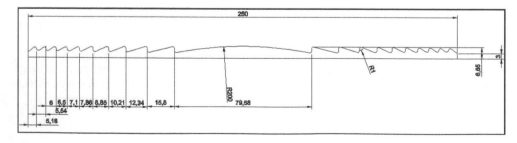

Figure 2. Auto-CAD design of Fresnel.

176

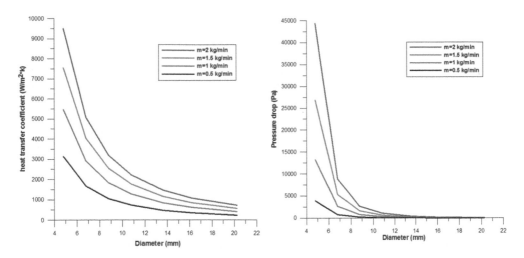

Figure 3. Variation of heat transfer coefficient and pressure drop with change in diameter.

2.2 *Design of absorber tube*

A selection of absorber tube diameter is carried out in such a way that we can get a maximum heat transfer with a low pressure-drop. We can find the required diameter using the trial and error method by fixing the mass flow rate and calculate the heat transfer coefficient and pressure drop for the different absorber tube diameter. We have selected an 8 mm copper tube, since it has optimum heat transfer coefficient and comparative low-pressure drop. The plot for change in heat transfer, pressure drop with respect to the change in diameter of a tube is shown in Fig. 5.

3 RESULTS AND DISCUSSION

The experimental setup is located at 23.7° N, 72.32° E (A-block) Nirma University, Ahmedabad. The orientation of experimental setup is in north to south direction; sun tracking is to be carried out from east to west. The experimental results are carried out with the Fresnel lens collector and evacuated tube collector for four different mass flow rates.

3.1 *Performance curves of Fresnel lens solar collector*

The purpose of this experiment is to find out the thermal efficiency of the Fresnel lens collector. During the test, water temperature inside the tank is measured using the MS-1208 sensor at an interval of 30 min. Beam radiation is measured by a digital pyranometer.

Fig. 4 shows a variation of water temperature, ambient temperature and beam radiation with time. Fig. 5 shows variation of hourly efficiency with time from result, it is found out that hourly efficiency decreases as time increases.

Fig. 6 shows that the heat loss increases as the water temperature increases. In the morning, at lower water temperature, heat loss is less and as the temperature increase heat loss also increases at the afternoon. Fig. 7 shows the variation in the average efficiency with the change in mass flow rate. Maximum efficiency 67.05% is found at 1 kg/min mass flow rate.

3.2 *Performance curves of evacuated tube collector*

The experimental result was also taken by removing Fresnel lens, which is called evacuated tube collector to validate the result of Fresnel lens collector.

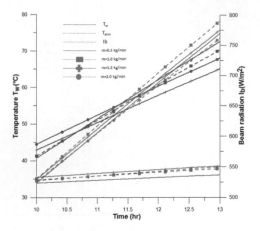

Figure 4.   Temperature vs. time.

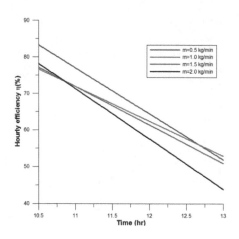

Figure 5.   Hourly efficiency vs. time.

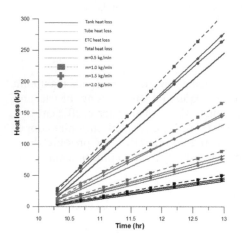

Figure 6.   Heat loss vs. time.

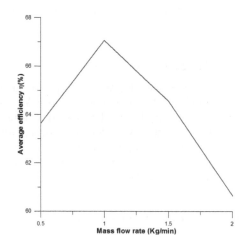

Figure 7.   Average efficiency vs. mass flow rate.

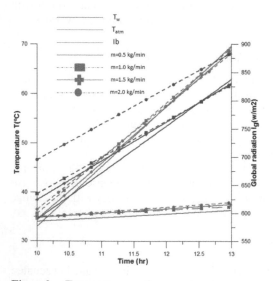

Figure 8.   Temperature vs. time.

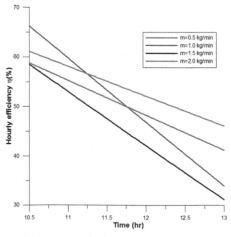

Figure 9.   Hourly efficiency vs. time.

178

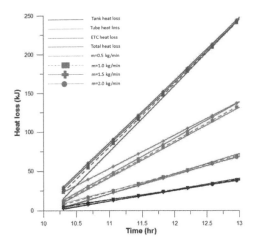

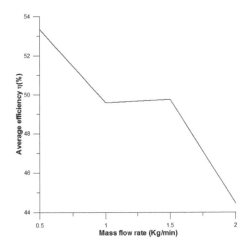

Figure 10. Heat loss vs. time.

Figure 11. Average efficiency vs. mass flow rate.

Maximum temperature achieved by the evacuated tube collector is 69°C as shown in Fig. 8. Maximum average efficiency archived using the evacuated tube collector is 53.34% at 0.5 kg/min mass flow rate as shown in Fig. 9.

4 CONCLUSION

The major conclusions drawn from the study are as follows:

- The efficiency of evacuated tube collector was found to be in the range of 44–53% at different flow rates.
- Due to the presence of Fresnel lens, the efficiency was increased to a range of 10–17% at different flow rates. The maximum efficiency of Fresnel lens collector was obtained as 67.05% at a mass flow rate of 1 kg/min.
- Four calculation results are under different mass flow rate showing that the flow rate does not impact the performance of solar collector, rather, it depends mainly on the intensity of solar radiation.
- The convective and radiation heat losses increased with the rise in water temperature which led to a decrease in efficiency at elevated temperatures. When the water temperature was raised from 38°C to 74°C, the heat losses increased from 25.56 kJ to 299.73 kJ in Fresnel lens collector.

REFERENCES

[1] H. Zhai, Y.J. Dai, J.Y. Wu, R.Z. Wang, L.Y. Zhang, "Experimental investigation and analysis on concentric solar collector using linear Fresnel lens", Energy Conversion and Management 51 (2010), 48–55.
[2] R.Z. Wang, K. Sumathy, W.T. Xie, Y.J. Dai, "Concentrated solar energy application using Fresnel lens: A review", Renewable and Sustainable Energy Review 15 (2011), 2588–2606.
[3] S. Sadhishkumar, T. Balusamy, "Performance improvement in solar water heating system: A review", Renewable and Sustainable Energy Review 37 (2014), 191–198.
[4] Fresnel technology, Inc. (1996–2014), www.fresneltech.com
[5] Yong Kim, Taebeom Seo, "Thermal performance comparisons of glass evacuated tube solar collector with shapes of absorber tube", Renewable Energy 32 (2007), 772–795.

Treatment of Gelatin Manufacturing Wastewater using anaerobic sequential batch reactor

D.K. Mistry & U.D. Patel

Department of Civil Engineering, Faculty of Technology and Engineering, The Maharaja Sayajirao University of Baroda, Vadodara, India

ABSTRACT: Gelatin Manufacturing Wastewater (GMW) containing Chemical Oxygen Demand (COD) and ammonical-nitrogen (NH_4-N) in the range of 1600–3000 mg/L and 150–250 mg/L, respectively, is conventionally treated by aerobic biological processes. Anaerobic degradation of GMW was attempted using AnSBR with an aim to evaluate COD removal efficiency and effect of NH_4-N concentration on the process. After acclimatization, COD removal efficiency close to 70% at Organic Loading Rate (OLR) of 0.05–0.35 g COD/g VSS/d and Volumetric Loading Rate (VLR) of 0.14–0.37 g COD/L/d was achieved for an operation cycle of 0.5 h feed-22 h reaction-1 h settling-0.5 h decanting. NH_4-N concentration in the range of 150–250 mg/L did not have significant adverse effect on COD removal. However, the COD removal efficiency was adversely affected due to sudden drop in ambient temperature. When ambient temperature dropped to 14–15°C, COD removal efficiency dropped to less than 20%.

1 INTRODUCTION

Gelatine is a heterogeneous mixture of water-soluble proteins derived from the collagen of animal hide or bone. It has diversity of uses; for example, in food from wine fining to confectionary, and in industry from matches to photography. Consumption of gelatine in India is expected to grow at a rate of 10–15% per year from the consumption of 13000 MT in year 2011 (Morrison, 2012). With increasing economic growth and use of gelatine in food and beverages, and neutraceuticals, global gelatin consumption is forecast to reach 3.96 lakh MT by the year 2017 from 3.71 lakh MT in 2012 (38th Annual Report, 2013–14, Nitta Gelatin India Ltd.).

Gelatin manufacturing process in India mainly involves selective hydrolysis followed by extraction of collagen from bovine bones. Each step of manufacturing process generates huge amounts of wastewater having high Biochemical Oxygen Demand (BOD), COD, suspended solids (SS), SO_4^{-2}, and ammonical nitrogen (NH_4-N) (Maree et al., 1990). High COD and BOD removal efficiencies (ca. 90%) have been reported by vermi-filter based treatment of GMW (Ghatnekar et al., 2010). Conventionally, this wastewater is treated by aerobic biological processes such as activated sludge process, trickling filter, etc. Aerobic treatment of GMW leads to generation of huge amount of waste biomass sludge which needs to be dewatered and disposed every day. In addition to this, it also demands huge amount of energy to run aerators to provide sufficient dissolved oxygen. Anaerobic biodegradation for GMW treatment is generally not explored much due to high concentrations of SO_4^{-2}, and NH_3-N. Moreover, GMW contains significant concentrations of proteins which are degraded more slowly than carbohydrates. Anaerobic degradation of proteins is a complex process involving various groups of microorganisms. Proteins are first hydrolyzed and degraded by proteolytic enzymes into peptides and individual amino acids. The peptides and amino acids are then acidified into Volatile Fatty Acids (VFA), hydrogen, ammonia, and reduced sulphur (Herbert et al., 2002). Thus, NH_4-N concentration in the effluent of anaerobic treatment may be more than that found in the influent. However, anaerobic treatment works in the absence of oxygen. Thus, energy consumption is substantially less as compared to aerobic treatment. On the contrary, methane is produced as one of the end products of anaerobic digestion which can be used as a source of energy. Besides this, the growth rate of anaerobic bacteria is very slow hence excess biomass sludge generated is 1/10th to 1/15th of that of aerobic biological treatment. Anaerobic treatment of GMW using Upflow Anaerobic Sludge Blanket (UASB) (Wurster et al., 2004), Upflow Anaerobic Reactor (Fang and Yu, 2002; Yu and Fang, 2003) has been reported. However, there are no reports on application of Anaerobic Sequential Batch Reactor (AnSBR) for the treatment of GMW. AnSBR has been reported to be suitable for treatment of organic wastewaters. Moreover, it offers some advantages over other systems such as, flexible operation, no recycling of liquid or solids, and use of the same vessel for reaction and settling (Cheong and Hansen, 2008).

Thus, the objectives of the present study were: (1) to study COD removal from GMW, and (2) to study the effect of NH_3-N on COD removal in AnSBR.

2 MATERIALS AND METHOD

2.1 *Gelatin manufacturing wastewater*

The wastewater after screening, oil and grease removal, equalization, and coagulation-flocculation-primary sedimentation, was collected from a gelatin manufacturing industry located at Vadodara, and used without any further adjustment (except pH adjusted to 6.9–7.1), as a feed to the AnSBR. Table 1 shows range of characteristics of wastewater used in the study.

2.2 *Anaerobic sludge*

The seed anaerobic sludge was collected from the UASB of Sewage Treatment Plant located at Atladra, Vadodara. The characteristics of sludge were as mentioned in Table 2.

2.3 *Anaerobic Sequential Batch Reactor (AnSBR)*

A closed cylindrical AnSBR of 10 L capacity having 40 cm height and 18 cm diameter was fabricated from Plexiglass (Figure 1). Sampling ports were provided at 3 L and 4 L levels, and one at the bottom of the reactor for sludge removal. A mechanical stirrer connected to a DC 12 V geared motor was operated at 30 RPM for proper mixing during the reaction time.

Table 1. Range of characteristics of GMW used in the study.

Characteristic	Values of parameter
pH	8–10.5
COD	1600–3000 mg/L
BOD_3[27]	900–1700 mg/L
Sulphate, SO_4^{-2}	285–606 mg/L
Ammonical nitrogen, NH_4-N	70–190 mg/L
TSS	120–230 mg/L

Table 2. Characteristic of anaerobic sludge used in study.

Characteristic	Values of parameter
TSS	26.4 g/L
Total fixed Suspended Solids, TFSS	13.2 g/L
Total volatile Suspended Solids, TVSS	13.2 g/L

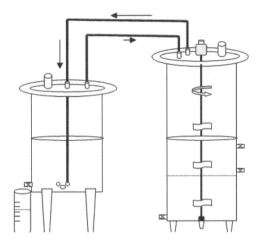

Figure 1.   Anaerobic Sequential Batch Reactor (AnSBR).

The biogas generated was passed through KOH solution and its volume was measured by volume displacement method. Due to some technical problem, the gas volumes measured were found to inconsistent and absurd and hence not reported here.

Throughout the study, the AnSBR was fed with 2.5 L of anaerobic sludge and 2.5 L of wastewater consisting of varying volumetric fractions of GMW and domestic sewage (during acclimatization period) or as-received GMW (after acclimatization period). The reactor operation cycle was: 0.5 h feed-22 h reaction-1h settling-0.5 h decanting.

Initially the reactor was fed with 250 mL of GMW mixed with 2.25 L of municipal sewage for 3 days. After three days, the feed consisted of 2 L of treated wastewater from the reactor mixed with 500 mL of GMW for the next 7 days. Then the fraction of GMW in the feed was gradually increased reaching to 2.5 L of real GMW after next nine days. Thus, after the acclimatization period of 19 days, the reactor was operated with 2.5 L of real GMW + 2.5 L of acclimatized sludge. After the completion of operation cycle of 24 h mentioned above, 2.5 L treated effluent was decanted out (keeping the sludge in the reactor) and replaced with new as-received GMW. After charging the feed, reactor contents were mixed for 10 min and a sample was withdrawn to analyze feed COD, alkalinity, VFA, and NH_4-N. The treated effluent was also analyzed for above parameters. Volumetric loading rate (g COD/L volume of reactor-d) and organic loading rate (g COD/g VSS-d) were calculated.

2.4   Analytical methods

2.4.1   Analysis of pH, TSS, VSS, COD, NH_4-N
pH of the feed and the effluent was measured by the pH meter. Total Suspended Solids, Volatile Suspended Solids, COD, NH_4-N of the sample were determined as per the Standard Method. (Eaton, A.D., 2012)

2.4.2   Analysis of alkalinity and volatile fatty acids
Titration procedure for measurements of VFA and Alkalinity according to Buchauer (1984) was used as under: Before analysis the sample was filtered through a 0.45 μm membrane filter. Filtered sample (50 mL) was taken into a titration vessel, the size of which was determined by the basic requirement to guarantee that the tip of the pH electrode was always below the liquid surface. Initial pH was recorded. The sample was titrated slowly with 0.05 N sulphuric acid until pH 5.0 was reached. The added volume A1 (mL) of the titrant was recorded. More acid was slowly added until pH 4.3 was reached. The volume A2 (mL) of added titrant was recorded. The latter step was repeated until pH 4.0 was reached, and the volume A3 (mL) of added titrant was recorded once more. The calculations for alkalinity and Volatile Fatty Acids (VFA) were made as under.

Alkalinity = (A1 + A2) * N * 1000/sample volume (mmol/L as CaCO₃)
VFA (mg/L as acetic acid equivalent) = {[131340 * N * (A2 + A3)]/volume of sample} – (3.08 * Alkalinity) – 10.9

Alkalinity to VFA ratio (A/V) was calculated by multiplying alkalinity value in mmol/L with 100 (MW of CaCO₃) and dividing by value of VFA.

3 RESULTS AND DISCUSSION

3.1 *COD removal in AnSBR*

It may be noted from Figure 2 that there was significant fluctuations in the feed quality.

The feed COD ranged from 800–1700 mg/L. AnSBR was able to tolerate these fluctuations and the COD of treated effluent remained close to 400 mg/L achieving COD removal efficiency of 70–80% at OLR of 0.05–0.35 g COD/g VSS/d and VLR of 0.14–0.37 g COD/L/d (Figure 3). It may further be noted that COD removal at 25th and 26th day dropped to less than 10%. It was observed that the normal ambient temperature on these two days dropped suddenly from 30–35 to 14–15°C. Ndon and Dague (1997) studied the effect of temperature and hydraulic retention time (HRT) on COD removal from low-strength wastewater using AnSBR. The authors reported that there was no COD removal at HRT of 12h and 16 h when temperature was dropped below 20°C.

During the acclimatization it was noticed that the reactor responded well to gradually increasing fraction of GMW and COD, achieving effluent COD close to 200 mg/L with the feed COD ranging between 600 to 1000 mg/L.

3.2 *A/V ratio*

Figure 4 shows variation of A/V ratio with respect to COD removal efficiency in AnSBR after acclimatization. The feed A/V ratio ranged from 2.0–5.0, whereas the A/V ratio in

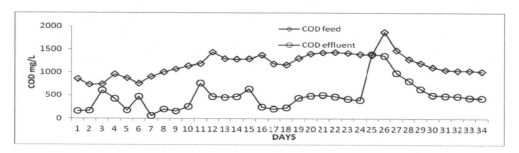

Figure 2.   COD removal in AnSBR after acclimatization.

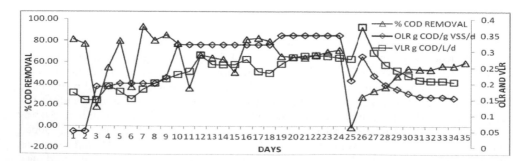

Figure 3.   % COD removal efficiency, OLR and VLR relationship in AnSBR.

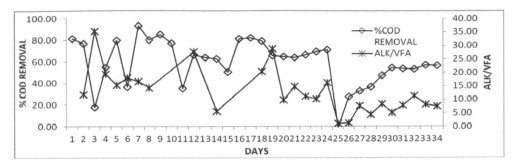

Figure 4.   % COD removal and Alkalinity to volatile fatty acid ratio of effluent.

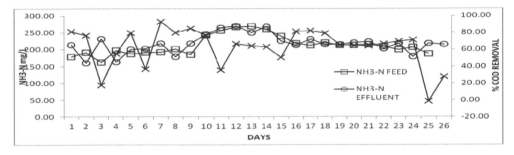

Figure 5.   Effect of NH$_3$-N on % COD removal.

effluent increased to about 5–10 with some of the days going as high as 30–35. This indicates that there was no accumulation of acids and that they were consumed by methanogens. The results of A/V are thus consistent with COD removal. It may further be noted that A/V ratio in effluent on 25th and 26th days dropped suddenly when temperature dropped to 14–15°C, as compared to other days with normal temperature of 30–35°C. The A/V ratio in the effluent, on these two days dropped to 1.16. It is known that the optimum temperature for methogens is around 35°C. Thus, reduction in A/V ratio may be due to accumulation of VFA at lower temperature when methanogenesis may be temporarily inhibited.

3.3   Effect of NH$_4$-N on COD removal

It may be noted from Figure 5 that the effluent NH$_4$-N exceeds the inlet NH$_4$-N concentration by 30–40 mg/L. The increase in NH$_4$-N may be attributed to ammonification of organic nitrogenous compounds (Shen, 2007) contained in GMW. Figure 5 also reveals that the range of NH$_4$-N (150–250 mg/L) observed in the study was not inhibitory to the anaerobic process yielding COD removal efficiency close to 70% irrespective of inlet NH$_4$-N concentration. Burak et al. (2012) studied effect of NH$_3$-N concentration on COD removal from a landfill leachate in a UASB reactor. It was reported that COD removal efficiency was maintained close to 90% when influent NH$_3$-N concentration was less than or equal to 150 mg/L. However when NH$_3$-N concentration was increased to ≥600 mg/L, the COD removal efficiency was drop to 10% due to free ammonia toxicity.

4   CONCLUSION

1. The results reported in the study showed that AnSBR can be successfully used for the treatment of GMW with inherent advantages such as low energy consumption, and less generation of excess biomass as compared to the conventional aerobic biological processes.

COD removal ranging from 50–70% could be achieved at OLR of 0.05–0.35 g COD/g VSS/d and VLR of 0.14–0.37 g COD/L/d.
2. COD removal was adversely affected by low ambient temperatures. On the other hand, NH_4-N concentration in the range of 150–250 mg/L did not have significant adverse effect on COD removal.

REFERENCES

Buchauer, K. 1984. A comparison of two simple titration procedures to determine volatile fatty acids in influents to wastewater and sludge treatment process. *Water SA* 24(1): 49–56.

Burak, Y., Deniz, A. & Bulent, M. 2010. Effect of high ammonia concentration on UASB reactor treating sanitary landfill leachate. *Fen Bilimleri Dergisi,* 24(2): 59–67.

Cheong, D.Y. & Hansen, L.C. 2008. Effect of feeding strategy on the stability of anaerobic sequential batch reactor responses to organic loading condition. *Bioresource Technology* 99: 5058–5068.

Eaton, A.D., Clesceri, L.S. & Greenberg, A.E. 2012. Standard method for the examination of water and wastewater, 20th Edition. *American Public Health Association.*

Fang, H.H.P. & Yu, H. 2002. Mesophilic acidification of gelatinaceous wastewater. *Journal of Biotechnology* 93(2): 99–108.

Ghantekar, S.D., Kavain, M.F., Sharma, S.M., Ghantekar, S.S., Gantekar, G.S. & Gantekar, A.V. 2010. Application of vermi-filter-based effluent treatment plant (pilot scale) for biomanagement of liquid effluents from the gelatin industry. *Global science book.*

Maree, J.P., Cole, C.G.B., Gerber, A. & Barnard, J.L. 1990. Treatment of gelatin factory effluent. *Water SA* 16(4): 265–268.

Morrison, N. 2012. Religion plays big part in pharma gelatin shortage for India. (http://www.in-pharmatechnologist.com/Ingredients/Religion-plays-big-part-in-pharma-gelatine-shortage-for-India-Beroe). (Site visited on 4th May 2015).

Ndon, U.J. & Dague, R.R. 1997. Effect of temperature and hydraulic retention time on anaerobic sequential batch reactor treatment of low-strength wastewater. *Water Research* 31(10): 2455–2466.

Shen, D.S., R. He., Liu, X.W., Zhang, Z. J. 2007. Characteristics of the bioreactor landfill system using an anaerobic-aerobic process for nitrogen removal. *Bioresource Technology* 98: 2526–2532.

Wurster, A., Merwe-botha, M. & Haycock, L. 2004. Anaerobic pretreatment of high strength gelatine production effluent. Proceedings of the 2004 Water Institute of Southern Africa (WISA) Biennial Conference.

Yu, H. & Fang, H.H.P. 2003. Acidogenesis of gelatin-rich wastewater in an upflow anaerobic reactor: influence of pH and temperature. *Water Research* 37: 55–66.

Multi-disciplinary Sustainable Engineering: Current and Future Trends – Tekwani, Bhavsar & Modi (Eds)
© *2016 Taylor & Francis Group, London, ISBN 978-1-138-02845-6*

Residual COD removal from OBA wastewater by improvement in biodegradability

S.S. Mapara & U.D. Patel
Department of Civil Engineering, Faculty of Technology and Engineering,
The Maharaja Sayajirao University of Baroda, Vadodara, India

ABSTRACT: The study was carried on conventionally treated Optical Brightening Agent manufacturing Wastewater (OBAW), for the removal of residual COD. Experiments revealed formation of large amount of fluffy, very slow-settling precipitates when wastewater was acidified to pH 3.0. About 30% COD reduction was obtained on removal of precipitates at pH 3.0. The settling rate of these precipitates improved when 4 g/L and higher doses of Waste Carbon Dust (WCD) were added as a weighting agent. Fenton treatment was explored with COD: H_2O_2 as 7.5:1 and H_2O_2: Fe^{+2} as 10:1 and 20:1, on acidic supernatant obtained after weighted sedimentation. Fenton's treatment did not yield any further COD removal. However, Zahn-Wellens test revealed that Fenton's treatment improved the biodegradability of wastewater from 30% in the absence of any treatment, to 52% and 88%, for H_2O_2: Fe^{+2} doses of 10:1 and 20:1, respectively.

1 INTRODUCTION

Optical Brightening Agents (OBA) are chemical compounds that absorb light in the ultraviolet and violet region (usually 340–370 nm) of the electromagnetic spectrum, and re-emit light in the blue region (typically 420–470 nm). These additives are often used to enhance the appearance of color of fabric and paper, causing a "whitening" effect, making materials look less yellow by increasing the overall amount of blue light reflected (Smulders et al, 2002).

OBAs are sophisticated compounds similar to synthetic dyes which usually possess several sulfonic acid groups conferring upon the compounds, high water solubility. In addition, the sulfonic groups also deactivate (by resonance effect) the OBA towards electrophilic attack. In turn, this indicates a marked resistance of such compounds to conventional biological treatment (Lemeune et al, 1996). OBAs are poorly biodegradable, thus pass through wastewater treatment plants without significant degradation, neither under aerobic nor anaerobic conditions. Only by adsorption onto biomass a partial removal of OBAs is achieved (Horsch et al, 2003). Poiger et al. (1998) studied behavior of Fluorescent Whitening Agents (FWAs) during sewage treatment and reported that 53–98% of FWAs were removed mainly due to adsorption to primary and activated sludge. The authors did not find any evidence for biodegradation of FWA in aerobic (activated sludge) biological process for sewage and anaerobic degradation of sludge. Moreover Kohler et al. (2006) reported that activated sludge treatment of FWA manufacturing wastewater increased the effluent toxicity which indicates that new ecotoxic metabolites might have formed during the biological treatment of FWA manufacturing wastewater.

Various alternatives to the conventional activated sludge treatment are employed for non biodegradable or toxic industrial wastewaters. Among these, chemical oxidative treatments; and especially, Advanced Oxidation Processes (AOP) are well known for their capacity for oxidizing and mineralizing almost any organic contaminant (Sirtori et al., 2009). Advanced oxidation technologies like Fenton's reagent (Fe^{+2}/H_2O_2) and the photo chemically enhanced Fenton process have been found useful in treatment of OBAW.

Fenton's reagent, which involves homogenous reaction and is environmentally acceptable is a system based on the generation of very reactive oxidizing free radicals, especially hydroxyl radicals, which have a stronger oxidation potential than ozone (2.8 V for •OH vs. 2.07 V for ozone).

The Fenton's reactions at acidic pH lead to the production of hydroxyl radicals and ferric ion and its recycling to ferrous ions as shown in equations 1–10.

$$Fe^{+2} + H_2O_2 \rightarrow Fe^{+3} + OH^- + \cdot OH \tag{1}$$

$$Fe^{+3} + H_2O_2 \rightarrow Fe^{+2} + HO_2 \cdot + H^+ \tag{2}$$

Hydroxyl radicals may be scavenged by reaction with another Fe^{2+} or with H_2O_2

$$\cdot OH + Fe^{+2} \rightarrow HO^- + Fe^{+3} \tag{3}$$

$$\cdot OH + H_2O_2 \rightarrow HO_2 \cdot + H_2O \tag{4}$$

Hydroxyl radicals may react with organics starting a chain reaction

$$\cdot OH + RH \rightarrow H_2O + R\cdot, RH = organic\ substrate \tag{5}$$

$$R\cdot + O_2 \rightarrow ROO\cdot \rightarrow product\ of\ degradation \tag{6}$$

Ferrous ions and radicals are produced during the reactions as shown below:

$$H_2O_2 + Fe^{+3} \rightleftarrows H^+ + FeOOH^{2+} \tag{7}$$

$$FeOOH^{2+} \rightarrow HO_2 \cdot + Fe^{+2} \tag{8}$$

$$HO_2 \cdot + Fe^{+2} \rightarrow HO^{2-} + Fe^{+3} \tag{9}$$

$$HO_2 \cdot + Fe^{+3} \rightarrow O_2 + Fe^{+2} + H^+ \tag{10}$$

It has also been reported that AOP such as Fenton's treatment can partially degrade refractory compounds and improve the biodegradability of wastewaters (Benatti and Tavares, 2012). Sirtori et al. (2009) observed that a particular pharmaceutical wastewater containing Nalidixic acid was tolerant to biodegradation. Subsequent to Fenton's treatment to completely remove Nalidixic acid, the biodegradability improved and ecotoxicity of wastewater decreased. Liu et al. (2011) observed that aniline manufacturing wastewaters is less biodegradable and very toxic to microorganisms, and the COD removal efficiency are low by direct biological treatment. The toxicity of the aniline wastewater was found to be reduced after pretreatment by the photo—Fenton oxidation. The photo Fenton pretreatment improved the BOD_5/COD ratio from 0.21 to 0.43 which showed the improvement in biodegradability.

The present study was carried out using a real OBAW after conventional primary (coagulation—flocculation—sedimentation) and secondary (activated sludge process) treatment from an OBA manufacturing industry located at Bharuch, Gujarat, India. The raw wastewater from the unit has COD in the range of 5500–6000 mg/L which is reduced to 3000–3400 mg/L after primary and secondary treatments. The treated wastewater is then sent to Multiple Effect Evaporator (MEE) plant for concentration. The condensate from MEE is either recycled or disposed and the solid slurry is dried and disposed into a landfill site. The high organic content of the feed to MEE results in frequent chocking of MEE. Thus, the aim of the study was to reduce the residual COD from 3000–3400 mg/L to less than 1000 mg/L so that the chocking in MEE can be avoided and/or the biodegradability of wastewater is enhanced to such an extent that MEE can be avoided. The Fenton's treatment was explored with an aim to improve the biodegradability.

During the initial experiments it was found that acidification of OBAW resulted in formation of a large volume of organic precipitates coupled with COD removal. However, these precipitates were very fluffy and light in weight, settling very slowly. Therefore another aim of the study was to improve the settling rate of the precipitates.

2 MATERIALS AND METHOD

2.1 *The OBA wastewater*

The wastewater was obtained from the outlet of secondary clarifier of an OBA manufacturing industry located at Bharuch, Gujarat, India. The company manufactures three different kinds of products (consisting of disulfonic, tetrasulfonic, and hexasulfonic acid groups) and there are three wastewater streams coming out of three manufacturing plants. The composite COD of raw wastewater is about 5500–6000 mg/L. The total flow is around 350 kL/day. The range of characteristics of secondary-treated wastewater collected from the OBA unit, used in the present study, is shown in Table 1.

2.2 *Experimental methods to improve settling rate*

2.2.1 *Use of carbon dust as a weighting agent*

In a typical experiment, 250 mL wastewater sample was taken. The Waste Carbon Dust (WCD, obtained from the Bag Filter system attached to the coal fired boiler plant in the unit and used as it is) addition and pH adjustment to 3.0 were carried out either by Method 1 or 2, as described below. Following the WCD addition and pH adjustment, the contents were stirred for 30 minutes on a magnetic stirrer at high speed. After stirring, each sample was transferred in a 250 ml Borosilicate Glass measuring cylinder, and the rate of settling and the final volume of sludge produced were noted down. The COD of supernatant was measured after filtering it through Whatman 42 No. filter paper.

Method 1: Sample pH was adjusted to 3.0 using 6 N H_2SO_4, precipitates were formed, and subsequently dose of WCD (2, 4, 6, 8, 10 g/L) was added.

Method 2: WCD dose (2, 4, 6, 8, 10 g/L) was first added and then pH was adjusted to 3.0 using 6 N H_2SO_4.

2.2.2 *Use of boiler ash as a weighting agent*

The Boiler ash was obtained from the coal-fired boiler plant and used as it is without any treatment as a weighting agent. The boiler ash was added as per Method 1 only and procedure as mentioned above for WCD was followed in all the experiments using boiler ash.

2.3 *Fenton treatment*

The supernatant at pH 3.0, after removal of precipitates, was used for Fenton's treatment. COD: H_2O_2 ratio of 7.5:1 and H_2O_2: Fe^{+2} molar ratios of 10:1 (1.25 mL/L of 24% H_2O_2 and 125 mg/L $FeSO_4.7H_2O$) and 20:1 (1.25 mL/L of 24% H_2O_2 and 250 mg/L $FeSO_4.7H_2O$) were used. In a typical procedure, $FeSO_4.7H_2O$ crystals (corresponding to required Fe^{+2} concentrations) were added first, followed by drop wise addition of H_2O_2 with stirring. The reaction was allowed to run for 3 hours. After the completion of reaction, pH was adjusted to 10.5 with hydrated lime followed by addition of alum to adjust pH to 7.0. Magnafloc 1011 (2 ppm) was added to improve the flocculation.

Table 1. Characteristics of OBAW used in current study.

Characteristics	Value of parameters
pH	6.45–8.55
TS (mg/L)	28000–29250
TDS (mg/L)	26500–27500
Chloride (mg/L)	12550–13250
BOD_5 (mg/L)	130–160
COD (mg/L)	3000–3400

2.4 Biodegradability analysis

The supernatant after carrying out Fenton treatment, was checked for improvement in the biodegradability by using modified Zahn—Wellens/EMPA test (OECD, 1992).

2.5 Analytical methods

COD (Closed reflux titrimetric method), BOD_5, Chloride, and Total Dissolved Solids (TDS) were determined as procedure given in Standard Methods (Rice et al., 2012). Residual H_2O_2 was measured using iodometric method. Since H_2O_2 is oxidized by $K_2Cr_2O_7$ during COD test, The COD values were corrected for residual H_2O_2 concentration as 0.47 mg/l COD/mg/l of residual H_2O_2 (Lee, 2011).

3 RESULTS AND DISCUSSION

3.1 Improvement in settling rate of precipitates at pH 3.0.using WCD as a weighting agent

About 30% COD removal was obtained due to separation of organic precipitates formed at pH 3.0 with or without the addition of WCD. Addition of WCD as Method 1 did not improve the settling rate. Figure 1 shows time course profiles of precipitate volume at varying WCD doses added as per Method 2. It may be noted that addition of WCD at concentrations of 4, 6, 8 and 10 g/L, significantly improved the rate of settling. At WCD doses of 0 g/L and 2 g/L no settling of precipitates was observed even after 24 hours. Based on these results WCD dose of 4 g/L was used for further studies. Figure 2 shows pictures of measuring cylinders at various WCD doses after 2 h of settling time as per Method 2.

Boiler Ash (BA) has alkaline characteristic due to the presence of metal oxides in it. When different doses of BA were added to OBAW at pH 3.0 the pH increased to 4.5–10 resulting in dissolution of organic precipitates. Thus, boiler ash was considered ineffective as weighting agent at acidic pH.

3.2 Fenton's treatment & improvement in biodegradability

Table 2 shows change in COD value after Fenton's treatment. It may be seen from Table 2 that there is no COD reduction due to Fenton's treatment; on the contrary COD increases slightly. Horsch et al. (2003) observed that there was COD reduction of just 5% in FWA wastewater when Dissolved Organic Carbon (DOC):H_2O_2 ratio of 1:1 and H_2O_2: Fe^{+2} ratio of 100:1 were employed. The lower COD reduction was attributed to the presence of high concentration of chloride ions. Siedlecka and Stepnowski (2005) studied degradation of phenol, 2-chlorophenol, and 2-nitrophenol by Fenton's reaction in the presence of chlorides and nitrates. The authors reported that the presence of anions reduced the degradation rate of all compounds due to the reduction in OH· generation and formation of much less reactive

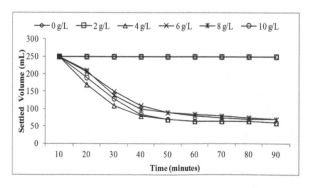

Figure 1. Rate of precipitate settlement at different WCD doses.

Figure 2.   Precipitate settlement at different WCD doses.

Table 2.   Reaction conditions and results of COD removal after Fenton's treatment.

$H_2O_2:Fe^{+2}$ molar ratio	H_2O_2 (24%) (mL)	$FeSO_4.7H_2O$ (mg)	Initial COD after WCD treatment (mg/L)	COD after Fenton's treatment after correction for residual H_2O_2 (mg/L)
10:1	1.25	250	2368	2436
20:1	1.25	125	2368	2274

Reaction conditions: Initial pH = 3.0, sample volume = 1 L, reaction time = 3 h, initial COD (Residual) after secondary treatment = 3264 mg/L

inorganic radicals such as SO_4^-, Cl^- and Cl_2^-. In a similar study Lu et al. (2005) reported that inhibition of aniline degradation by Fenton's reagent could be overcome by extending the reaction time when chloride concentration was low; however, at higher concentration of chloride the oxidation of aniline was inhibited completely due to the complexation of Fe-Cl. The OBAW treated in the current study contain high chloride ion concentration in the range of 12550–13250 mg/L which could have been the reason for lower COD reduction.

3.3   *Biodegradability assessment*

Figure 3 compares plots of % COD removal vs. time for (1) OBAW, (2) neutral supernatant obtained from method 2, (3) standard glucose solution, and (4) OBAW after Fenton's treatment. 100% biodegradability of glucose solution indicates that the microbial culture used is live and active. It may be further noted that while biodegradability of OBAW and supernatant after Method 2 is 30–35%, biodegradability improved to 52% and 88% for OBAW after Fenton treatment for H_2O_2: Fe^{+2} dose of 10:1 and 20:1, respectively. Horsch et al. (2003) observed that about 25% biodegradability is obtained after Fenton's treatment. The higher extent of biodegradability in the present study could be due to inherent difference in the nature of OBA compounds used by Horsch et al. (2003) and those contained in OBAW used by us.

4   CONCLUSION

1. Lowering the pH of OBAW to 3.0 resulted in formation of precipitates, removal of which yielded 30% C.O.D removal with or without addition of WCD. The precipitates were fluffy in nature with negligible settling rate. Addition of WCD doses of ≥ 4 g/L as a weighting agent prior to the adjustment of pH to 3.0, significantly improved the rate of settling of the precipitates.
2. The biodegradability of the OBAW after conventional primary and secondary treatments is 30% and after removal of organic precipitates is 35% which could be improved by

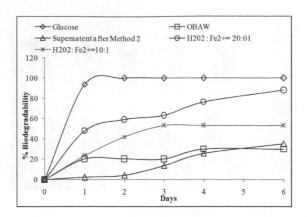

Figure 3.   Time course profiles for biodegradability of OBAW with and without Fenton's treatment.

Fenton's treatment to 52% and 88% for COD: H_2O_2 ratio of 7.5:1 and H_2O_2: Fe^{+2} of 10:1 to 20:1, respectively.

3. Thus a treatment sequence consisting of: WCD addition (4 g/L), further followed by pH adjustment to 3.0, hence followed by Fenton's treatment, and finally by biodegradation, can achieve up to 90% removal of residual COD from OBAW.

REFERENCES

Benatti, C.T. & Tavares, C.R.G. 2012. Fenton's Process for the Treatment of Mixed Waste Chemicals. *Organic pollutants ten years after the Stockholm Convention-Environmental and Analytical Update*: 247–268.

Horsch, P., Speck, A. & Frimmel, F.H. 2003. Combined advanced oxidation and biodegradation of industrial effluents from the production of stilbene–based fluorescent whitening agent. *Water Research* 37: 2748–2756.

Kohler, A., Hellweg, S., Escher, B.I. & Hungerbuhler, K. 2006. Organic pollutant removal versus toxicity reduction in industrial wastewater treatment: The example of wastewater from fluorescent whitening agent production. *Environmental Science & Technology* 40: 3395–3401.

Lee, E., Lee, H., Kim, Y.K., Sohn, K. & Lee, K. 2011. Hydrogen peroxide interference in chemical oxygen demand during ozone based advanced oxidation of anaerobically digested livestock wastewater. *International Journal of Environmental Science and Technolog* 8(2): 381–388.

Lemeune, S., Barbe, J.M., Trichet, A. & Guilard, R. 2008. Fluorescent Brightener 28 Removal by Ozonation or Advanced Oxidation Processes (O_3/H_2O_2) *Ozone: Science & Engineering: The Journal of the International Ozone Association* 19(2): 129–144.

Liu, Q.Y., Liu, Y.X. & Lu, X.J. 2012. Combined photo Fenton and biological oxidation for the treatment of aniline wastewater *Procedia Environmental Sciences* 12: 341–348.

Lu, M., Chang, Y., Chen, I. & Huang, Y. 2005. Effect of chloride ions on the oxidation of aniline by Fenton's reagent. *Journal of Environmental Management* 75(2): 177–182.

Oecd Guidelines for Testing of Chemicals 1992. Zahn-Wellens/EMPATest 302 B.

Poiger, T., Field, J.A., Field, T.M. & Siegrist, H. 1998. Behavior of fluorescent whitening agents during sewage treatment. *Water Research* 32 (6): 1939–1947.

Rice, E.W., Baird, R.B. & Eaton, A.D. 2012. Standard Methods for the Examination of Water and Wastewater-22nd edition.

Siedlecka, E.M. & Stepnowski, P. 2005. Phenols Degradation by Fenton Reaction in the presence of Chlorides and Sulfates. *Polish Journal of Environmental Studies* 14(6) 823–828.

Sirtori, C., Zapata, A., Oller, I., Gernjak, W., Aguera, A. & Malato, S. 2009. Decontamination industrial pharmaceutical wastewater by combining solar photo Fenton and biological treatment. *Water Research* 43: 661–668.

Smulders, E., Rybinski, W., Sung, E., Rahse, W., Steber, J., Wiebel, F. & Nordskor, A. 2002. Laundry Detergents. *Ullmann's Encyclopedia of Industrial Chemistry.*

Photocatalytic degradation of reactive turquoise blue dye using MFe_2O_4 (M = Ni, Zn) spinels by reactive grinding and sol-gel methods

J.M. Pardiwala, Femina Patel & Sanjay Patel
Institute of Technology, Nirma University, Ahmedabad, Gujarat, India

ABSTRACT: In this study, the photo catalytic degradation of reactive turquoise blue (RB21) dye was investigated by using MFe_2O_4 (M = Ni, Zn) spinel catalysts prepared by reactive grinding and sol-gel methods. The samples characterized by TG-DTA, XRD, particle size, and UV-NIR. XRD analysis revealed that $NiFe_2O_4$ spinel prepared by reactive grinding using improved planetary ball mill with jar speed: 350 rpm (clockwise), sun wheel speed: 200 rpm (anti-clockwise), ball to powder weight ratio: 15:1, milling time 11 h was of a single phase spinel. Traces of other phases were detected in addition to spinel phase in other catalysts. $NiFe_2O_4$ catalyst prepared by reactive grinding showed high photocatalytic activity compared with other spinels which completely remove the RB21 dye; TOC and COD removal efficiency was 81 and 75% after 120 min. The best photocatalytic performance was achieved after 120 min of UV light irradiation with a first order kinetic rate constant k of $0.0168 \ min^{-1}$.

Keywords: spinel; dye; photocatalytic degradation; reactive grinding; sol-gel

1 INTRODUCTION

Water pollution is a serious problem faced by developing and developed countries of the world (Bagheria et al. 2013, Borhan et al. 2013, Lu et al. 2013). Textiles, paper, plastics, leather, food, and the cosmetic industry uses different types of dyes to color the products (Casbeer et al. 2012, Verma et al. 2012, Borhan et al. 2013, Ismail et al. 2013, Borhan et al. 2014). Generally, 30–40% of these dyes that are coming from these industries remain in the waste water (Ismail et al. 2013).

Dyes are classified according to the chemical structure and type of application (Fan et al. 2009, Bagheria et al. 2013). Most of the dyes used in industry usually have a synthetic origin and complex aromatic molecular structures which make them very stable and relatively difficult to biodegrade (Fan et al. 2009, Casbeer et al. 2012, Verma et al. 2012, Borhan et al. 2013, Borhan et al. 2014). Varieties of methods such as physical-chemical (coagulation, adsorption, filtration, sedimentation, and catalytic action), biological (activated sludge), and advanced oxidation (ozonation, membrane separation, electrochemical, photo-catalysis, and ultrasonic techniques) are used for the removal of dyes from their effluents (Fan et al. 2009, Casbeer et al. 2012, Verma et al. 2012, Ismail et al. 2013, Lu et al. 2013, Borhan et al. 2014). Among these methods, the photocatalysis process is the one that is most widely used because it is a low cost, simple, and efficient process that creates low secondary pollution, no sludge production, and reduces the foul odors (Borhan et al. 2013, Lu et al. 2013, Borhan et al. 2014). Various semiconductors (TiO_2, ZnO, Fe_2O_3, CdS, GaP, and ZnS) and multi component oxides are used as photocatalysts to degrade dye (Casbeer et al. 2012, Verma et al. 2012, Borhan et al. 2013, Lu et al. 2013, Borhan et al. 2014). Semiconductors are used widely in industries as photo catalysts for the degradation of dyes (Marinca et al. 2011, Lu et al. 2013, Vanaamudan et al. 2013). However, the separation of these catalysts from treated water, especially from a large volume of water, is expensive to produce, also it creates secondary pollution and is time consuming, which limits their application in the industrial fields (Verma et al. 2012, Borhan et al. 2014). Another limitation of semiconductor use as a photocatalyst is that photocatalytic processes can be activated by ultraviolet light, which utilizes only 4% of the incoming solar energy (Borhan et al. 2013, Borhan et al. 2014). Hence, another class of oxides, spinel (AB_2O_4) is used as a novel catalytic material with a narrow band gap and is capable of utilizing a large portion of the solar spectrum for the degradation of dyes from its effluents (Verma et al. 2012, Bagheria et al. 2013, Borhan et al. 2013, Vanaamudan et al. 2013).

The general formula of the spinel type oxides is $(A)[B_2]O_4$, where the tetrahedral sites (A) are occupied by divalent cations (Zn^{2+}, Ni^{2+}, Co^{2+}, Cu^{2+} etc.) and octahedral sites (B) by trivalent cations (such as Cr^{3+}, Al^{3+}, In^{3+}, La^{3+} etc.) in a cubic structure lattice (Verma et al. 2012, Bagheria et al. 2013, Borhan et al. 2013). Spinel can be recovered easily and reused from other degradation processes due to their magnetic properties making them less expensive compared with semiconductors (Borhan et al. 2013, Borhan et al. 2014).

Spinels are mainly prepared by sol-gel, co-precipitation, hydrothermal, micelle technique, micro-emulsion, solid state reaction, ceramic method, and mechano-chemical route (Marinca et al. 2011, Verma et al. 2012, Vanaamudan et al. 2013, Zhang & Yuanxin 2013). The conventional methods require handling of large amounts of organic salt, solvent or surfactant, which usually makes them expensive and also creates an environmental pollution (Vanaamudan et al. 2013). The novel synthesis method called reactive grinding also known as mechano-chemical synthesis that using high energy ball mill is developed which is simple, low in production cost, eliminating heating step to synthesize spinels, and also overcomes the limitations of conventional methods used for preparing spinels (Zhang & Yuanxin 2013). To the best of our knowledge, limited work has been carried out for the development of spinels by reactive grinding method using high energy ball mills mainly planetary ball mill for photocatalytic degradation of reactive dyes. Thus, in this study, MFe_2O_4 (M = Ni, Zn) spinels were synthesized by reactive grinding and sol-gel methods and its photocatalytic properties (under ultraviolet light irradiation for degradation of Reactive Turquoise Blue (RB21)) were investigated.

2 MATERIAL AND METHODS

2.1 Material

All chemicals used were analytical reagents with more than 98.0% purity except Fe_2O_3 and NiO (95.0%), which were supplied by CDH, Finar, and HPLC grade. Commercial grade reactive turquoise blue dye was obtained from International Welspun India Ltd without further purification.

2.2 Catalyst preparation

Sol-gel method: A typical procedure for preparing spinel $NiFe_2O_4$ by sol-gel method is as follows: metal nitrates $Fe(NO_3)_3 \cdot 9H_2O$ (0.1279 mol, 51.69 g) and Ni $(NO_3)_2 \cdot 6H_2O$ (0.064 mol, 18.56 g) were first dissolved in 50 ml of distilled water separately. The molar ratio of citric acid to total metal cations was fixed at unity. Subsequently, 15 ml of aqueous solution containing 10 wt% excess (30.931 g) of citric acid, to ensure complete complexation of the metal ions, was slowly added to the precursor solution under vigorous stirring at room temperature for 15 min. To make a gel, 40 ml Ethylene Glycol (EG) was added by stirring. The resulting solution was heated at 80°C in an oil bath to form a viscous gel which finally yields a solid precursor upon slow solvent evaporation at 110°C overnight in hot air oven; it was finally calcined in muffle furnace under air at 700°C (ramp = 10°C/min) for 5 h in order to achieve the corresponding spinel structure. The same procedure was followed for the preparation of $ZnFe_2O_4$ spinel.

194

Reactive grinding method: spinel $NiFe_2O_4$ prepared by a reactive grinding method using an improved planetary ball mill. NiO (4.70 g) and Fe_2O_3 (10.20 g) were premixed by hand grinding for 5 min. The mixed powder was then loaded into a stainless steel jar (250 mL) with stainless steel balls (4 balls of 19 mm size, 7 balls of 14 mm size, and 7 balls of 10 mm size) under an air atmosphere. The ball to powder weight ratio was 15:1. The jar was closed with a thick cover and sealed with O-ring. The revolution per minute (rpm) of the sun wheel was kept at 200 (anti clock wise) and that of the jar at 350 (clock wise). The milling was carried out for 11 h in the improved planetary ball mill. During milling, temperature was increased due to impact of balls within the jar which was cooled using fans in order to maintain 40°C temperature. The milling atmosphere in the jar could be controlled by allowing air through air pipe (oxidizing atmosphere). After milling, the as-ground material was separated from the balls and no heat treatment was given. The same procedure was followed for the preparation of $ZnFe_2O_4$ spinel catalyst.

2.3 Catalyst characterization

The thermal behavior of the samples was investigated by TG-DTA using SF 752 (Metpler Toledo) with the heating rate of 10°C/min. Phase analysis was determined by X-ray diffraction using X'pert-MPD system (Philips) at 40 kV and 30 mA using Cu Kα radiation ($\lambda = 1.5406$ Å). The particle size was determined by laser diffraction method using Zeta sizer (Malven). Band gap was measured using UV-vis-NIR spectrophotometer (Agilent, Carry 5000) in the wavelength range of 190–2500 nm at room temperature. The concentration of RB21 during the degradation was monitored by UV–visible spectrophotometer (Shimadzu, Japan). Total Organic Carbon (TOC) and COD in the solution was measured with a TOC analyzer (Shimadzu, Japan) and (HACH, US Canada), respectively, to investigate the mineralization degree of RB21. A 450 W UV-Visible lamp photocatalytic reactor (Lelesil Innovative system Pvt Ltd, Mumbai, India) was used as the illumination source. Solar power meter (Tenmars TM 207) was used to measure the intensity of light.

2.4 Preparation of dye solutions

Stock solutions of RB21 (1 g/L) were prepared by dissolving an accurately weighed amount of RB21 in the double-distilled water and subsequently diluting it to the required concentration. The structure of the dye is shown in Figure 1.

2.5 Photocatalytic activity of dyes solutions

The photocatalytic activity of the prepared spinel catalysts was evaluated by degradation of RB21 dye under UV-light irradiation. In each experiment, the reaction suspension was prepared by adding 0.8 g of spinel catalyst powder into 500 mL RB21 solution with an initial concentration of 50 mg/L and pH 7. The suspension was magnetically stirred in the dark for 30 min to reach adsorption–desorption equilibrium at a constant temperature prior to irradi-

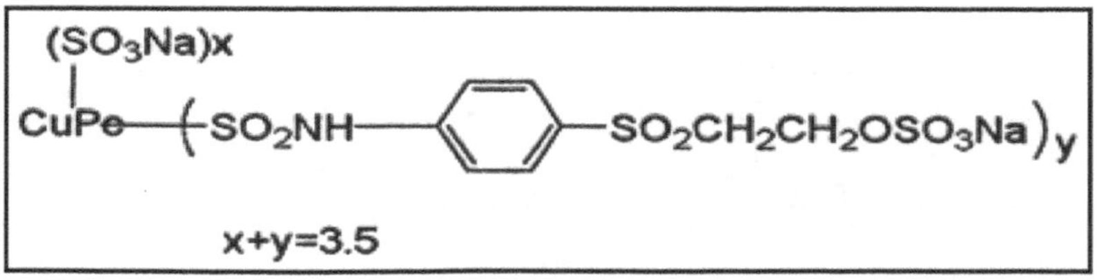

Figure 1. Structure of Reactive Turquoise Blue (RB21).

ation. After irradiation, each experiment was conducted for 90 min and samples were drawn at the time intervals of 15, 30, 45, 60, 75, and 90 min for measuring the dye degradation. The absorbance of the solution at λ_{max} = 620 nm with time has been measured to monitor RB21 concentration. All experiments were performed at room temperature. The dye removal efficiency was calculated based on the following equation:

$$\text{Dye removal efficiency (\%)} = \frac{C_0 - C_t}{C_0} \times 100$$

where C_0 is the initial concentration of RB21 (mg/L) and C_t the concentration of RB21 at time t (mg/L).

3   RESULTS AND DISCUSSION

3.1   *Thermo gravimetric and differential thermal analysis*

In order to understand the thermal decomposition of the RG 1(NiFe$_2$O$_4$ spinel catalysts prepared by reactive grinding) thermo gravimetric measurements (TG) have been done. The weight loss of the samples have been obtained by milling for 11 h from the mixture of NiO$_2$ and Fe$_2$O$_3$ powders in a measured temperature range from 25 to 800°C. A weight loss region from 25°C to 400°C with a weight loss of 0.7673% for precursors of RG 1 catalyst was observed in the TG curves of the precursors of RG 1 catalyst after it was milled for 11 h in the planetary ball mill. The weight loss region from 25°C to 400°C resulted due to desorption of moisture. DTA curve shows a large endothermic peak around 100°C to 150°C which is assigned to the evaporation of moisture. No further peak or weight loss appeared after the sample was given heat treatment above 400°C indicating that all oxides components were decomposed completely and the spinel oxide of NiFe$_2$O$_4$ began to form.

3.2   *X-ray diffraction pattern*

The X-ray diffraction patterns of MFe$_2$O$_4$ (M = Ni, Zn) spinel catalysts are shown in Figure 2. The spectra show that spinel phase was developed in all the samples. As shown in Figure 2, after 11 h of milling, the typical patterns of the two starting oxides (NiO and Fe$_2$O$_3$) and other phases were vanished and formed NiFe$_2$O$_4$ spinel structure (JCPDS card 98-018-2237) without any traces of other impurities in case of RG 1. No secondary phase was detected in the XRD pattern which ensures the phase purity of the final products. The results show that improved planetary ball mill imparted sufficient energy at 350 rpm and caused the mech-

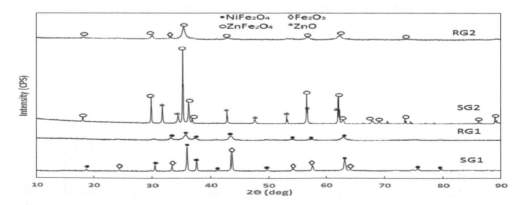

Figure 2.   XRD patterns of MFe$_2$O$_4$ (M = Ni, Zn) spinel catalysts.

Table 1. Physical properties of MFe_2O_4 (M = Ni, Zn) spinel catalysts.

Catalyst composition	Catalyst designation	Calcination temp/time	XRD phases	Crystallite size d_{XRD} (nm)	Particle size (nm)	Band-gap (eV)
$NiFe_2O_4$	SG 1	700°C, 5 h	$NiFe_2O_4$ Fe_2O_3	42.19	447.7	1.84
$NiFe_2O_4$	RG 1	w/o heat treatment	$NiFe_2O_4$	21.11	136.9	1.63
$ZnFe_2O_4$	SG 2	700°C, 5 h	$ZnFe_2O_4$ ZnO	59.21	544.8	2.10
$ZnFe_2O_4$	RG 2	w/o heat treatment	$ZnFe_2O_4$ Fe_2O_3	24.66	475.4	1.65

anochemical reaction between NiO and Fe_2O_3 which caused the spinel phase to occur after 11 h milling. Other minor phases such as Fe_2O_3 (JCPDS card 98-002-2505) were detected in addition to the major AB_2O_4 spinel phase in case of SG 1 (JCPDS card 98-005-2387), which indicates that the calcinations temperature is not sufficient to form a pure spinel phase. Spectra of SG 2 show the presence of AB_2O_4 spinel structure (JCPDS card 98-017-0914) with other phase ZnO (JCPDS card 04-016-6648). Other minor phases such as Fe_2O_3 (JCPDS card 00-002-1047) were detected in addition to the major AB_2O_4 spinel phase (JCPDS card 01-077-0011) in RG 2. Due to intense milling, iron contamination dislodged from the grinding balls and the jar during the milling. This is the reason for the presence of Fe_2O_3 impurities in case of RG 2. From XRD spectra, crystal sizes of the catalysts (d_{XRD}) were calculated using Scherrer equation ($d_{XRD} = 0.9\lambda/\beta\cos\theta$) with Warren's correction for instrumental broadening where d_{XRD} is the crystal size, λ is the wavelength of Cu-Kα radiation ($\lambda = 1.5406$ Å), β is the effective line width of the X-ray reflection calculated by the expression $\beta^2 = B^2 - b^2$ (where B is the Full Width at Half Maximum (FWHM), b the instrumental broadening), and θ is the diffraction angle. Physical properties of MFe_2O_4 (M = Ni, Zn) spinel catalysts are reported in Table 1. XRD spectra shown in Figure 2 exhibited some differences in the intensity of the spinel peaks at the same angles which confirmed the higher crystallinity for spinel prepared by sol-gel compared with reactive grinding.

Table 1 shows the band gap energies of MFe_2O_4 (M = Ni, Zn) spinel catalysts. It can be seen that each of the MFe_2O_4 (M = Ni, Zn) spinel catalysts have a band gap of approximately 2 eV or lower making them effective under UV and visible light irradiation.

3.3 *Evaluation of photocatalytic activity*

For photo degradation experiments, the influence of two different pH conditions (1 and 12) was studies using RG 1 (initial concentration of RB21 dye (50 mg/L) and contact time (120 min)) in addition to the pH = 7. RB21 shows slight adsorption at alkaline pH and is more strongly adsorbed at acidic and neutral pH. The photo degradation of the dye is found to be at a maximum at neutral pH. At acidic and alkaline pH, the photocatalytic activity is not comparable with the results of neutral pH. Table 2 shows the influence of initial pH of dye solution on the photocatalytic activity using RG 1 spinels.

Reactive turquoise blue dye generates negatively charged ions when dissolved in aqueous media, it can therefore be concluded that the rate of adsorption of such a dye is higher in acidic media but there was negligible removal of TOC and COD. The results clearly indicate that the best activity is observed for RG 1 at neutral pH. Therefore, pH 7 has been selected for other experiments. In general, the natural pHs of the RB21 dye solution was 7. Therefore, these solutions were used directly without any pH adjustment.

To evaluate the photocatalytic performance of the MFe_2O_4 (M = Ni, Zn) spinel catalysts, the catalytic decomposition process of the RB21 molecules in aqueous solution was conducted (the initial pH is 7) under UV light irradiation at a given interval time. After 120 min

Table 2. Influence of initial pH of dye solution on the photocatalytic activity using RG 1 spinels.

pH	Dye removal %	% TOC removal	% COD removal
1	90	42	Not detect
7	90	80	75
12	40	46	30

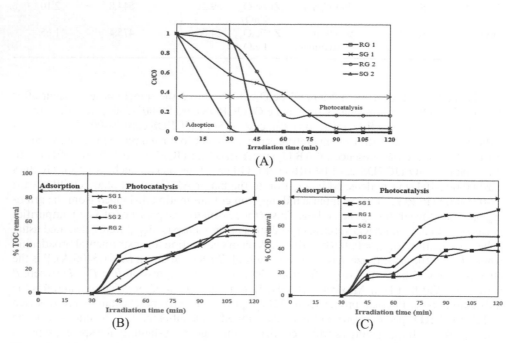

(A)

(B)

(C)

Figure 3. Photocatalytic degradation of RB21 in the presence of different catalysts.

of reaction, RB21 removal efficiency (Figure 3A) was 100, 83, 96, and 99%, TOC removal efficiency (Figure 3B) was 81, 49, 53, and 56% and COD removal efficiency (Figure 3C) was 75, 40, 45, and 52% for RG 1, RG 2, SG 1, and SG 2, respectively. It should be noted that all tests were performed using the same amount of catalyst and the characteristic RB21 absorption peak at λ_{max} = 620 nm decreased gradually and almost no color was observed after 120 min indicating that RB21 was degraded by RG 1, SG 1, and SG 2. There are no new peaks observed which means that RB21 molecules are mainly degraded into CO_2 and H_2O.

RG 1 ($NiFe_2O_4$ catalyst prepared by reactive grinding) showed high photocatalytic activity compared with other spinels which completely degrade RB21 dye and remove 81% of total organic carbon and 75% COD from the RB21 solution after 120 min due to a narrow band gap (1.63 eV). Under UV light irradiation, some electrons (e^-) in the Valence Band (VB) were excited to the Conduction Band (CB) causing the generation of holes (h^+) in the VB simultaneously. A portion of the photo generated electrons would recombine with holes in the VB, while others transferred to the surface and reacted with the adsorptive oxygen molecule to yield $\bullet O_2^-$. The generated $\bullet O_2^-$ would further combine with H^+ to produce $\bullet HO_2$. The reactive species, such as $\bullet O_2^-$, $\bullet HO_2$, and h^+_{VB}, all could oxidize the RB21 dye. Under UV light irradiation, RG 1 photocatalyst can generate e^-/h^+, which in turn produces reactive oxygen species through redox processes for the degradation of the RB21. RG 1 has a relatively narrow band gap 1.63 eV) making them capable of such processes.

Figure 4 shows the RG 1 ($NiFe_2O_4$) spinel catalyst attracted to a magnet after use in RB21 photo degradation, illustrating its complete removal from the solution. However, the mag-

Figure 4. Photograph the RB21 catalyst solution before (left) and after (right) photo degradation, and RG 1 being attracted by a magnet (right).

netic catalysts have the advantage of being easily removed from the solution by applying a magnetic field, thus, reducing time and costs of their separation/recovery. Two cycles of RB21 photo degradation were performed with the RG 1 ($NiFe_2O_4$) catalyst in order to evaluate its stability. There was a small reduction in RB21 degradation upon reuse of RG 1 ($NiFe_2O_4$) catalyst with photocatalytic activities of 80% (TOC removal efficiency) after two cycles, suggesting a good potential for catalyst recycling.

4 CONCLUSION

In summary, MFe_2O_4 (M = Ni, Zn) spinel catalysts were prepared by reactive grinding and sol-gel methods. The catalysts were successfully used for the photocatalytic degradation of RB21 dyes under UV light irradiation. XRD confirmed the presence of a single phase in $NiFe_2O_4$ spinel prepared by a reactive grinding method. The highest RB21 (100%), TOC (81%), and COD (75%) removal after 120 min of reaction were obtained with the $NiFe_2O_4$ spinel prepared by a reactive grinding method due to narrow band gap and phase purity. Moreover, the catalyst could easily be separated from the treated solution for reuse by simply applying an external magnetic field. Spinel catalysts are promising photo catalyst for the degradation of reactive dyes with high photocatalytic activity. Further studies are necessary to put all the new materials into practice.

REFERENCES

Bagheri, Gh., A., Ashayeri, V. & Mahanpoor, K. 2013. Photocatalytic efficiency of $CuFe_2O_4$ for photodegradation of acid red 206. *International Journal of Nano Dimension* 4(2): 111–115.

Borhan, A.I., Samoila, P., Hulea, V., Iordan, A.R. & Palamaru, M.N. 2013. Photocatalytic activity of spinel $ZnFe_{2-x}Cr_xO_4$ nanoparticles on removal Orange I azo dye from aqueous solution. *Journal of the Taiwan Institute of Chemical Engineers.* 6.

Borhan, A.I, Samoila, P., Hulea, V., Iordana, A.R., & Palamaru, M.N. 2014. Effect of Al^{3+} substituted zinc ferrite on photocatalytic degradation of Orange I azo dye. *Journal of Photochemistry and Photobiology A: Chemistry* 279: 17–23.

Casbeer, E., Sharma, V.K. & Li, X.Z. 2012. Synthesis and photocatalytic activity of ferrites under visible light: A review. *Separation and Purification Technology* 87: 1–14.

Fan, G., Zhijun, G., Yang, Lan., Li Feng. 2009. Nanocrystalline zinc ferrite photocatalysts formed using the colloid mill and hydrothermal technique *Chemical Engineering Journal* 155: 534–541.

Ismail, B., Hussain, S.T. & Akram, S. 2013. Adsorption of methylene blue onto spinel magnesium aluminate nanoparticles: Adsorption isotherms, kinetic and thermodynamic studies. *Chemical Engineering Journal* 219: 395–402.

Lu, D., Zhang, Y., Lin, S., Wang, L. & Wang, C. 2013. Synthesis of magnetic $ZnFe_2O_4$/graphene composite and its application in photocatalytic degradation of dyes. *Journal of Alloys and Compounds* 579: 336–342.

Marincaa, T.F., Chicinas, I., Isnard, O., Pop, V., Popa F. 2011. Synthesis, structural and magnetic characterization of nanocrystalline nickel ferrite-$NiFe_2O_4$ obtained by reactive milling. *Journal of Alloys and Compounds* 509: 7931–7936.

Vanaamudan, A., Pathan, N. & Pamidimukkala, P. 2013. Adsorption of Reactive Blue 21 from aqueous solutions onto clay, activated clay, and modified clay. *Taylor and Francis*: 1–11.

Verma, A.K., Dash, R.R., & Bhunia, P. 2012. A review on chemical coagulation/flocculation technologies for removal of colour from textile wastewaters. *Journal of Environmental Management* 93: 154–168.

Zhang, L., & Wu, Y. 2013. Sol-Gel Synthesized Magnetic $MnFe_2O_4$ Spinel Ferrite Nanoparticles as Novel Catalyst for Oxidative Degradation of Methyl Orange, *Journal of Nanomaterials*. 6: 1–6.

Multi-disciplinary Sustainable Engineering: Current and Future Trends – Tekwani, Bhavsar & Modi (Eds)
© *2016 Taylor & Francis Group, London, ISBN 978-1-138-02845-6*

Analysis of linear parallel plate regenerator for active magnetocaloric regenerative refrigeration using CFD as a tool

Alzubair A. Saiyed
B.H. Gardi College of Engineering and Technology, Rajkot, Gujarat, India

V.J. Lakhera
Nirma University, Ahmedabad, Gujarat, India

Jayesh J. Barve
Adani Institute of Infrastructure Engineering, Ahmedabad, Gujarat, India

ABSTRACT: Active Magnetocaloric Regenerative Refrigeration (AMRR) system is a potentially attractive alternative to Vapor Compression Refrigeration (VCR) technology. To investigate the performance of an AMRR system, it is important to understand the heat transfer and pressure drop phenomena taking place within the regenerator. With this objective, a transient three dimensional model of a linear parallel plate regenerator for AMRR system for cold blow process has been developed and solved by theoretical as well as numerical method. The magnetocaloric effect is neglected particularly for the present analysis. The numerical model is flexible and used to accomplish several parametric studies on linear parallel plate regenerator for different design selections and operating conditions. The heat transfer coefficient and pressure drop phenomenon were explored numerically for four fluid channel thickness (0.2, 0.4, 0.6, and 0.8 mm) at the five Reynolds numbers of Heat Transfer Fluid (HTF) (25, 50, 100, 500, and 1000) involving two different linear parallel plate materials (Cu and Gd). Finally the numerical model of the linear parallel plate regenerator is validated by comparing it with theoretical results. The validation shows the good agreement between numerical results and theoretical results.

Keywords: linear parallel plate regenerator; active magnetocaloric regenerative refrigeration; heat transfer; pressure drop; cold blow; numerical method

1 INTRODUCTION

Magnetocaloric Refrigeration (MR) is a cooling technology, which works on the principle of "magnetocaloric effect". Generally MR system is used to achieve extremely low temperatures, nearly about below 1 K, and also for the ranges used in the commercial refrigeration system which mainly depend on the design of the refrigeration system. When the magnetic field is applied to a suitable material (magnetocaloric material), the temperature of the material changes, which is easily noticeable. This magneto-thermodynamic phenomenon is also known as a magnetocaloric effect. The chemical element gadolinium and some of its alloys give the notable change in temperature whenever the magnetic field applied, which is the best example of the magnetocaloric effect. Gadolinium's temperature is detected to rise when it arrives at certain magnetic fields. When the magnetic field is removed, the temperature of gadolinium returns to normal. References [1–4] provide more information regarding the MR technology.

MR system appears to be a promising technology for cryogenic application as well as room temperature refrigeration. Compactness and efficiency are the main advantages of

201

MR system. From the past three decades, the experimental prototype and numerical modeling analysis developed by the researcher from academia, industries, and Government R&D organization across the globe. Results have proven that the technology has potential, but understanding its principles and predicting its performance have been extremely difficult. This technology has been successfully demonstrated in lab-scale prototypes by several researchers/technologists from the academia and the industry that is globally and technologically maturing fast. But, viable and commercial products are still awaited due to several systems and materials challenges. The objective of the present study is as follows:

1. Develop a validated model of a linear parallel plate regenerator for AMRR system in ANSYS FLUENT.
2. Simulation of the regenerator model with various materials of parallel plates and fluid channel thickness in ANSYS FLUENT.

2 NUMERICAL SIMULATION

2.1 Geometry of solution domain

To optimize the performance of AMRR system, it is important to understand the heat transfer phenomena taking place in the regenerator. With this objective in mind, 40 cases of regenerator simulated in ANSYS FLUENT 14.5, which including two parallel plate material (Cu and Gd), four fluid channel thickness (0.2, 0.4, 0.6, and 0.8 mm) and five Heat Transfer Fluid (HTF) flow Reynolds number (25, 50, 100, 500, and 1000). The various combinations of simulations are listed in Table 1.

In the linear parallel plate regenerator, the characteristic scales of the length and flow section is in a millimeter scale. Therefore, it is very difficult to simulate the entire heat exchanger with the existing computing power. For the present CFD analysis, only two parallel plates with one fluid channel (sandwich structure model) is selected as shown in Figure 1.

Table 1. The various combinations of simulations for the present study.

Case	Parallel plate material	Fluid channel thickness (mm)	Reynolds numbers
1	Copper (Cu)	0.2, 0.4, 0.6, 0.8	25, 50, 100, 500, 1000
2	Gadolinium (Gd)	0.2, 0.4, 0.6, 0.8	25, 50, 100, 500, 1000

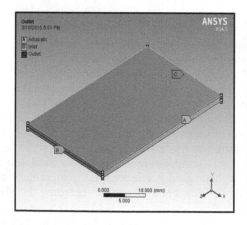
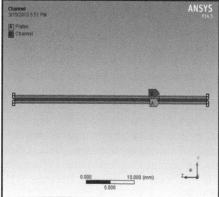

Figure 1. Simulation object: sandwich structure model.

Table 2. Simulation boundary conditions.

At inlet of HTF	Velocity inlet
At outlet of HTF	Outflow
Fluid cell zone condition	Water
Solid cell zone condition	Cu or Gd (depends on analysis)
Interface between solid and fluid	Coupled
The outer wall of the both parallel plates	Adiabatic wall (heat flux is equal to zero)

2.2 Simulation assumptions and boundary conditions

Generally 1 to 1.5 T magnetic fields are applied to the present dimensional geometry of regenerator leading to a temperature increase up to 4–5°C. The outer walls of the regenerator take an adiabatic wall because there is no heat transfer from regenerator to the atmosphere. The cycle time for the cold blow process of the AMRR cycle is 1 s while there is a HTF flow from cold heat exchanger side (inlet) to hot heat exchanger side (outlet).

The computational domain of parallel plate regenerator model contains boundary conditions which are listed in Table 2.

3 THEORETICAL PERFORMANCE PARAMETER FORMULATION

3.1 Heat transfer in linear parallel plate regenerator through HTF

The heat transfer between the regenerator and fluid is determined from the Nusselt number. Here the theoretical correlation suggested by Nickolay and Martin [4] for heat transfer between parallel plates is used to determine the local Nusselt number as;

$$Nu_l = Nu_a - \frac{1}{3}\left(\frac{Nu_2 - b}{Nu_a}\right)^{n-1} \cdot Nu_2$$

Gz is the Graetz number and Nu_a is the average Nusselt number defined as;

$$Nu_a = (Nu_1{}^n + b^n + (Nu_2 + b)^n)^{1/n}$$

With $Nu_1 = 7.541$ $Nu_2 = 1.841 \, Gz^{1/3}$ b = 0 and n = 3.592

3.2 Pressure drop of HTF flow

The pressure determines the amount of pumping power needed to run a regenerator. It is, therefore, important to characterize the pressure drop for design. The pressure drop (ΔP) is determined by the Hagen–Poiseuille law.

4 RESULTS AND DISCUSSION

4.1 Effect of fluid channel thickness on convective heat transfer coefficient characteristics

Convective heat transfer coefficient characteristics in the regenerator are predicted as a function of variation in plate spacing. For all Reynolds number, the heat transfer coefficient value is increased by decreasing the channel thickness and represented by graphs as shown in Figure 2 (for copper material) and in Figure 3 (for gadolinium material). As the convective heat transfer coefficient is inversely proportional to the hydraulic diameter of the fluid

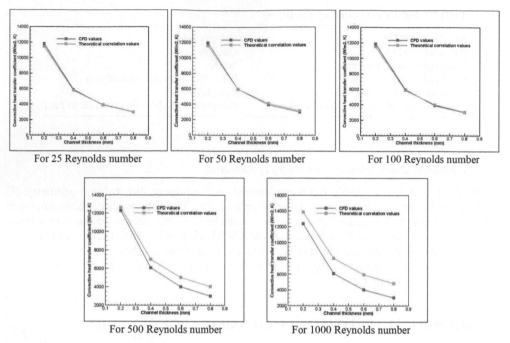

For 25 Reynolds number   For 50 Reynolds number   For 100 Reynolds number

For 500 Reynolds number   For 1000 Reynolds number

Figure 2. Convective heat transfer coefficient ($W/m^2 \cdot K$) vs. channel thickness (mm) for copper material. A) For 25 Reynolds number, B) For 50 Reynolds number, C) For 100 Reynolds number, D) For 500 Reynolds number, and E) For 1000 Reynolds number.

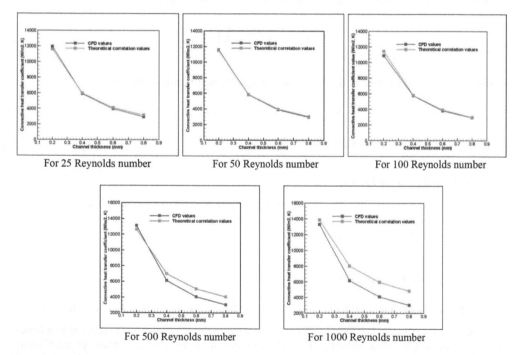

For 25 Reynolds number   For 50 Reynolds number   For 100 Reynolds number

For 500 Reynolds number   For 1000 Reynolds number

Figure 3. Convective heat transfer coefficient ($W/m^2 \cdot K$) vs. channel thickness (mm) for gadolinium material. A) For 25 Reynolds number, B) For 50 Reynolds number, C) For 100 Reynolds number, D) For 500 Reynolds number, and E) For 1000 Reynolds number.

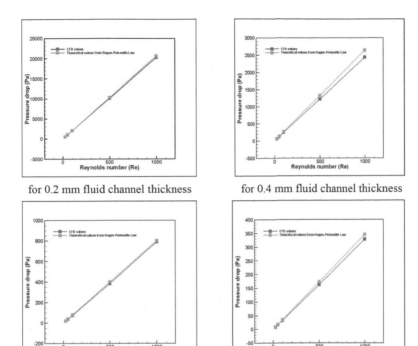

for 0.2 mm fluid channel thickness   for 0.4 mm fluid channel thickness

for 0.6 mm fluid channel thickness   for 0.8 mm fluid channel thickness

Figure 4. Pressure drop *vs.* Reynolds number for various fluid channel thickness. A) For 0.2 mm fluid channel thickness, B) For 0.4 mm fluid channel thickness, C) For 0.2 mm fluid channel thickness, D) For 0.6 mm fluid channel thickness, and E) For 0.8 mm fluid channel thickness.

channel, hence, the small thickness fluid channel has high a Nusselt number comparatively with the large thickness fluid channel.

For a high density of heat flux transportation application, as much as possible, a small hydraulic diameter fluid channel is selected. But due to the manufacturing limitation, the very small hydraulic diameter (nearer to macro and mini channel) regenerator is costly.

4.2 *Pressure drop characteristics*

Due to small hydraulic diameter of fluid channel, a large pumping work is required to overcome the pressure drop that will decrease the effectiveness of the cooling system (by increasing the power required for pump). For a fixed fluid channel thickness, the pressure drop increased with increasing Reynolds number as shown in Figure 4. The result data derived from the Hagen–Poiseuille law match well with the CFD numerical result data [5].

4 CONCLUSION

Based on the present analysis of the parallel plate regenerator behavior, the following conclusions can be drawn:

1. The results of numerical CFD simulations using the continuum model are in good agreement with the correlation suggested by Nickolay and Martin for heat transfer coefficient value in the three dimensional channels.

2. For a particular given size of the fluid channel thickness (0.2, 0.4, 0.6, and 0.8 mm), the continuum approach cannot be applied. Even when the channel dimension is between 200 microns to 0.9 mm, some deviation of the results can be found. This is the probable reason for the deviation of some numerical results of heat transfer coefficient value derived from FLUENT ANSYS not matching with the theoretical results.
3. For a given thickness, the relation between fluid channel pressure drop and Reynolds number is linear. The fluid channel pressure drop increased with the increasing Reynolds number.

REFERENCES

[1] Gomez, J.R., Garcia, R.F., Carril, J.C. & Gomez, M.R. 2013. A review of room temperature linear reciprocating magnetic refrigerators. Renewable and Sustainable Energy Reviews 21: 1–12.
[2] Bouchard, J., Nesreddine, H. & Galanis, N. 2009. Model of a porous regenerator used for magnetic refrigeration at room temperature. International Journal of Heat and Mass Transfer 52: 1223–1229.
[3] Gschneidner, K.A., Jr. & Pecharsky, V.K. 2008. Thirty years of near room temperature magnetic cooling: Where we are today and future prospects. International Journal of Refrigeration 31: 945–961.
[4] Petersen, T.F. 2007. Numerical modelling and analysis of a room temperature magnetic refrigeration system. Ph.D. thesis, The Technical University of Denmark.
[5] Saiyed, A.A. 2015. Investigations on Regenerator Designs for Magnetocaloric Refrigeration. M.Tech Thesis, Institute of Technology, Nirma University, Ahmedabad.

Multi-disciplinary Sustainable Engineering: Current and Future Trends – Tekwani, Bhavsar & Modi (Eds)
© *2016 Taylor & Francis Group, London, ISBN 978-1-138-02845-6*

Combined catalytic ozonation and UV treatment for removal of reactive red 120 dye from water

S. Sharma, T. Patel & J. Ruparelia

Department of Chemical Engineering, Nirma University, Ahmedabad, Gujarat, India

ABSTRACT: Treatment of effluents from textile and tannery industries containing dyes has not been very successful using conventional methods like biological treatments. To overcome this problem various Advanced Oxidation Processes (AOPs) are being suggested in literature. In this paper an attempt has been made with the focus on the degradation of reactive red 120 dye in water by catalytic ozonation in the presence of titanium oxide catalyst and UV light. The removal of dye was monitored by the removal of TOC and color employing all the three processes, ozonation, ozonation/catalyst, ozonation/catalyst/UV have been investigated, and degradation efficiency order observed as ozonation/catalyst/UV > ozonation/catalyst > ozonation. The results proved that the catalyst may be activated by UV + ozone and increases the dye degradation efficiency.

1 INTRODUCTION

Dye and textile industries are major producers of wastewater. Dyes are an important class of pollutants, even if they are present in trace quantities. Dye and textile industries discharges large amount of synthetic wastewater containing dyes (Robinsion 2001). A loss of 1–2% in production and 1–10% in use are estimated. Due to their large-scale production and extensive application, synthetic dyes can cause considerable non-aesthetic pollution and pose a serious health risk factor (Martinez-Huitle 2009). Conventional methods are not effective for degradation of dye in textile effluents because they contain high concentration of toxic pollutants. Chemical oxidation with ozone is a current technology for the removal of organic pollutants and disinfection (Beltrain et al. 2002, Moussavi et al. 2009, Sui et al. 2012, Ikhlaq et al. 2012, Moussavi et al. 2012). Advanced oxidation processes can provide effective technological solution for such effluents.

Ozonation generates hydroxyl radical (•OH), a powerful oxidizing agent, which can completely degrade or mineralize the pollutants into harmless products. Ozone is a very powerful oxidizing agent and it has highly unstable molecules. Ozone degrades pollutants by two mechanism, either by direct reaction in which pollutants react with dissolved molecular ozone or indirect reaction in which pollutants react with hydroxyl radical (•OH) (Baig et al. 2001). Catalyst is also used in the ozonation process to improve the efficiency of the process and to reduce the ozone consumption. The combination of catalytic ozonation process along with UV light further enhances the efficiency of the process and considerably reduces the ozone consumption rate.

For the simple ozonation process, ozone reacts with water and it gives difficult chemical reaction and mechanism like the direct or indirect reaction with ozone.

Direct reaction of ozone with dye molecule occurs as follows (Hordem et al. 2003):

$$O_{3+} \text{ dye} \rightarrow \text{dye}_{oxid} \text{ (oxidation product of dye)} \tag{1}$$

Indirect reaction occurs with the reaction:

$$O_3 + H_2O \xrightarrow{OH\bullet} \bullet OH + O_2 \tag{2}$$

O_3/UV process is an advanced water treatment method for the effective oxidation and destruction of toxic and refractory organics in wastewater. Basically, aqueous systems saturated with ozone are irradiated with UV light of 254 nm in a reactor convenient for such heterogeneous media.

O_3/UV oxidation process is more complex than the other ones, since •OH radicals are produced through different reaction pathways. There is a general agreement about involved reactions (Galindo et al. 1999).

$$O_3 + H_2O + hv \rightarrow H_2O_2 + O_2 \tag{3}$$

$$H_2O_2 + hv \rightarrow 2HO \bullet \tag{4}$$

Photocatalytic processes make use of a semiconductor metal oxide as a catalyst and of oxygen as an oxidizing agent (Konsawa 2003). Many catalysts have been so far tested, although, only TiO_2 in the anatase form seems to have the most interesting attributes such as high stability, good performance, and low cost (Kiwi 1993).

$$TiO_2 + hv \rightarrow e^- + h^+ \tag{5}$$

$$TiO_2^{(h+)} + H_2O^{ad} \rightarrow TiO_2 + HOad \bullet + H^+ \tag{6}$$

$$TiO_2^{(h+)} + HO^{ad} \rightarrow TiO_2 + HOad \tag{7}$$

$$TiO_2^{(h+)} + RX^{ad} \rightarrow TiO_2 + RXad^+ \tag{8}$$

2 EXPERIMENTAL INVESTIGATIONS

2.1 *Material*

In the present study, dye sample of reactive red 120 was taken from the local vendor and used without any further purification. The molecular formula of reactive red 120 dye is $C_{44}C_{12}H_{24}N_{14}Na_6O_{20}S_6$. These dyes are called azo dye as the—N = N—group is present. Distilled water was used for the preparation of the synthetic wastewater for all the experiments.

Ozone was generated from the oxygen by using corona–discharge method. The applied ozone dose was controlled by using flow meter which had the capacity of 0–60 LPH. The semi batch pyrex glass bubble column reactor was used for conducting all the experiments. The unreacted ozone gas was trapped into the two impinger bottles which was filled with the 2% KI solution.

2.2 *Method (preparation of titanium dioxide catalyst)*

The Titanium Tetra Isopropoxide (TTIP) solution (97% synthesis) is directly purchased from the local vendor. TTIP solution is mixed with the 8 ml of the deionized or demineralized water. The solution is stirred up to 1 h for the proper mixing. 1 M of HNO_3 acid was added to maintain the pH of the solution upto 3 pH. The preparation of 1 M of HNO_3 is done by taking 3.2 ml solution of the concentrated HNO_3 and by adding 46.8 ml of deionized water. After mixing the solution mixture is then calcined at 400°C for 6 h in muffle furnace. The calcined mixture is then converted into the anatase phase TiO_2 catalyst, which is used for the degradation of the synthetic wastewater.

3 EXPERIMENTAL SET-UP

Ozonation studies were done using an experimental setup consisting of an ozonator, oxygen cylinder, bubble column reactor, two impinge bottles, and circular sintered glass gas

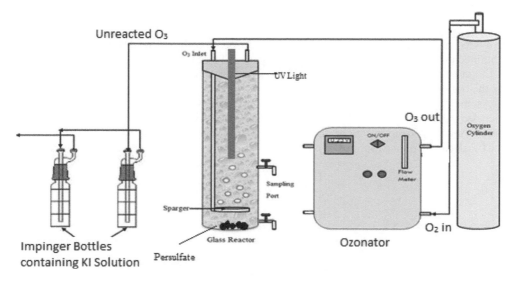

Figure 1. Structure of RR 120.

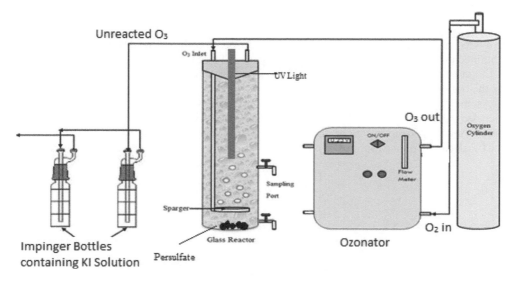

Figure 2. Schematic diagram.

distribution tube as depicted in Figure 1. Silicon tubing was used for the connection between ozone generators, reactor column, and impinge bottle.

The UV light was kept in the reactor, and the catalyst was stirred by using the magnetic stirrer, which was located at the bottom of the reactor.

4 RESULTS AND DISCUSSION

4.1 *Decolorization of reactive red 120*

The synthetic dye waste water was prepared at two different concentrations of 500 and 1000 ppm, and ozone gas flow-rate of 40 LPH. There are three different operating conditions namely, ozonation, ozone/UV, and Ozone/UV/TiO$_2$ catalyst. The color intensity of the dye is measured in the UV spectrophotometer. After the completion of the process, complete color removal from the wastewater was observed.

The complete decolorization of reactive red 120 dye at 1000 ppm and 500 ppm takes 120 min and 80 min for the simple ozonation process, and for the catalytic ozonation process it requires 110 and 75 min, respectively, for color removal. For the catalytic ozonation process 3 g/2 L of the catalyst is used for the reaction. Whereas in case of ozonation/catalyst/UV process it requires 80 min and 60 min for the complete color removal of reactive red 120 having a dye concentration of 1000 ppm and 500 ppm, respectively.

4.2   Effect on TOC

The experiments were also done to see the effect of the ozone concentration in the gas. Unfortunately the ozonation apparatus does not allow one to change the concentration easily without the flow-rate. Three initial ozone concentrations 30, 40, and 50 LPH were used to decolorize the dye and to remove the total organic carbon from the RR 120 dye, respectively.

4.3   Effect on pH

The experiments were carried out at the different pH values of 2, 4, 6, 8, and 10 for the single dye RR 120. The concentration of the dye is 500 ppm and the ozone dosage is 40 LPH. The obtained experimental data is shown in the Figure 3. The Figure 3 indicates that the pH value is the most important factor in the ozonation process since pH determines the dissociation of the organic compounds.

As mentioned above, ozone decomposes partly in •OH-radicals. When the pH value increases, the formation of •OH radicals increases. In a solution with a high pH value, there are more hydroxide ions present, see formulas below. These hydroxide ions act as an initiator for the decay of ozone (Song 2007):

$$O_3 + OH^- \rightarrow HO_2^- + O_2 \qquad (9)$$
$$O_3 + HO_2\text{-} \rightarrow {}^\bullet OH + O_2^{\bullet-} + O_2 \qquad (10)$$

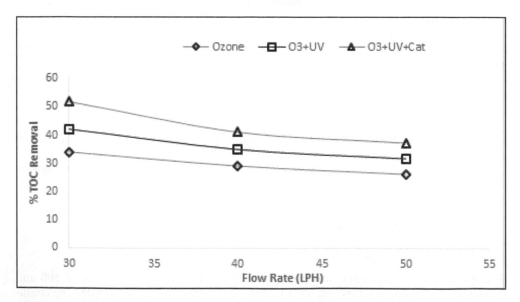

Figure 3.   % TOC reduction on RR 120 at different operating condition [ozone dose: 30, 40, and 50 LPH, and UV light range: 8 watt, TiO_2 anatase catalyst dose: 2 g/L and dye concentration: 500 mg/L].

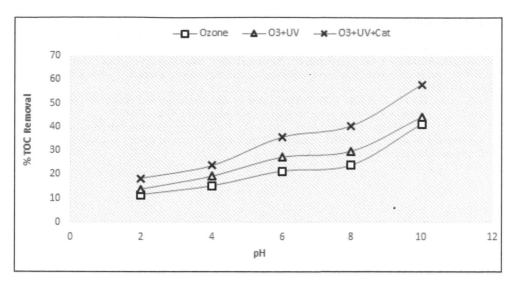

Figure 4.   Effect of pH on decolorization and TOC [ozone dose: 40 LPH, pH dose: 2, 4, 6, 8, and 10, and dye concentration: 500 mg/L].

5   CONCLUSION

This experimental work reveals that TOC removal for O_3/UV/TiO_2 catalyst process was almost double as compared with the ozonation process and 10% higher than that of the ozone/UV process. It also provides a valuable comparison between O_3, O_3/UV, and O_3/UV/ TiO_2. The UV light and TiO_2 anatase phase catalyst is more effective for wastewater treatment since it generates •OH radicals, which help to degrade the recalcitrant organic matter. It can also be concluded from the given result at higher pH, O_3/UV/TiO_2 gives almost 60% TOC reduction.

REFERENCES

Baig, S., Liechit, P.A. 2001. Ozone treatment for biological COD removal. *Water Sci. Technol* 43 (2): 197–204.
Beltrain, F.J., Rivas, F.J., Espinosa, R.M. 2002. Catalytic ozonation of oxalic acid in aqueos TiO_2 slurry reactor. *Applied catalysis B: Environmental* (39): 221–231.
Galindo, G., Kalt, A., 1999. UV/H_2O_2 oxidation of azo dyes in aqueous media: evidence of a structure-degradability relationship. *Dyes Pigments* (42), 199–207.
Hordem, B.K., Ziolek, M., Nawrocki, J. 2003. Catalytic ozonation and methods of enhancing molecular ozone reactions in water treatment. *Applied Catalysis B: Environmental* (46), 639–669.
Ikhlaq, A., Brown, D, R., Kasprzyk-Hordern, B. 2012. Mechanisms of catalytic ozonation on alumina and zeolites in water: Formation of hydroxyl radicals Applied Catalysis B: Environmental 123–124 (2012) 94–106.
Kiwi, J., Pulgarin, C., Peringer, P. and Gra"ztel, M. 1993. Beneficial effects of heterogeneous Photocatalysis on the biodegradation of antraquinone sulfonate observed in water treatment. *New Journal of Chemistry*, 17(7), 487–494.
Konsawa, A.H. 2003., Decolorizaton of wastewater containing direct dye by ozonation in a batch bubble column reactor. *Desalination* 158 (2003), 233–240.
Martinez-Huitle, C.A., Brillas, E. 2009. Decontamination of wastewaters containing synthetic organic dyes by electrochemical methods: A general review. *Appl. Catal. B: Environ.* 87, 105–145.
Moussavi, G., Mahmoudi, M. 2009. Degradation and biodegradability improvement of the reactive red 198 azo dye using catalytic ozonation with MgO nanocrystals. *Chemical Engineering Journal* 152: 1–7.

Moussavi, G., Khosravi, R. 2012. Preparation and characterization of a biochar from pistachio hull biomass and its catalytic potential for ozonation of water recalcitrant contaminants, *Bioresource Technology* 119: 66–71.

Robinson, T., McMullan, G., Marchant, R., Nigam, P. 2001. Remediation of dyes in textile effluent: a critical review on current treatment technologies with a proposed alternative. *Bioresource. Technol* (77), 247–255.

Song, S., Ying, H., He, Z., Chen, J. 2007. Mechanism of decolorization and degradation of CI Direct Red 23 by ozonation combined with sonolysis. *Chemosphere* (66), 1782–1788.

Sui, M., Xing, S., Sheng, L., Huang, S., Guo, H. 2012. Heterogeneous catalytic ozonation of ciprofloxacin in water with carbon nanotube supported manganese oxides as catalyst, *Journal of Hazardous Materials* 227–228: 227–236.

Multi-disciplinary Sustainable Engineering: Current and Future Trends – Tekwani, Bhavsar & Modi (Eds)
© 2016 Taylor & Francis Group, London, ISBN 978-1-138-02845-6

Remote sensing application to study ambient air quality over Ahmedabad

Y.S. Rami
Department of Environmental Engineering, L.D. College of Engineering, Ahmedabad, Gujarat, India

A. Chhabra
Space Applications Center, ISRO, Ahmedabad, Gujarat, India

M.J. Pandya
Department of Environmental Engineering, L.D. College of Engineering, Ahmedabad, Gujarat, India

ABSTRACT: Currently, air pollution resulting from various aerosols is of great research interest owing to its contribution for deteriorating ambient air quality and serious impacts on human health etc. Thus, it is essential to understand Spatio-temporal and seasonal distribution of aerosols; in particular for urban environments using satellite-derived remote sensing data. This paper reports a recent study carried out to understand Spatio-temporal variations in Aerosol Optical Depth (AOD) in Ambient Air Quality of Ahmedabad using MODIS satellite sensor-derived data of the recent period 2012–2014. The results show that AOD varies from 0.2–0.5 over Ahmedabad district. However, it is reported higher in the range of 0.5–0.9 over urban Ahmedabad; indicating Ahmedabad as "Fairly Polluted Area". This study provides important results which may be useful inputs towards planning Greener and Eco-friendly alternative technologies for controlling air pollution and improving the ambient air quality of urban Ahmedabad.

1 INTRODUCTION

Aerosols are suspended particulate matter with few nanometers to micron size originated from various natural and anthropogenic sources. By mass, 90% aerosols have natural origins which include volcanic eruption, desert dust, forest fires etc. and remaining 10% aerosols have an anthropogenic origin, viz. industrial emissions, vehicular exhaust, incineration, mining etc. Aerosols are important components of earth's climate system as it affect radiation budget of the earth, climate, hydrological cycle, cloud formation. Anthropogenic aerosols are responsible for 'Radiative Forcing' (RF) of climate through multiple processes which can be grouped into two types: aerosol–radiation interactions (ari) and aerosol–cloud interactions (aci) (IPCC, 2013). Aerosols over the Southern Asia cause negative forcing at the surface (cooling) and positive effect at 'Top of Atmosphere' (Satheesh et al. 2000). Ramanathan et al. (2001) reported that additional heating and cooling affect tropical rainfall patterns and disturb hydrological cycle. Aerosols have a large effect on air quality, in particular in densely populated areas where high concentrations of fine Particulate Matter (PM) are associated with premature death and the decrease of life expectancy (Kokhanovsky et al. 2009). Based on the source, chemical composition, and their shape, aerosols have various impacts over the environment, human health and agriculture etc.

It is essential to understand the Spatio-temporal variation of aerosols over local, regional and global scales. Satellite Technology through spaceborne remote sensing of aerosols has an advantage of providing information at spatial scales with frequent repetitive observations. Aerosol Optical Depth (AOD) defined as an integral of the aerosol light extinction over the vertical path through the atmosphere is a fundamental parameter for studying aerosols. Satellite data is increasingly used to obtain information on aerosol properties. Various satellites have been used to retrieve AOD over ocean and land, for example; Advanced Very High Resolution Radiometer (AVHRR), Total Ozone Mapping Spectrometer (TOMS) series was extended with Ozone Mapping Instrument (OMI) to deliver 'Aerosol Index', Polarization and Directionality of the Earth Reflectance (POLDER), MODerate Imaging Spectroradiometer (MODIS), Multi-angle Imaging Spectroradiometer (MISR). Recently, Cloud-Aerosol Transport System (CATS) was launched on January 10, 2015, to understand about aerosol and aerosol-cloud interaction.

MODerate Imaging Spectroradiometer (MODIS) sensor onboard NASA's Terra and Aqua Satellites provides valuable information on aerosols. Terra's orbit around the Earth is timed so that it passes from North to South across the equator in the morning at 10:30 AM, while Aqua passes South to North over the equator in the afternoon at 1:30 PM. Terra/Aqua MODIS are viewing the entire Earth's surface every 1 to 2 days, acquiring data in 36 spectral bands. Out of the 36 bands, MODIS give AOD at 7 bands i.e. 0.47, 0.55, 0.65, 0.85, 1.25, 1.64, 2.1 μm with a wide swath of ~2330 km. MODIS AOD products are generated using aerosol retrieval algorithms. These algorithms are developed for vegetated land surfaces and remote ocean regions i.e. denoted as Dark Target (DT) algorithm because they operate best on regions that are "dark" visually. Another algorithm denoted as Deep-Blue algorithm designed for surface appear bright visually and they are fairly dark in the near–UV wavelength (Levy et al. 2013).

Ahmedabad is the largest city in state of Gujarat and the fifth-largest urban agglomeration in India with a population of 55.71 lakhs (Census, 2011). In recent times, Ahmadabad has witnessed rapid growth in population, industries, transport amenities, automobiles, road networks, besides installation of many new industrial projects, skyscraper buildings and construction projects in and around Ahmadabad city. This study was undertaken with an objective to understand the Spatio-temporal variability of Aerosol Optical Depth (AOD) in the ambient air quality of Ahmedabad for the recent period 2012–2014.

2 STUDY AREA AND METHODOLOGY USED

The study area is Ahmedabad; an urban, industrialized and highly populated city of Gujarat in Western India. Ahmadabad is the fifth largest city of India with an estimated population of 55.71 lakhs as per Census, 2011. It is located between 22.0°–23.30°N Latitude to 71.30°-73.0°E longitude (Figure 1).

For the purpose of the study, Aqua/MODIS satellite L2 C006 AOD product (MYD04_3 K) was collected from NASA's Goddard Space Flight Center Earth Sciences Distributed Active Archive Center (DAAC, *http://ladsweb.nascom.nasa.gov*) for the study period January 2012-December 2014. MYD04 have a nominal spatial resolution of 3 X 3 Km² at nadir. Generally, MODIS data are organized into 5 minutes swath segments called granules. Out of all seven channels provided in AOD product, AOD at 550 nm was used for the purpose of this study.

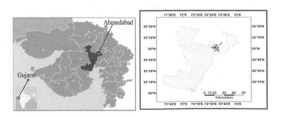

Figure 1.   Map of sampling points in Ahmedabad.

A detailed spatial database of daily granules over the Ahmedabad district was generated. The image analysis was carried out using ENVI (Environment for Visualizing Images) Version 4.4 software.

ENVI is ideal software for the visualization, analysis, and presentation of all types of digital imagery. The administrative boundary coverage of Ahmedabad was collected from the whole India coverage using ARCGIS ver 9.2. Further, the Ahmedabad district coverage was overlaid to extract the study area. The Spatio-temporal patterns of AOD over Ahmedabad were generated and detailed analysis was done.

3 RESULTS AND DISCUSSION

The Spatio-temporal analysis of MODIS-derived Aerosol Optical Depth (AOD) over Ahmedabad district indicates AOD in the range of 0.2–0.5 during the study period 2012–2014 (Figure 2). However, the AOD values varied from 0.5–0.9 over Ahmedabad city during the study period 2012–2014. Higher AOD values were observed over Ahmedabad city than Ahmedabad district as a whole during the study period. Also, there are seasonal variations in AOD with a minimum during December-January in contrast to maximum values during July-August. The higher MODIS-derived AOD values observed during monsoon season may be attributed to erroneous and overestimation due to cloud cover and limitation of optical sensor during monsoon season.

The monthly Spatio-temporal patterns for each of the study year 2012–2014 were studied in detail. However, for the purpose of this paper, only spatial analysis for the recent year 2014 is illustrated here in Figure 3. During 2014, mean AOD varies over Ahmedabad district ranged from 0.29 ($\pm$ 0.15) in December to 1.2 ($\pm$ 0.59) in August month. Using the MODIS data, Prasad et al. (2004) studied the AOD distribution over India with seasonal variations. His study reported very high AOD over Gujarat region as observed in summer which may due to dust transport from the Thar Desert and also from the Middle East. However, in the monsoon season, estimated value of AOD is 4.35 which may be erroneous due to clouds and relative humidity. Srikanth et al. (2013) also reported higher AOD values during monsoon season in the month of July over Bangalore. 6. The remote sensing data analysis indicate that Ahmedabad may be classified under "Fairly Polluted Area" (AOD 0.6 to <1).

Month–wise minimum, maximum, mean and standard deviation of AOD over Ahmedabad district for the study period of the year 2012–2014 in given in Table 1.

Wang and Christopher (2003) presented an inter-comparison between satellite-derived AOD and PM$_{2.5}$ mass in the South-eastern United States. Their study showed a good

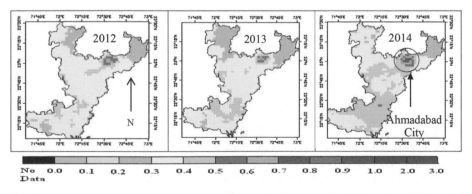

Figure 2. Spatio-temporal variations in satellite derived mean AOD over Ahmedabad during 2012–2014.

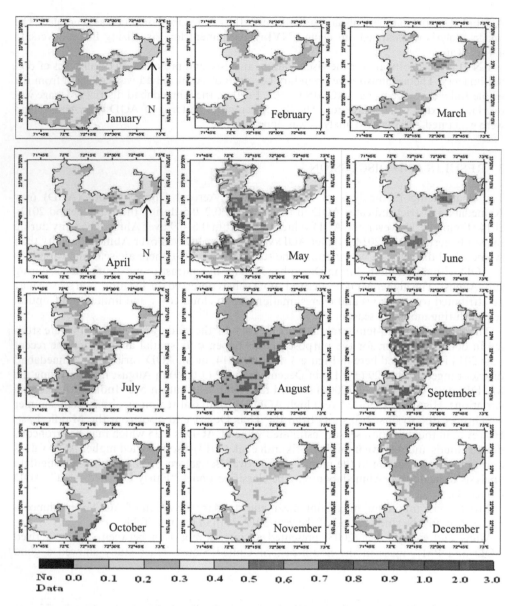

Figure 3. Mean monthly AOD distribution over Ahmedabad district for the year 2012.

positive correlation of remote sensing derived MODIS AOD with $PM_{2.5}$ mass (R = 0.7) and demonstrated that satellite-derived AOT is a useful tool for air quality studies over large spatial domains to track and monitor aerosols. Kumar et al. (2007) studied an empirical relationship between $PM_{2.5}$ and AOD in Delhi metropolitan a significant positive association between AOD and $PM_{2.5}$. They found that average AOD gradually declines with increasing distance from the city's center. The average concentration of AOD in the north-eastern parts of Delhi was more than 0.6, as compared to the outside Delhi boundaries with AOD less than 0.5. This may be attributed to local sources of emissions in the urban areas. Thus, AOD provides a good correlation with Particulate Matter concentrations as reported through various studies in the literature.

Table 1. Monthly summary of MODIS-derived AOD over Ahmedabad district for the year 2012–2014.

Years	2012				2013				2014			
Months	Min	Max	Mean	Std. Dev.	Min	Max	Mean	Std. Dev.	Min	Max	Mean	Std. Dev.
January	0.12	0.56	0.28	0.06	0.23	1.11	0.32	0.05	0.19	0.83	0.42	0.35
February	0.17	0.67	0.31	0.06	0.32	0.89	0.47	0.08	0.12	0.69	0.33	0.24
March	0.16	0.67	0.36	0.07	0.18	0.64	0.37	0.09	0.14	0.73	0.30	0.10
April	0.20	0.62	0.38	0.06	0.23	1.07	0.57	0.16	0.12	0.71	0.34	0.18
May	0.12	1.31	0.46	0.15	0.15	2.71	0.57	0.31	0.19	0.91	0.40	0.22
June	0.18	0.80	0.37	0.09	0.55	4.35	1.29	0.75	0.23	1.01	0.42	0.17
July	0.01	1.83	0.43	0.20	0.13	2.11	0.61	0.34	0.09	2.69	1.19	0.70
August	0.43	4.35	1.20	0.55	0.15	0.81	0.34	0.09	0.28	2.97	1.28	0.59
September	0.07	1.38	0.43	0.21	0.17	0.64	0.32	0.07	0.20	1.51	0.63	0.26
October	0.22	0.71	0.39	0.008	0.09	0.65	0.27	0.07	0.26	0.83	0.46	0.19
November	0.23	0.65	0.36	0.005	0.16	0.64	0.30	0.06	0.26	0.85	0.40	0.24
December	0.15	0.52	0.29	0.06	0.12	0.52	0.30	0.06	0.18	0.56	0.29	0.15

4 CONCLUSION

This study was aimed at understanding Spatio-temporal variations in Aerosol Optical Depth; an important parameter of aerosol studies in Ambient Air Quality of Ahmedabad using MODIS satellite sensor-derived data of the recent period 2012–2014. The detailed spatial analysis indicates AOD varies from 0.2–0.5 over Ahmedabad district during the study period. However, the AOD values varied from 0.5 to 0.9 over urban Ahmedabad indicating "Fairly Polluted Area". Higher AOD observed in summer season than the winter season. This is in contrast to low values in post–monsoon season due to the washout of the aerosols through rainfall. This study provides important results towards ambient air quality of urban Ahmedabad that calls for proper planning and implementation of air pollution control measures to improve the quality of life of local residents.

REFERENCES

Census, 2011 http://www.census2011.co.in/census/city/314-ahmedabad.html.
DAAC, http://ladsweb.nascom.nasa.gov.
IPCC, 2013: Climate Change 2013: The Physical Science Basis. The contribution of Working Group I to the Fifth Assessment Report of the Intergovernmental Panel on Climate Change [Stocker, T.F., D. Qin, G.K. Plattner, M. Tignor, S.K. Allen, J. Boschung, A. Nauels, Y. Xia, V. Bex and P.M. Midgley (eds.)]. Cambridge University Press, Cambridge, United Kingdom and New York, NY, USA, pp 1535.
Kokhanovsky, A. & Leeuw, G. 2009. Satellite Aerosol Remote Sensing over Land, Praxis Publishing (Springer).
Kumar, N., Chu, A., Foster, A. (2007). An empirical relationship between PM2.5 and aerosol optical Depth in Delhi Metropolitan. Atmospheric Environment, 1, 41(21): 4492–4503.
Levy, R., Mattoo, S., Munchak, L., Remer, L., Sayer, A., Patadia, F. & Hsu N., 2013. The Collection 6 MODIS aerosol products over land and ocean, Atmospheric Measurement Techniques, 6, 2989–3034.
Prasad, A. & Singh, R. & Singh, A. 2004. Variability of Aerosol Optical Depth Over Indian Subcontinent Using MODIS Data, Journal of Indian Society of Remote Sensing 32,4: 313–316.
Ramanathan, V., Crutzen, P., Kiehl, J. & Rosenfeld, D. 2001. Aerosols, Climate, and the hydrological cycle. Science, 294: 2119–2124.
Satheesh, S & Ramanathan, V. 2000. Large differences in tropical aerosol forcing at the top of the atmosphere and Earth's surface. Nature, 405: 60–63.
Sreekanth, V. 2013. Satellite-derived aerosol optical depth climatology over Bangalore, India. Advances In Space Research 51, 12: 2195–2404.
Wang J. and Christopher S. (2003). Intercomparison between satellite-derived aerosol optical thickness and PM2.5 mass: Implications for air quality studies. Geophysical Research Letters, 30, No. 21, 2095, doi:10.1029/2003GL018174.

Electrochemical oxidation as an alternative treatment for real industry wastewater

P.P. Saxena, J.P. Ruparelia & N. Mehta
Department of Chemical Engineering, Nirma University, Ahmedabad, Gujarat, India

ABSTRACT: Industrial processes generate a variety of complex, toxic and non-biodegradable compounds which are difficult and expensive to treat by conventional wastewater treatment processes. Real wastewater containing refractory organics from a Common Effluent Treatment Plant (CETP) functional in one of the local industrial estates was used for the present work. In this work, electrochemical oxidation of wastewater with and without prior coagulation with ferric chloride ($FeCl_3.6H_2O$) was investigated using indigenously prepared Dimensionally Stable Anode (DSA). The treatment performance was evaluated in terms of Chemical Oxygen Demand (COD) and Total Organic Carbon (TOC) reduction. It was found that overall COD and TOC reduction were higher for electrooxidation combined with coagulation. But it resulted in higher specific energy consumption along with sludge generation in comparison to only electrooxidation.

1 INTRODUCTION

A variety of complex, toxic and non-biodegradable compounds are generated from industrial processes which are difficult and expensive to treat by conventional wastewater treatment processes. Electrochemical Technology (electrooxidation) as an advanced refractory wastewater treatment alternative provides adaptability, energy competence, amenability to automation, cost effectiveness and environmental compatibility [Chen 2003]. Electrooxidation is highly robust in degrading these pollutants directly at the electrode surface with no secondary pollution generation in terms of sludge since electron that brings about mineralization of pollutants is a clean reagent for this process [Trasatti 2000, Comninellis et al. 2010]. Small-scale industries (SSIs) play a pivotal role in overall industrial growth and development. Wastewater of individual industries often contains significant concentration of pollutants; and to reduce them by individual treatment up to the desired concentration, becomes techno-economically difficult [Chatzisymeon et al. 2006, Anglada, et al. 2009, Zayas et al. 2011]. Concept of Common Effluent Treatment Plant (CETP) not only helps the industries in easier pollution control, but also acts as a step towards cleaner and greener environment and service to the society at large.

In the present work, source has been identified for wastewater from one of the operational CETPs in a local industrial estate that employs physico-chemical treatment followed by conventional activated sludge based biological process for wastewater treatment with the treated sludge ultimately disposed to secured landfill disposal site. The CETP treats a mix of effluent resulting from pharmaceutical, dyes and intermediates, pigments, fine chemicals, other organic and inorganic chemicals and textile processing. Details of this source are given in table 1.

In the present work, the CETP wastewater was subjected to electrooxidation treatment as an alternative method and its performance was assessed in terms of observable quantifiable indicators that include Chemical Oxygen Demand (COD), Total Organic Carbon (TOC), turbidity and specific energy consumption. In one mechanism, only electrooxidation was used while in other, electrooxidation was used after the wastewater was subjected to

Table 1. Details of CETP wastewater source.

Member units	>600
Inlet norms for member units	*BOD 1200 mg/L
	*COD 3000 mg/L
	*TSS 600 mg/L
	pH 6.5–8.5
Solid waste disposal	Secured landfill disposal site
Final treated effluent discharge	Mega pipeline carrying treated effluent mixed with treated sewage of one of the Sewage Treatment Plants (STP) before disposal into a river source
Operational cost	Rs. 15–18 per m^3 of effluent
	Rs. 250 per MT of solid waste

*BOD = Biochemical Oxygen Demand, COD = Chemical Oxygen Demand, TSS = Total Suspended Solids.

coagulation using $FeCl_3.6H_2O$ as coagulant to assess preliminary reduction in turbidity and COD. Indigenously prepared Dimensionally Stable Anode (DSA) was used for experimental work in batch mode at room temperature.

2 MATERIALS AND METHODS

2.1 Chemicals

Solution of $FeCl_3.6H_2O$ (98% CDH) coagulant was prepared for dosing varying amounts of coagulant solution in jar test apparatus. NaCl (CDH make LR grade) was used to increase the conductivity of the solution. $RuCl_3.3H_2O$ (99.9% Aldrich), $ZrOCl_2.8H_2O$ (98% CDH), $SbCl_3$ (98.5% CDH), $SnCl_4.5H_2O$ (98% Aldrich) were mixed to prepare a precursor solution. Isopropanol (99.8% CDH) and hydrochloric acid (37% CDH) were used as solvents to prepare the precursor solution. Pretreated titanium metal petal was used as a substrate to coat the precursor solution.

2.2 Experimental setup for coagulation and electrooxidation

Known quantities of ferric chloride coagulant were added to the wastewater in jar test apparatus to find the optimal coagulant dose. It was thoroughly stirred for one minute at 150 rpm and thereafter slowly stirred for two minutes at 40 rpm. The wastewater sample was left steady for 40 min for sedimentation [Klimiuk, et al. 1999, Verma et al. 2012, Sahu et al. 2013]. Figure 1 shows electrooxidation laboratory unit of 1.0 L capacity. Titanium substrate with mixed metal oxide coating was employed as DSA type anode and stainless steel (SS 304) as cathode with an electrode spacing of 8 mm. A constant potential was applied by means of a D.C. power supply (Aplab India, LD-3205).

2.3 Electrode preparation

Ti/ $ZrO-RuO_2-Sb_2O_5-SnO_2$ used as working electrode was fabricated indigenously by thermal decomposition process. Titanium metal piece (65 mm × 75 mm × 3 mm) was polished with sic paper and etched in 10% oxalic acid at 80°C for 1 hr. It was then rinsed with deionized water before applying the coating. Precursor solution was prepared by dissolving 0.196 g $RuCl_3.3H_2O$, 0.241 g $ZrOCl_2.8H_2O$, 0.114 g $SbCl_3$ and 0.452 g $SnCl_4.5H_2O$ in a solvent mixture consisting of 10 mL isopropanol and 0.5 mL hydrochloric acid. Precursor solution was

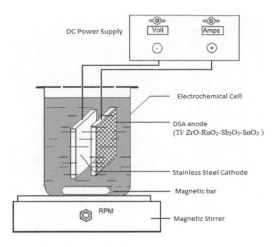

Figure 1.   Experimental setup for electrooxidation.

applied by brush technique on the pretreated titanium substrate at room temperature, dried at 80°C for 5 min to allow the solvents to vaporize and then baked at 550°C for 5 min. This procedure was repeated about 12–15 times to provide a proper coating layer. Finally the electrode was heated at 550°C for an hour. The total substrate oxide loading was about 2–3 mg/cm^2 [Chen et al. 2002, Scialdone 2003, Adams et al. 2009, Soni et al. 2012, Nordin et al. 2013].

2.4   Methodology

Experiments were conducted by two different mechanisms. In mechanism-I, only electrooxidation was used while in mechanism-II, electrooxidation was used with prior coagulation. Experimental conditions of mechanism I and II are shown in Table 2. Table 3 shows overall percentage COD and TOC reduction for only electrooxidation and combined coagulation and electrooxidation.

2.5   Analysis

For analytical determination, sampling was done at periodic time interval. COD was determined using standard closed reflux titrimetric method (APHA Standard 5220 C) using COD reactor (Hach DRB200, COD reactor, USA). TOC was determined using TOC analyzer (TOC-VCSH, Shimadzu, Japan). Percentage COD removal, percentage TOC removal and energy consumption are calculated using Equations 1, 2, 3 and 4 respectively [Martınez-Huitle et al. 2009, Wu et al. 2014, Martınez-Huitle et al. 2015].

$$\%COD\,Removal = \left(\frac{COD_0 - COD_t}{COD_0}\right) \times 100 \qquad (1)$$

where, COD_0 and COD_t represent COD values at initial and electrolysis time t respectively.

$$\%TOC\,Removal = \left(\frac{TOC_0 - TOC_t}{TOC_0}\right) \times 100 \qquad (2)$$

where, TOC_0 and TOC_t represent TOC values at initial and electrolysis time t respectively.

Energy consumption based on COD and TOC mass is calculated using Equations 3 and 4.

Table 2. Experimental conditions for mechanisms I & II.

	Mechanism I (only electrooxidation)	Mechanism II (electrooxidation after coagulation)
Anode	$Ti/RuO_2 - CeO_x - Sb_2O_5 - SnO_2$	$Ti/RuO_2 - CeO_x - Sb_2O_5 - SnO_2$
Cathode	Stainless Steel (SS 304)	Stainless Steel (SS 304)
Voltage (V)	4–4.5	4–4.5
Current Density (mA/cm²)	40	40
Volume (L)	1.0	1.0
Time (min)	60	60
Initial COD (mg/L)	3424	2465
Initial TOC (mg/L)	723	615
NaCl added (g/L)	4	4
pH	7–8	7–8
Rpm for electrooxidation unit	200	200
Coagulant dose (mL)	–	60
Turbidity before electrooxidation (NTU)	127	12
Coagulation conditions	–	40 rpm 40 min settling time

Table 3. Overall percentage COD and TOC reduction.

	Only electrooxidation	Coagulation + Electrooxidation		
		Only coagulation	Electrooxidation after coagulation	Overall
% COD Reduction	41	28	38	66
% TOC Reduction	29	15	24	39

$$E = \left(\frac{UI\Delta t}{COD_t - COD_{t+\Delta t}} \right) \times V \qquad (3)$$

$$E = \left(\frac{UI\Delta t}{TOC_t - TOC_{t+\Delta t}} \right) \times V \qquad (4)$$

where $(COD)_t$, $(COD)_{t+\Delta t}$, $(TOC)_t$, and $(TOC)_{t+\Delta t}$ are the COD and TOC values at t and t+Δt time (hr) in (g O_2 dm^{-3}), respectively, U is the average cell potential(V), I is the applied current (A), and V is the volume of the electrolyte solution (L), to get energy consumption in kWh/g of COD reduction.

3 RESULTS AND DISCUSSION

3.1 *Variation in COD and TOC removal efficiency with electrolysis time*

Figure 2 and Figure 3 show the variation of COD and TOC removal efficiency with treatment time for both mechanism I and mechanism II respectively. Ferric chloride acting as coagulant reduced the initial COD by 28%, initial TOC by 15% and turbidity by 91% along with sludge generation. Both COD and TOC removal were found to increase with the electrolysis time. After 60 min of electrolysis, the COD and TOC removal efficiencies, respectively, reached 41% and 29% for mechanism I and 38% and 24% for mechanism II.

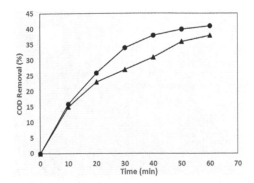

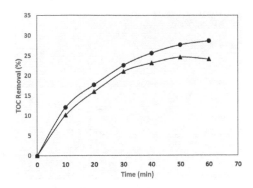

Figure 2. COD removal efficiency (%) with treatment time for mechanism I (•) and mechanism II (▲).

Figure 3. TOC removal efficiency (%) with treatment time for mechanism I (•) and mechanism II (▲).

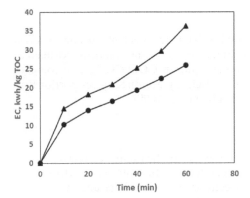

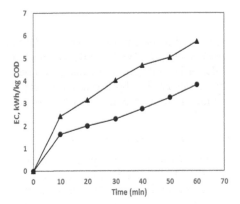

Figure 4. Energy Consumption (EC) per unit TOC mass with treatment time for mechanism I (•) and mechanism II (▲).

Figure 5. Energy Consumption (EC) per unit COD mass with treatment time for mechanism I (•) and mechanism II (▲).

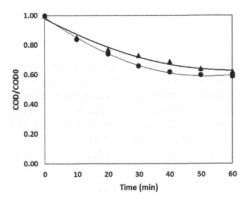

Figure 6. Normalized COD against electrolysis time for mechanism I (•) and mechanism II (▲).

3.2 Variation in specific energy consumption with electrolysis time

Figure 4 and Figure 5 show the variation of Energy Consumption (EC) per unit TOC and COD mass with treatment time for both mechanism I and mechanism II respectively. After 60 min of electrolysis, the energy consumption per unit COD and TOC mass, respectively,

Figure 7.　SEM images of unused (a) and used (b) Ti/ZrO-RuO_2-Sb_2O_5-SnO_2 electrocatalytic coating.

were found to be almost 4 kWh/kgCOD and 26 kWh/kgTOC for mechanism I and 6 kWh/kgCOD and 37 kWh/kgTOC for mechanism II. Figure 6 shows the trend of normalized COD for both the mechanisms. It can be observed that trends of COD/COD_0 ratio coincide. This indicates that both oxidation rate and process efficiency are directly proportional to organic matter concentration [Panizza et al. 2008, Panizza et al. 2009].

3.3 *Coating surface morphology*

The morphology of catalytic coating surface was observed by Scanning Electron Microscopy (SEM). The SEM images of unused and used catalytic coating surface are depicted in Figure 7(a) and Figure 7(b) respectively. Unused electrode showed a tinier and even crystal structure resulting in improved current distribution on the surface thereby promoting the conductivity, stability and catalytic activity of the electrode. Cracks were clearly visible on the used electrode due to reactions taking place at the surface. Similar observations for used and unused plates were reported by Soni et al. (2012) and Weng et al. (2013).

4　CONCLUSION

The present study reported work on actual wastewater. It was observed that the overall COD and TOC removal efficiencies were 41% and 29% respectively for only electrooxidation and 66% and 39% for electrooxidation combined with coagulation. Specific energy consumption per unit COD and TOC mass was found to be lower by 33% and 30% respectively for electrooxidation without coagulation. Although overall COD and TOC reduction were higher for electrooxidation combined with coagulation but the process of coagulation added to the cost of operation. In addition, it also leads to sludge generation and associated sludge disposal problems.

REFERENCES

Adams, B., Tian, M & Chen, A. 2009. Design and electrochemical study of SnO_2-based mixed oxide electrodes. *Electrochimica Acta* 54: 1491–1498.
Anglada, A., Urtiaga, A. & Ortiz, I. 2009. Contributions of electrochemical oxidation to wastewater treatment: fundamentals and review of applications. *Journal of Chemical Technology and Biotechnology* 84 (12): 1747–1755.

Chatzisymeon, E., Xekoukoulotakis, N.P., Coz, A., Kalogerakis, N. & Mantzavinos, D. 2006. Electrochemical treatment of textile dyes and dye house effluents. *Journal of Hazardous Materials* 137 (2) 998–1007.

Chen, G. 2003. Electrochemical technologies in wastewater treatment. *Separation and Purification Technology* 38 (1): 11–41.

Chen, G., Chen, X. & Yue, P.L. 2002. Electrochemical Behavior of Novel Ti/IrOx-Sb$_2$O$_5$-SnO$_2$ Anodes. *J. Phys. Chem. B 106:* 4364–4369.

Comninellis, C. & Chen, G. (ed.) 2010. *Electrochemistry for the Environment.* Berlin: Springer.

Klimiuk, E., Filipkowska, U., Libecki, B. 1999. Coagulation of Wastewater Containing Reactive Dyes with the Use of Polyaluminium Chloride (PAC). *Polish Journal of Environmental Studies* 8 (2): 81–88.

Martınez-Huitle, C.A. & Brillas, E. 2009. Decontamination of wastewaters containing synthetic organic dyes by electrochemical methods: A general review. *Applied Catalysis B: Environmental* 87 (3–4): 105–145.

Martınez-Huitle, C.A. & Brillas, E. 2015. Decontamination of wastewaters containing synthetic organic dyes by electrochemical methods. An updated review. *Applied Catalysis B: Environmental* 166–167: 603–643.

Nordin, N., Siti Fathrita, M.A., Riyanto, Othman, M.R. 2013. Textile industries wastewater treatment by electrochemical oxidation technique using metal plate. *International Journal of Electrochemical Science* 8: 11403–11415.

Panizza M. & Cerisola, G. 2008. Removal of colour and COD from wastewater containing acid blue 22 by electrochemical oxidation. *Journal of Hazardous Materials* 153: 83–88.

Panizza, M. & Cerisola, G. 2009. Direct and mediated anodic oxidation of organic pollutants, *Chemical Reviews* 109: 6541–6569.

Sahu, O.P. & Chaudhari, P.K. 2013. Review on Chemical treatment of Industrial Waste Water. *Journal of Applied Sciences and Environmental Management* 17 (2): 241–257.

Scialdone, O. 2009. Electrochemical oxidation of organic pollutants in water at metal oxide electrodes: A simple theoretical model including direct and indirect oxidation processes at the anodic surface. *Electrochimica Acta* 54: 6140–6147.

Soni, B.D. & Ruparelia, J.P. 2012. Studies on effects of electrodes for decontamination of dyes from wastewater. *Journal of Environmental Research and Development* 6 (4): 973–980.

Soni, B.D. & Ruparelia, J.P. 2012. Application of Ti/RuO$_2$-SnO$_2$-Sb$_2$O$_5$ Anode for Degradation of Reactive Black-5 Dye. *World Academy of Science, Engineering and Technology* 6: 11–22.

Trasatti, S. 2000. Electrocatalysis: understanding the success of DSA®. *Electrochimica Acta* 45: 2377–2385.

Verma A.K., Dash, R.R. & Bhunia, P. 2012. A review on chemical coagulation/flocculation technologies for removal of colour from textile wastewaters. *Journal of Environmental Management* 93 (1): 154–168.

Weng, M., Zhou, Z. & Zhang, Q. 2013. Electrochemical Degradation of Typical Dyeing Wastewater in Aqueous Solution: Performance and Mechanism. *International Journal of Electrochemical Science* 8: 290–296.

Wu, W., Huang, Z.H. & Lim, T.T. 2014. Recent development of mixed metal oxide anodes for electrochemical oxidation of organic pollutants in water. *Applied Catalysis A: General* 480: 58–78.

Zayas, T., Picazo, M. & Salgado, L. 2011. Removal of organic matter from paper mill effluent by electrochemical oxidation. *Journal of Water Resource and Protection* 3: 32–40.

Multi-disciplinary Sustainable Engineering: Current and Future Trends – Tekwani, Bhavsar & Modi (Eds)
© 2016 Taylor & Francis Group, London, ISBN 978-1-138-02845-6

Studies on copper chlorine thermochemical cycle for hydrogen production

Dhaval S. Andadia, Y.B. Solanki & R.K. Mewada
Institute of Technology, Nirma University, Ahmedabad, Gujarat, India

ABSTRACT: Hydrogen is green fuel for future. Currently, major hydrogen sources are based on the crude oil. The Concept of hydrogen production from water splitting using electrolysis is very old. But it is not economically feasible. If hydrogen can be produced from water through the economically feasible process, then it is best sustainable fuel option and one of the potential solutions for the current energy and environmental problems. Many papers have been published about thermochemical cycles, but overall problems associated with the various techniques are not discussed. The five-step copper chlorine cycle consists of copper chlorination (hydrogen production), electrolysis, drying, hydrolysis and decomposition step. The highest temperature required within all five steps is 530°C. This paper discusses a setup for all five steps which have been carried out separately and products of all steps have been confirmed with basic properties and stoichiometric calculations. Recovery and recycle of $CuCl_2$ and HCl have been done successfully and reuse of the same within the process is demonstrated.

1 INTRODUCTION

Hydrogen is a clean energy carrier that has the potential to solve the problem of climate change. The Major challenging task for hydrogen as a future energy carrier is a workable, low-cost method of generating hydrogen from the water splitting. In the world, hydrogen is currently derived by reforming of fossil fuels, such as coal or methane, i.e., Steam-Methane Reforming (SMR). Hydrogen is mostly used in Industrial such as an agricultural, petrochemical, food processing, plastic, manufacturing, and lightly used in industries such as the transportation sector.

Large scale sustainable methods of hydrogen production require an energy source such as nuclear or solar energy. Electrolysis is a commercially available technology that uses electricity for water splitting and hydrogen production. In past, 200 possible techniques have been proposed, but very few techniques and theories have been developed that work and show the practical feasibility of the process (Orhan, M. F., Dincer, I. & Rosen, M. A., 2010). Few of the noticeable cycles are sulphur-based cycles, copper-chlorine (Cu-Cl), iron-chlorine (Fe-Cl), cerium-chlorine (Ce-Cl), vanadium-chlorine (V-Cl), Most of these cycles needed very high temperature above 850°C that are not obtainable from existing reactor. Most important advantages of copper chlorine cycle are low operating temperatures, the low demand of material of construction, common chemical agents and a moderately low electrochemical cell voltage (Z.L. Wang, G.F. Naterer, et al. 2010).

In the 1960s, the concept of thermochemical production of hydrogen from water was proposed. Compared to other thermochemical cycles, Cu-Cl cycle has a high efficiency, to produce hydrogen at a low cost and smaller impact on the environment by dropping air-bone releases, solid waste and energy supplies (G.F. Naterer, S. Suppiah, L. Stolberg, et al. 2010).

2   STEPS FOR COPPER-CHLORINE THERMOCHEMICAL CYCLE

Five steps for the Cu-Cl thermochemical process are discussed below:

1. Hydrogen production: It is an exothermic reaction where hydrogen chloride gas reacts with copper metal to produce cuprous chloride (CuCl).
2. Electrolysis: This reaction takes place when cuprous chloride is dissolved in hydrochloric acid and a potential difference is applied across the solution. Copper is deposited on the cathode and copper (II) chloride remains in the HCl solution.
3. Drying: Drying converts aqueous $CuCl_2$ from electrolysis to solid $CuCl_2$.
4. Hydrolysis Process: In this step copper (II) chloride reacts with steam to form copper (II) oxychloride and produces hydrochloric acid as gas form.
5. Decomposition: During decomposition reaction, copper (II) oxychloride is converted into molten CuCl and oxygen is released.

3   EXPERIMENTAL PROCEDURE

Experiments were carried out for all five steps individually to optimize the various parameter like molar ratio, temperature, concentration, time, current etc.

3.1   *Chemicals used*

All chemicals are procured from CDH, Mumbai and were used without purification as follows: $CuCl_2$, Cu, NaOH, Methyl Orange, Silver Nitrate. All solutions were prepared in distilled water.

3.2   *Experimental setup*

Figure 1 shows experimental setup used for processing of all five steps.
    Individually and Table 1 shows all the reactions taking place in all five steps.

3.2.1   *Hydrolysis process*
Borosilicate glass packed bed reactor was filled with rasching ring packing and 11 gram $CuCl_2$ for hydrogen production. The Temperature was maintained and controlled by PID

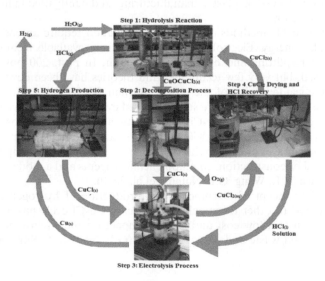

Figure 1.   Experiment setup for five step Cu-Cl cycle.

Table 1. Five step Cu-Cl cycle reaction.

Process steps	Reactions	Temp cond.
1) Hydrogen Production	$2Cu_{(s)} + 2HCl_{(g)} \rightarrow 2CuCl_{(l)} + H_{2(s)}$	500°C
2) Electrolysis Reaction	$2CuCl_{(aq)} \rightarrow Cu_{(s)} + CuCl_{2(aq)}$	20–80°C
3) Drying Process	$CuCl_{2(aq)} \rightarrow CuCl_{2(s)} + H_2O_{(g)}$	100–150°C
4) Hydrolysis Reaction	$2CuCl_{2(s)} + H_2O_{(g)} \rightarrow CuOCuCl_{2(s)} + 2HCl_{(g)}$	375–400°C
5) Decomposition process	$CuOCuCl_{2(s)} \rightarrow 2CuCl_{(l)} + 0.5O_{2(g)}$	530°C

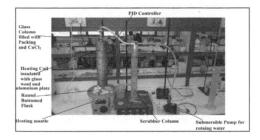

Figure 2. Hydrolysis process setup.

Figure 3. Raw Material and product at different stages.

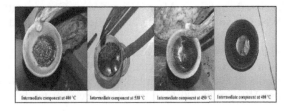

Figure 4. Intermediate and product at different stages.

Figure 5. Electrolysis process setup.

Figure 6. Remaining solution and deposited Cu.

controller with +/– 1°C accuracy. Distilled water was taken in a Round Bottom Flask (RBF) placed on a heating mantle. One end of the column was placed in the RBF and the other one was connected to the scrubber column for absorption and recovery of HCl gas using water as a solvent. The Column was heated up to 400°C within 20 min. During the reaction, generated HCl gas was absorbed in absorption column using water circulation. The Reaction was carried for 2 hours. After the reaction, intermediate product (CuOCuCl₂) was collected and weight of the product mixture was measured.

3.2.2 *Decomposition process*
Intermediate product mixture, CuOCuCl₂, was taken in the crucible and exposed to burner for heating at 530°C. At reaction temperature, decomposition reaction took place and oxygen gas was released. Molten CuCl was taken in petri dish for quench cooling, then after the solid CuCl was crushed, weighed and stored in closed box.

3.2.3 *Electrolysis process*

Molten CuCl was added in 11.75 N HCl and 0.5 N CuCl solution was made. 60 ml (0.5 N) CuCl solution was used in the electrolyzer. Graphite rod was used as a cathode and as well as anode electrode. DC current supply with constant ampere was started. After the reaction, deposited copper was filtered from the solution and dried in the oven. The Remaining solution was taken for further recovery of HCl and $CuCl_2$. As shown in Figure 5, Fabricated Glass Electrolyzer was used for Electrolysis.

Operating conditions:
– Pressure: 1 atm
– Temperature: 25–80°C
– Current supply: 0.5–2 A

Electrodes:
– Anode: Catalyst-free graphite
– Cathode: Catalyst-free graphite

Electrolyzer notation:
– Graphite | $CuCl_{(s)}$, $HCl_{(l)}$, 0.5–1 N || CuCl(s), $HCl_{(l)}$, 0.5–1 N | Graphite

3.2.4 *Drying process*

In the Drying process, HCl solution was recovered from produced aqueous solution of $CuCl_{2\,(aq)}$ in electrolysis process and moisture content was removed from the $CuCl_2$.

Procedure:
– First take a remaining aqueous solution of $CuCl_{2\,(aq)}$ in HCl recovery system and placed on a heating mantle. The Outlet of reactor is connected with spiral condenser and stored HCl solution in the closed container. After HCl recovery, Concentration of HCl solution is checked with 5 N NaOH solution and $CuCl_2$ is collected at the bottom of reactor.

3.2.5 *Hydrogen production process*

In Hydrogen Production step, Copper is reacted with HCl gas in pipe reactor at high pressure and 500°C and produced hydrogen gas and CuCl as a solid form. Nickel alloy material is sustainable for handling HCl gas at 500°C so that whole reactor made by this material.

Procedure:
– Reactor was filled with rasching ring packing and $Cu_{(s)}$ 11 gram (as per required) for H_2 production. The heating coil controlling the temperature was woven around to the reactor. Reaction was stopped after 3 hours. The Reaction mixture (gas mixture) was passed through Water-trap to remove any HCl and hydrogen was collected in the gas collecting

Figure 7.   Drying and HCl recovery setup.

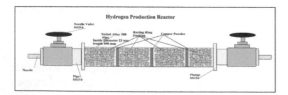
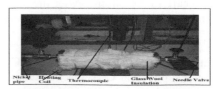

Figure 8.   Schematic diagram of H_2 production reactor.   Figure 9. H_2 pipe reactor setup.

bottle by replacement method. Water replacement method provides the quantity of gas produced during the reaction. The Gas sample was collected and GC analysis was carried out and CuCl was sent for copper recovery.

4   RESULTS AND DISCUSSION

4.1   *Hydrolysis reaction*

Results:
Steam/$CuCl_2$ ratio was varied. With the decrease in molar ratio, the quantity of $CuOCuCl_{2(s)}$ increases and that of $HCl_{(g)}$ decreases (which is not preferable).

4.2   *Decomposition process*

Results:
From experiment result, we conclude that increasing the retention time of the intermediate in glassware column during the reaction, formation of CuO increases with increasing the retention time and formation of CuCl decrease (which is not preferable).

4.3   *Electrolysis process and drying process*

4.3.1   *Change of parameter: current*
Results:
Take, CuCl = 3 gram, $HCl_{(l)}$ = 60 ml, $CuCl_{(aq)}$ = 3 gram CuCl dissolved in 60 ml of 11.75 N HCl solution.

Experiment	Current (A)	Cu (gram)	$CuCl_2$ (gram)	Volume of HCl recover (ml)	Hydrogen produce (gram)
1	0.5	0.32	2.51	55	0.0255
2	1	1.21	1.55	55	0.0122
3	1.5	1.35	1.15	56	0.0101
4	2	1.63	0.56	56	0.006

From the material balance, $Cu_{(s)}$ = 0.95 gram, $CuCl_{2(s)}$ = 2.05 gram
As current supply increases copper deposition increases but $CuCl_2$ production in solution decreases. It means some other reaction like chlorine production might have taken place. Confirmation for chlorine presence could not be done.

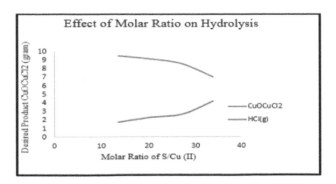

Figure 10.   Effect of Molar ratio on Hydrolysis reaction.

5 CONCLUSION

Five step Cu-Cl thermochemical cycle is selected for hydrogen production from water. Cu-Cl cycle is having relatively lower reaction temperature compared to other cycles like a sulfur—iodine cycle. In the present study of the Cu-Cl thermochemical cycle, all five steps were carried out separately. Various parameters were studied. Various intermediate products were identified using their basic properties. Currently, exact quantification of an intermediate product like CuCl or $CuCl_2$ could not be identified and quantified. But based on the copper weight collected and water displacement during the reaction, the quantities of the products formed were calculated. Hydrogen product was confirmed with GC analysis. Further optimization and suitable reactors are required for all the steps. A Selection of the suitable material of construction for high pressure and high temperature with resistance to hydrochloric acid is a major challenge.

REFERENCES

[1] G.F. Naterer, S. Suppiah, et al. 2011, clean hydrogen production with the Cu-Cl cycle—progress of international consortium, I: Experimental unit operations, *International Journal of Hydrogen Energy* 36 (2011) 15472–15485.
[2] G.F. Naterer, S. Suppiah, et al. 2013, Progress of international hydrogen production network for the thermochemical Cu-Cl cycle, *International Journal of Hydrogen Energy* 38 (2013) 740–759.
[3] Orhan, M.F., Dincer, I. & Rosen, M.A. 2010, An exergy-cost-energy-mass analysis of a hybrid copper-chlorine thermochemical cycle for hydrogen production, International journal of hydrogen energy 35 (2010) 4831–4838.
[4] Mehmet F. Orhan, Ibrahim Dincer, Marc A. Rosen, 2011, Design of systems for hydrogen production based on the Cu-Cl thermochemical water decomposition cycle: Configurations and performance, *International Journal of Hydrogen Energy* 36 (2011) 11309–11320.
[5] G.F. Naterer, S. Suppiah, L. Stolberg b, M. Lewis, et al., Progress of international program on hydrogen production with the copper-chlorine cycle, *International Journal of Hydrogen Energy* 39 (2014) 2431–2445.
[6] Z.L. Wang, G.F. Naterer, et al. 2010,Comparison of different copper chlorine thermochemical cycles for hydrogen production, *International Journal of Hydrogen Energy* 34 (2009) 3267–3276.
[7] G.F. Naterer, S. Suppiah, L. Stolberg, et al. 2010,Canada's program on nuclear hydrogen production and the thermochemical Cu-Cl Cycle, *International Journal of Hydrogen Energy* 35 (2010) 10905–10926.

Multi-disciplinary Sustainable Engineering: Current and Future Trends – Tekwani, Bhavsar & Modi (Eds)
© 2016 Taylor & Francis Group, London, ISBN 978-1-138-02845-6

Feasibility study on degradation of RR120 dye from water by O_3, O_3/UV and O_3/UV/Persulfate

S. Sharma, S. Patel & J. Ruparelia
Department of Chemical Engineering, Nirma University, Ahmedabad, Gujarat, India

ABSTRACT: An attempt has been made, in the present study, to do a comparative investigation of the efficiency of ozone based oxidation processes viz., O_3, O_3/UV, O_3/UV/Persulfate, for the removal of Reactive Red 120 dye in synthetic wastewater. All the adopted ozone based oxidation processes for the study were tested for the most appropriate conditions concerning varied parameters like initial pH, ozone flow rate, persulfate concentration. Removal of the dye was observed to be dependent on the type of ozone based oxidation process and their process parameters as well. The maximum TOC removal was 75% and was achieved by O_3/UV/Persulfate process.

1 INTRODUCTION

The available water resources in the world are decreasing whereas the industrial wastewater volume is constantly increasing. Wastewater generated from textile industries possesses high potential to pollute environment, which could have execrable effects on the ecology (Shu 2006). Textile wastewater is considered to be very tricky to treat, since it contains significant amount of recalcitrant substances such as sizing agents, dyeing agents, surfactants, volatile organic compounds, wetting agents, and bleaching agents, etc., and thus troublesome to treat using conventional methods (Arslan & Balcioglu, 2000; Sundarajan et al. 2007). Advanced Oxidation Processes (AOPs) are emerging as a substitute to conventional methods and it holds high potential to be the best option in the near future. AOPs oxidize the dye and organic pollutants which are present in the wastewater. Various researches have been carried out to determine the effect of the ozone based oxidation technology for treating the effluents of textile and dyes & dye intermediate industries, and they have revealed that it leads to considerable color removal, degrades organic compounds, increases biodegradability, and provides effective disinfection without any residuals (Arslan & Balcioglu, 2000). In addition to ozone to be employed alone as oxidizing agent, persulfate ($S_2O_8^{2-}$) is used as an alternative oxidant in the oxidation of contaminants (Tan et al. 2013). Persulfate has drawn attention as it has several advantages such as high oxidizing potential ($E_0 = 2.01$ V), non-selectively reactive, ease of storage and transportation, pH independence, reasonably stable at room temperature and low cost. It was reported that, performance of persulfate can further be improved by homolysis of the peroxide bond using heat (Ghauch et al. 2012), transition metal ions (Me^{n+}) such as Fe^{2+}, microwave (Qi, Liu, Lin, Zhang, Ma, Tan, & Ye, 2014) and UV light (Fang & Shang, 2012) to form sulfate radical ($SO_4^{-\cdot}$), a stronger oxidant ($E_0 = 2.60$ V) than persulfate, to significantly enhance the oxidation of contaminants. Research shows that sulfate radicals are more efficient than hydroxyl radicals for the degradation of compounds like 2, 4-dichlorophenol, atrazine, and naphthalene(Rao et al. 2014).

The foremost objective of this study was to determine the degradability of Reactive Red 120 using ozone based AOPs in a lab scale glass reactor to analyze various parameters like decolorization, COD and TOC removal with different operating conditions.

2 MATERIAL AND METHODS

2.1 Chemicals

In the present study, RR120 bearing solution was used as simulated wastewater in the ozone-assisted AO process. The RR120 was taken from local supplier and used to prepare synthetic wastewater without further purification process. Chemicals like Potassium Iodide (LR grade, Ranbaxy Laboratory Limited), Sodium Thiosulphate ($Na_2S_2O_3$) (LR grade, High Purity Laboratory Chemical), Sodium persulfate ($Na_2S_2O_8$), NaOH and H_2SO_4 (LR grade, S.D. Fine Chem. Limited) were purchased and used without further purification. Phillips (TUV G5 T5 8 W $\lambda = 254$ nm) UV Light was used as UV light source.

2.2 Instruments and analytical procedures

The experiments were carried out in a semi batch mode. 3.0 Lit. of 500 mgL^{-1} dye solution was taken into the reactor. The reactor configuration is as shown in Figure 1. Ozone flow rate was set in the range of 30 to 60 LPH. Initial pH was adjusted using 1 N NaOH and 1 N H_2SO_4 solution. COD to persulfate(PS) ratio was set in the range of 1:5 to 1:12. The excess ozone was trapped in 2% KI solution, and was measured by idometric method (Clesceri et al. 1998).

In this study, the effectiveness of the ozone based AOPs were determined by analyzing various performance indicators like Total Organic Carbon (TOC) (mg L^{-1}), Chemical Oxygen Demand (COD) (mg L^{-1}), and color at different process operating conditions. TOC was measured by TOC Analyzer (TOC-VCPH/CPN, Shimadzu Corporation, Japan). COD of the samples were determined by using HACH DRB 200 reactor. The UV scans of the samples were done using UV-visible spectrophotometer (UV-1800 Shimadzu, Japan) at 200 nm to 800 nm. The Ozone gas concentration was measured using ozone analyzer (ELTECH 200, Eltech Engineers, Mumbai). However, appropriate standard methods for the examination of water and wastewater (APHA) were used to analyze raw and treated wastewater samples.

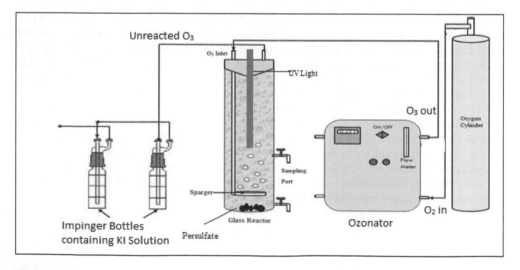

Figure 1. Reactor system.

3 RESULTS AND DISCUSSION

3.1 *Effect of ozone flowrate*

The experiments were conducted to observe the effect of the ozone concentration on dye degradation under fixed initial dye concentration, pH and temperature. The ozone flowrate was maintained at 30, 40, 50 & 60 LPH to study its effect on decolorization, total organic carbon and chemical oxygen demand removal from the RR 120 dye respectively.

Figure 2 depicts that the percentage removal of TOC was increased with increasing the ozone flow rates with match with the previous research (Matheswaran et al. 2009). The percentage of TOC removal efficiency was achieved from up to 70% at 1.25 h by the ozone flow rate of 60 LPH.

It was also observed that O_3/UV/Persulfate gives better results for removal of TOC as compared to other two processes.

3.2 *Effect of initial pH*

As it was reported that O_3 is a strong chemical oxidant and it can directly react with the unsaturated compounds. Further, in alkaline pH, Ozone (O_3) can be decomposed by producing $^\bullet$OH (Wei et al. 2011). Therefore, O_3 and $^\bullet$OH both could react and decompose RR120 into simpler end products. However, the direct reaction of O_3 is selective and is slow (Singh 2008).

TOC removal efficiency of RR120 solution varied from 32% to 76% depending upon type of treatment given as shown in Figure 3. At higher pH, destruction showed better dye degradation and TOC removal efficiency.

In O_3/UV/Persulfate system, result reveals that RR120 was decomposed more rapidly at a high pH due to two reasons.

It was reported that persulfate activated under under alkaline condition as well as in UV. Thus, chemical reaction representing the phenomena may be given by (Tan et al. 2013, Lin et al. 2013).

$$S_2O_8^{2-} \xrightarrow{OH^-} 2SO_4^{\bullet-} \tag{1}$$

In addition, it may also be proposed that, part of $SO_4^{\bullet-}$ was transformed to OH$\bullet$ in alkaline condition and reaction showing the phenomena may be given by:

$$SO_4^{\bullet-} + OH^- \rightarrow SO_4^{2-} + OH^\bullet \tag{2}$$

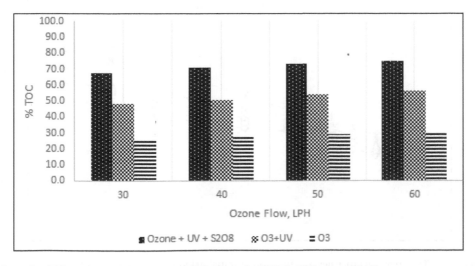

Figure 2.   Effect of ozone flowrate on TOC removal of TOC. (500 ppm; pH:10).

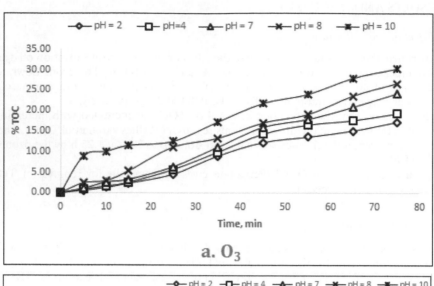

a. O_3

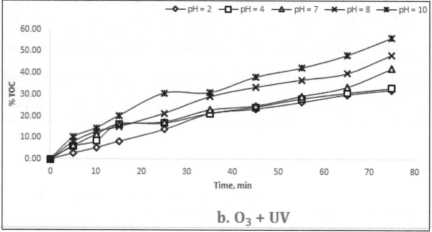

b. O_3 + UV

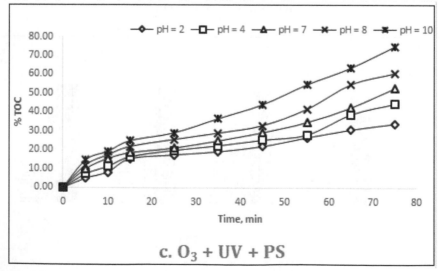

c. O_3 + UV + PS

Figure 3.   TOC removal efficiency as a function of initial pH and ozone based AOP's.

3.3 *Effect of persulfate concentration*

It is very essential to decide optimum persulfate dosage. Thus, Persulfate dosage was decided by COD to persulfate ratio and it was set in the range of 1:5 to 1:12. Different concentrations of persulfate were investigated in this study and the results are displayed in Figure 4. The initial dye concentration was fixed at 500 mg L^{-1} and initial pH value was 10.

It can be observed from Figure 4 that TOC reduction of RR120 increased from 67% to 75% by changing COD to persulfate ratio in the said range. It was observed that after 1:10 ratio TOC removal was marginally around 0.3%.

3.4 *Effect on ozone consumption*

Further quantity of ozone consumed by all the three processes for removal a gram of TOC was calculated. As shown in Figure 5, it was found that, in O$_3$/UV/Persulfate process quantity of ozone consumed for per gram removal of TOC was 15.8 mg which is less by 51.46% and 44.56% as compared to O$_3$ and O$_3$/UV process respectively.

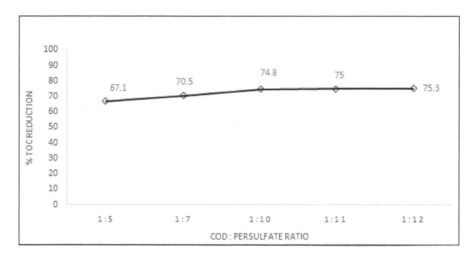

Figure 4.   Effect of persulfate dosage on TOC reduction.

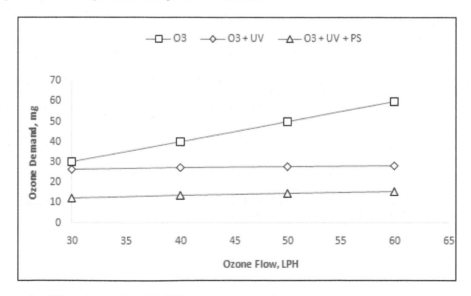

Figure 5.   Effect of ozone based AOP's on ozone demand.

4 CONCLUSION

This study explores a reliable and promising way to use O_3/UV/Persulfate for efficient removal of recalcitrant compounds such as dyes and intermediates. Higher efficiency of TOC removal also suggests that the treatment technique provide mineralization of dyes. It also provides valuable information regarding comparison of O_3, O_3/UV, and O_3/UV/Persulfate treatments for organic compound. In addition, the removal of RR120 dye in wastewater, the removal rate observed for O_3/UV/Persulfate, O_3/UV and O_3 was 75%, 52% & 30% respectively at 60 LPH of flow of ozone and 10 pH. It was also observed that COD to persulfate ratio was optimum around 1:10, further increase in persulfate concentration has shown negligible effect on the removal of TOC. Thus, the study proved the importance of various key factors in RR120 dye removal rate. Therefore, to maximize the treatment efficiency, key factors should be further carefully optimized, such as pH, ozone flowrate and persulfate concentration for the targeted compounds. The result demonstrates that O_3/UV/Persulfate has potential to replace conventional treatment methods when recalcitrant organic compounds in wastewater creates the issues of treatability.

REFERENCES

APHA, AWWA and WEF, Standard Methods for the Examination of Water and Wastewater (20th ed) 1998. Washington, DC: American Public Health Association.

Arslan, I. & Balcioglu, I.A. 2000. Effect of common reactive dye auxiliaries on the ozonation of dyehouse effluents containing vinylsulphone and aminochlorotriazine dyes. *Desalination* 130 61–71.

Clesceri, L.S., Greenberg, A.E. & Eaton, A.D. Standard methods for examination of water and wastewater (20) 1998. USA.

Fang, J.Y. & Shang, C., 2012. Bromate Formation from Bromide Oxidation by the UV/Persulfate Process. *Environ. Sci. Technol* 46: 8976–8983.

Ghauch, A., Tuqan, A.M. & Kibbi, N. 2012. Ibuprofen removal by heated persulfate in aqueous solution: A kinetics study. *Chemical Engineering Journal* 197: 483–492.

Lin, Y.T., Liang G. & Chen, J.H. 2011. Feasibility study of ultraviolet activated persulfate oxidation of phenol. *Chemosphere* 82: 1168–1172.

Matheswaran, M. & Moon, S. 2009. Influence parameters in the ozonation of phenol wastewater treatment using bubble column reactor under continuous circulation. *Journal of Industrial and Engineering Chemistry* 15: 287–292.

Oha, S.Y, Kima, H.W, Parka, J. M, Parka, H.S. & Yoon C. 2009. Oxidation of polyvinyl alcohol by persulfate activated with heat, Fe^{2+}, and zero-valent iron. *Journal of Hazardous Materials* 168: 346–351.

Qi, C., Liu, X., Lin, C., Zhang, X., Ma, J., Tan, H. & Ye, W. 2014. Degradation of sulfamethoxazole by microwave-activated persulfate: Kinetics, mechanism and acute toxicity. *Chemical Engineering Journal* 249: 6–14.

Rao, Y.F., Qua, L., Yanga, H. & Chu, W. 2014. Degradation of carbamazepine by Fe(II)-activated persulfate process. *Journal of Hazardous Materials* 268: 23–32.

Shilpi Singh, Maohong Fan, & Robert C. Brown. 2008. Ozone treatment of process water from a drymill ethanol plant. *Bioresource Technology* 99: 1801–1805.

Shu, H.Y. 2006. Degradation of dyehouse effluent containing C.I. Direct Blue 199 by processes of ozonation, UV/H_2O_2 and in sequence of ozonation with UV/H_2O_2. *Journal of Hazardous Materials B* 133: 92–98.

Sundrarajan, M., Vishnu G. & Joseph, K. 2007. Ozonation of light-shaded exhausted reactive dye bath for reuse. *Dyes and Pigments* 75: 273–278.

Tan, C., Gao,. N, Deng, Y., Zhanga, Y., Sui, M., Deng, J. & Zhou, S. 2013. Degradation of antipyrine by UV, UV/H_2O_2 and UV/PS. *Journal of Hazardous Materials* 260: 1008–1016.

Multi-disciplinary Sustainable Engineering: Current and Future Trends – Tekwani, Bhavsar & Modi (Eds)
© *2016 Taylor & Francis Group, London, ISBN 978-1-138-02845-6*

Performance prediction of a small size parabolic trough solar collector for water heating application

J.R. Mehta
The Maharaja Sayajirao University of Baroda, Vadodara, India

V.M. Patel
Sir Bhavsinhji Polytechnic Institute, Bhavnagar, India

ABSTRACT: A small size parabolic trough solar collector is fabricated and analyzed for water heating applications. The collector aperture area is 3 m² and has a concentration ratio of 18.5. The performance of this collector is predicted by heat transfer analysis, which involves finding flow regimes and applying relevant correlations. The given collector can provide a heating rate of up to 1.7 kW for a water outlet temperature of 90°C at 1000 W/m² solar irradiance. The effect of a lower irradiance on the efficiency and water flow rate is found for the collector with and without glass cover. The reduction in efficiency is moderate while that in the heating rate is drastic when the irradiance value reduces.

1 INTRODUCTION

1.1 Solar thermal energy

The growing demand for energy in emerging economies of the world and climate change concerns due to the combustion of fossil fuels have compelled us to look for sustainable energy sources like solar energy (IEA, 2014; Ahn and Graczyk, 2012). Solar PV technology is entering into the mainstream of electrical power generation technologies. On the other hand, solar thermal collectors can cater to the need of energy in form of heat. There is a huge demand for heat in temperature range of 80°C to 240°C for industrial processes like sterilizing, pasteurizing, drying, hydrolyzing, distillation, evaporation, washing, cleaning, and polymerization (Kalogirou, 2004). Some innovative and economical collectors are being investigated by researchers and developers in this field (Mehta and Rane, 2013, Rane and Mehta, 2014, Mehta and Gandhi, 2014, Mehta, 2014).

1.2 Parabolic Trough Collector (PTC)

Parabolic trough type solar collectors have been used over decades for power production, earliest one being SEGS plant in California, USA. Such collectors are large in size and cater to need of central electrical power plants. Recently, small size parabolic trough collectors are being investigated for fulfilling local needs of the commercial and industrial sector. Multiple small PTCs can be combined in series or parallel to get required amount of heat and temperature for the given application.

2 DEVELOPMENT OF THE PARABOLIC TROUGH COLLECTOR

A small size PTC was investigated in this work for water heating application. A trough with 2 m length and 1.5 m width has been fabricated, a size which is easy to handle. Number of such collectors can be put in series or in parallel to cater to required heating rate and

Figure 1. Parabolic trough solar collector analyzed in this work.

temperature. The outside diameter of absorber tube is taken 25.4 mm. If the size is kept very small, there will be more chances of reflected radiation missing the target (absorber tube). The concentration ratio is 18.5. The reflector sheet is a mirror finish acrylic sheet with reflectivity equal to 0.9. This material is much lighter and lower in cost as compared to an aluminum sheet. On the other hand, it is brittle and not scratch resistant. It is mounted on a painted mild steel structure.

The rim angle is kept 90°, a value which is the result of trade-off between optical efficiency and trough surface area. Tracking systems are costly and need regular maintenance. The axis of PTC may be kept horizontal and oriented in East-West direction. The advantage here is that only seasonal tracking is required. But, this type of tracking leads to the variation of heat collection over a day and annual output is low as compared to other tracking schemes.

Absorber tube material is copper and is painted with black chrome selective paint with absorptivity equal to 0.96. A 58 mm OD, 1.6 mm thick borosilicate glass tube is put over absorber tube to reduce convective losses. Transmissivity of the glass tube is 90%. Adding the glass tube increases the surface area for heat as well as transmittance loss; though temperature difference between tube surface and ambient reduces. The performance of collector is evaluated with and without the glass cover tube. Considering the geometrical accuracy of 0.8, the optical efficiency of the PTC comes out to be 69.1% without glass cover and 62.2% with cover.

3 SIMULATION: CONDITIONS AND METHODOLOGY

Water outlet temperature of the PTC is taken 90°C, slightly higher than the normal flat plate collectors can deliver. Water flow rate needs to be adjusted to achieve 90°C temperature at PTC outlet. The ambient temperature is taken as 37°C, which is the average of ambient temperature for the month of April for time period 9:00 to 15:00 solar time for Ahmedabad (Tyagi, 2009). Sky temperature is taken to be 6°C lower than the ambient temperature and radiant heat exchange is considered only with sky (Sukhatme and Nayak, 2008). Inlet temperature of the water is taken to be 3°C lower than the ambient temperature. The performance of the PTC is evaluated at 1000, 850, 700, and 550 W/m² solar irradiance values. Specific heat of water is taken a constant at 4,175 J/kg K. Other properties of water, necessary to calculate heat transfer coefficient, are calculated as a function of temperature.

Constant heat flux boundary conditions are assumed as the heat flux falling on the absorber tube from the reflector is ideally constant throughout the length. The PTC size is small, so it can heat only little quantity of water. The velocity of water through the tube and the Reynolds number are low and the flow is laminar. The value of Prandtl number is around two. It is found that the flow is thermally developing for the almost full length of the absorber tube.

The flow is also hydrodynamically developing for significant length of the absorber tube. Thus, the flow is either simultaneously developing or thermally developing and hydrodynamically developed for the major part of absorber tube. Following correlations are used for simultaneously developing and hydrodynamically developed but thermally developing flow (Shah and London 1978, Stephen and PreuBer, 1979):

$$Nu_{sd} = 1.953(\text{Re}\,\text{Pr}\,D/L)^{\frac{1}{3}} \tag{1}$$

$$Nu_{td} = 4.364 + \frac{0.086\left(\dfrac{\text{Re}\,\text{Pr}\,D}{L}\right)^{1.33}}{1 + 0.1\text{Pr}\left(\dfrac{\text{Re}\,D}{L}\right)^{0.83}} \tag{2}$$

A constant value of Nusselt number is considered throughout the length and this number is the weighted average of the two numbers for above two flow regimes.

3.1 *Performance of PTC without glass cover on absorber tube*

Following are the steps for finding the performance of the PTC:

a. Initially, a value of efficiency was assumed and flow rate of water was found for outlet temperature equal to 90°.
b. Properties of water should be evaluated at the mean temperature of surface of absorber tube and water. As the surface temperature is not known beforehand, it is first assumed and then corrected based on the calculated value.
c. Heat transfer coefficient of water over inner surface of tube was found and the surface temperature at inlet or outlet of absorber tube can be found with following equation:

$$T_{s.abt} = \frac{q''}{h_{i.abt}} + T_w \tag{3}$$

The average of inlet and outlet temperature is used for further calculations.
d. The outside heat transfer coefficient, which is dependent on the wind velocity was found using properties of the air at the mean film temperature. The convective and radiative heat losses from absorber tube surface can be evaluated as:

$$q_{c.abt} = A_{abt.o}h_{abt.o}(T_{s.abt.avg} - T_{amb}) \tag{4}$$

$$q_{r.abt} = \varepsilon\sigma A_{abt.o}(T^4_{s.abt.avg} - T^4_{sky}) \tag{5}$$

Following correlation is used to find Nusselt number on the outside surface of absorber tube (Hilpert, 1993).

$$Nu_{o.abt} = 0.615\,\text{Re}^{0.466} \tag{6}$$

e. Finally, the useful heat is found by deducting the losses from the energy absorbed by the absorber tube. Dividing useful heat with incident solar energy on reflector surface gives efficiency of the collector:

$$\eta_{th} = \frac{Q_u}{I_b A_{ap}}, \text{ where } Q_u = \dot{m}_w c_{pw}(t_{wo} - t_{wi}) \tag{7}$$

f. Above value of efficiency is compared with the assumed value. The assumed value is adjusted as required to match it with the calculated value. The parameters like water flow rate and surface temperature may also be required to be readjusted.

3.2 Performance with glass cover on absorber tube

Following would be the additional steps when the glass cover is put over the absorber surface:

g. The combined conduction and natural convection heat transfer coefficients between the absorber tube and glass cover are calculated as follows:

$$h_{c,abt-gc} = \frac{2k_{eff.air}}{d_{abt.o} \ln(d_{i.gc.}/d_{o.abt})} \tag{8}$$

$k_{eff.air}$ is found with following correlation (Raithby and Hollands, 1975):

$$\frac{k_{eff.air}}{k_{air}} = 0.317(Ra*)^{\frac{1}{4}} \tag{9}$$

Modified Rayleigh number is derived from Rayleigh number and geometry of the system (Sukhatme and Nayak, 2008).

Radiative heat transfer coefficient between glass cover and absorber tube can be found as:

$$h_{r,abt-gc} = \frac{\sigma(T_g^2 + T_a^2)(T_g + T_a)}{\left(\dfrac{1}{\varepsilon_r} + \dfrac{A_o}{A_{co}}\left(\dfrac{1}{\varepsilon_g} - 1\right)\right)} \tag{10}$$

h. The temperature of glass cover needs to be assumed first and later on corrected on basis of calculated value.
i. Convective heat transfer coefficient from the surface of the glass cover is calculated on basis of wind velocity and mean film temperature. Radiative heat transfer coefficient between glass cover and sky is also found.
j. The overall heat transfer coefficient from absorber surface to glass cover is found.

$$U_L = \left(\frac{A_{abt.o}}{A_{gc.o}\left(h_{c.gc-a} + h_{r.gc-a}\right)} + \frac{1}{\left(h_{r.abt-gc} + h_{c,abt-gc}\right)}\right)^{-1} \tag{11}$$

4 RESULTS AND DISCUSSION

4.1 PTC without glass cover

When solar irradiance is lower, the flow rate of water needs to be reduced to maintain constant outlet temperature at 90°C. The flow is laminar in all cases and the entry lengths are proportionate to Reynolds number, which reduces as flow rate reduces. Heat transfer coefficients in cases with lower solar irradiance are lower due to low value of Nusselt number. Figure 2 shows the reduction in heat transfer coefficient on the inner surface of tube with reduction in solar irradiance. The heat flux transported from absorber tube surface to water also reduces as solar irradiance reduces. The surface temperature of the absorber tube is seen to be reducing, but the reduction rate of surface temperature with reduction in solar irradiance is very low.

Outside heat transfer coefficient mainly depends on wind velocity and is not affected by irradiance value. So, the rate of reduction of the absolute value of heat loss from the surface of absorber with reduction in solar irradiance is quite low; while percentage heat loss is high. Percentage optical losses remain the same in all cases. The final effect is reduction of efficiency with reduction in solar irradiance. Radiation loss is found to be insignificant in all cases (less than 1% of incidence energy) due to smaller surface area for heat loss.

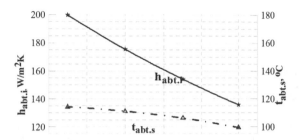

Figure 2.   Variation of heat transfer coefficient and surface temperature with solar irradiance.

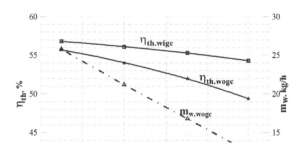

Figure 3.   Variation of efficiency and mass flow rate with solar irradiance.

The thermal efficiency of the PTC without glass cover at 1000 W/m² solar irradiance value is 55.7%, which reduces to 49.4% at 550 W/m² (Figure 3). The reduction in mass flow rate of water that can be heated to 90°C is quite sharp as heat flux as well, as efficiency of collector goes down.

4.2   *PTC with glass cover*

When a glass tube cover is used over the absorber tube, high temperature absorber tube is not directly exposed to wind. There are two thermal resistances acting in parallel between absorber tube and glass cover, namely, radiative and convective cum conductive, through the air gap. So, the glass surface remains at a much lower temperature (43 to 45°C) as compared to the absorber tube (102 to 105°C). The heat transfer area of glass cover is around 2.25 times that of absorber tube. Final effect is still to reduce heat losses and improve efficiency of the collector. It is seen from Figure 3 that improvement in efficiency due to glass cover is higher when solar irradiance is lower. Improvement in efficiency due to glass cover at 1000 W/m² is 2%, while that at 550 W/m² is 10% as compared to no glass cover. It is advisable to use glass cover as it is one time investment and not very costly; and protects the absorber tube paint. Evacuating the space between glass tube and receiver tube may not help much in increasing the efficiency.

5   CONCLUSION

A parabolic trough solar collector with an aperture area of 3 m² and a concentration ratio of 18.5 is analyzed for water heating up to 90°C. The collector can deliver around 1.8 kW heat at 1000 W/m² solar irradiance, which is reduced to around 0.9 kW at 550 W/m². The effect of lower irradiance on efficiency and water flow rate is found for the collector, with and without glass cover. Efficiency of the PTC is expected to be 57% at 1000 W/m² and 54% at 550 W/m² irradiance values respectively, when used with glass cover. Improvement in efficiency due to glass cover at 1000 W/m² is 2%, while that at 550 W/m² is 10% as compared to no glass cover. So, it is advisable to use glass cover, which will also protect the absorber tube.

REFERENCES

Ahn, S.J. & Graczyk D. 2012. Understanding energy challenges in India: Policies, players and issues. OECC/IEA. France.

Hilpert, R. 1993. Warmeabgabe von gelhizen Drahten und Rohren, Forsch Geiete Ingenieurw, 4:220.

IEA 2014. Key world energy statistics. Paris, France.

Kalogirou, S.A., 2004. Solar thermal collectors and applications. Progress in Energy and Combustion Sciences 30: 231–295.

Mehta, J.R. 2014. Regeneration of liquid desiccant in solar passive regenerator with enhanced performance. International Journal of Modern Engineering Research 4(7): 63–65.

Mehta, J.R. & Rane M.V. 2013. Liquid desiccant based solar air conditioning system with novel evacuated tube collector as regenerator. Procedia Engineering 51: 688–693.

Mehta, J.R. & Rane M.V. 2014. Two stage regeneration of liquid desiccant using solar energy for fresh air dehumidification. Proc. Intern. Sorption Heat Pump Conf., College Park, Washington, 31 Mar–3 Apr 2014.

Mehta, J.R. & Gandhi S.M. 2013. Investigation of solar still as liquid desiccant regenerator. Proc. Nirma University Int. Conf. on Engineering, Ahmedabad, India, 28–30 November 2013.

Raithby, G.D. & Hollands, K.G.T. 1975. A general method of obtaining approximate solutions to laminar and turbulent free convective problems, Advances in Heat Transfer, 11: 265.

Shah, R.K. & London, A.L. 1978. Laminar flow forced convection in ducts, Academic Press, New York.

Stephan, K. & PreuBer, P. 1979. Heat transfer and critical heat flux in pool boiling of binary and ternary mixtures. German Chemical Engineering, 2(3): 161–169.

Sukahtme S.P. & Nayak J.K., 2008. Solar Energy: Principles of thermal collection and storage. 3rd Ed, Tata McGraw-Hill Pub. Co. Ltd. New Delhi.

Tyagi, A.P. 2009. Solar radiant energy over India. India Meterological Department, Ministry of Earth Sciences, New Delhi.

Sustainable manufacturing processes

Experimental evaluation and optimization of warpage in injection molded Nylon 66 bevel gear using numerical simulation and neural network

Dattatray Chopade & Rajesh Metkar
Department of Mechanical Engineering, Government College of Engineering, Amravati, Maharashtra, India

Subhash Hiwase
SCM Data Engineering Services, Uppal, Hyderabad, Telangana, India

ABSTRACT: Uneven cooling of injection molded part sets up uneven distribution of stresses. These stresses warp the part and affect its dimensional accuracy. The lack of injection molding machine control leads to this defect in plastic parts. So as to keep the warpage minimum, accurate prediction of optimum process parameters is very important. Finite element analysis software, Moldflow Plastic Insight, Taguchi statistical design of experiment and artificial neural network model are used to find optimum process parameter values. The optimum combination of process parameters that can minimize the warpage is found out. Most effective factor contributing to warpage is the melt temperature for this problem. The value of warpage obtained by using recommended process setting is reduced by 5.3% after optimization. Neural network model, formulated on the basis of finite element simulation data, gives results with acceptable accuracy and can be used as an alternative to the costly and time consuming finite element simulation process.

1 INTRODUCTION

Injection molding is one of the most versatile, efficient, and widely used manufacturing processes for the manufacturing of plastic products in various shapes, sizes, and dimensions. A number of defects may occur during the injection molding process such as sink marks, warpage, weld lines, air traps, short shots, etc. These defects are caused mainly due to improper process parameter settings, mold design, and geometry of part as well as plastic material used to manufacture the part. Among these defects, warpage is the result of unequal stress in the molded part, when the stress is strong enough to strain or distort the piece.

Chen et al. (2009) used Design Of Experiments (DOE) approach to determine optimal process parameter settings. ANOVA and regression models were constructed for the results of simulation and experiment. Hakimian and Sulong (2012) discussed about warpage and shrinkage properties of injection molded micro gears polymer composites using Taguchi method. Kramschuster et al. (2005) investigated the effects of processing conditions on shrinkage and warpage behavior of a box-shaped, polypropylene part using conventional and microcellular injection molding. They used fractional factorial design approach of experiments. Wang et al. (2013) investigated the reduction of sink mark and warpage of the molded part in rapid heat cycle molding process. They used design of experiments via Taguchi methods to investigate the effect of processing parameters on warpage.

Taghizadeh et al. studied the warpage in plastic injection molded part using Artificial Neural Network (ANN). Back propagation neural network modeling for warpage prediction and optimization of plastic products during injection molding was used by Fei Yin et al. (2011). Back propagation neural network trained by the data from finite element simulations, which are conducted on Moldflow software. Farshi et al. (2011), optimized the injection molding process parameters by using Sequential Simplex Algorithm. Using more than one method for solving the problem of warpage is also popular. Hybrid optimization method was proposed by Deng et al. (2010), that combines a Mode-Pursuing Sampling (MS) method with the genetic algorithm for minimizing the warpage of injection molded plastic parts. Ozcelik and Erzumlu (2005) discussed the dimensional parameters affecting the warpage of thin shell plastic parts using Integrated Response Surface Method (RSM) and Genetic Algorithm. Chiang and Chang (2006) used an approach of orthogonal array with grey rational analysis and fuzzy logic to determine the optimal parameter setting for PC/ABS cell phone shell. Various researchers found different process parameters as most significant varying according to the problem under consideration.

2 EXPERIMENTAL DETAILS

A bevel gear (shown in Figure 1) having diameter of 16.42 mm, gear thickness of 3.52 mm, and a teeth number of 20, is used as a model for this study. The gear is made up of Nylon 66 A125 (Manufacturer-Unitika) and its material properties are given in Table 1. These values are taken from Moldflow Plastic Insight (MPI) material database. An 8-cavity mold layout is selected in this study and 3D mesh is used for meshing the model. The model consists of 521984 tetrahedral elements. It is created in MPI software which is a commercial software that is based on hybrid finite element method for solving pressure, flow, and temperature fields. FE model with feeding system and cooling channels is shown in Figure 2.

The parameters used for the warpage analyses are the melt temperature, mold temperature, packing pressure, injection time, and cooling time, and are represented by the letters "A," "B," "C," "D," and "E", respectively. Three levels of these parameters are selected taking reference from material database available in MPI and are given in Table 2. Taguchi L_{27} (3^5) orthogonal array are selected for the experimental design.

The FE analysis of this system is performed for all the 27 sets of process settings and is given in Table 3 with their warpage values obtained by FE analysis. The analysis sequence used in this case is "Fill + Pack + Warp". The packing time is kept constant at 5 s for all the cases. Other machine and process parameters are kept as default. The result of warpage for

Scale (20 mm)

Figure 1.   Gear used for this study.

Table 1.   Properties of plastic material (Nylon-66) used for bevel gear.

Properties	Values
Trade name	Unitika Nylon 66 A125
Recommended mold surface temperature (°C)	90
Recommended melt temperature (°C)	290
Ejection temperature (°C)	158
Elastic modulus (MPa)	2690

Figure 2.   FE model with feed system and cooling channels.

Table 2.   Setting of the processing parameters.

| Level | Experimental factors | | | | |
	A:Melt temperature (°C)	B:Mold temperature (°C)	C:Packing pressure (MPa)	D:Injection time (s)	E:Cooling time (s)
1	280	80	30	0.5	15
2	290	90	40	1	20
3	300	100	50	1.5	25

Table 3.   Process settings with their warpage values obtained by FE simulation.

| Test no | Process setting | | | | | Simulated warpage value (mm) | S/N ratio |
	A	B	C	D	E		
1	280	75	30	0.5	15	0.1252	18.04
2	280	75	30	0.5	20	0.1252	18.04
3	280	75	30	0.5	25	0.1252	18.04
4	280	85	40	1.0	15	0.1248	18.07
5	280	85	40	1.0	20	0.1251	18.05
6	280	85	40	1.0	25	0.1244	18.10
7	280	95	50	1.5	15	0.1255	18.03
8	280	95	50	1.5	20	0.1255	18.03
9	280	95	50	1.5	25	0.1255	18.03
10	290	75	40	1.5	15	0.1272	17.91
11	290	75	40	1.5	20	0.1272	17.91
12	290	75	40	1.5	25	0.1272	17.91
13	290	85	50	0.5	15	0.1285	17.82
14	290	85	50	0.5	20	0.1285	17.82
15	290	85	50	0.5	25	0.1289	17.79
16	290	95	30	1.0	15	0.1325	17.55
17	290	95	30	1.0	20	0.1322	17.57
18	290	95	30	1.0	25	0.1318	17.60
19	300	75	50	1.0	15	0.1318	17.60
20	300	75	50	1.0	20	0.1305	17.68
21	300	75	50	1.0	25	0.1318	17.60
22	300	85	30	1.5	15	0.1352	17.38
23	300	85	30	1.5	20	0.1352	17.38
24	300	85	30	1.5	25	0.1352	17.38
25	300	95	40	0.5	15	0.1343	17.44
26	300	95	40	0.5	20	0.1344	17.43
27	300	95	40	0.5	25	0.1344	17.43

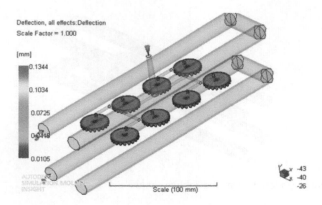

Figure 3.   Warpage analysis result of the 8 cavity bevel gear mold.

one analysis is shown in Figure 3. The maximum warpage value of the part and its location can be seen directly from the picture.

The objective of this study is to minimize the warpage to the greatest extent possible and develop a prediction model for warpage so as to reduce the time and cost involved in repetitive FE analyses. Hence, a neural network model is developed using MATLAB neural network toolbox.

3   RESULTS AND DISCUSSION

3.1   *Taguchi method*

The Taguchi method is applied to design the experiment and to determine the effect of process parameters on warpage. The measured warpage values and their S/N ratios are given in Table 3. The S/N ratio is used to evaluate the effect of altering a particular parameter value on the desired output.

"The smaller, the better" quality characteristic is selected while calculating the S/N ratios as the goal of this study is to minimize the warpage of the part. The response table of S/N ratios is given in Table 4. The S/N response graphs are plotted using the data given in Table 4 and are shown in Figure 4. The optimum combination of process parameters can be determined by selecting the level with highest value for each factor. Hence, the best combination of process parameters for this study is A1, B1, C3, D3, and E1. We have selected E1 instead of E2 because there is a negligible difference between S/N ratios of two levels, and selection of E1 causes reduction of cooling time from 20 s to 15 s. Thus, the cycle time of the process gets reduced by large amounts without compromising the quality of the part. The difference values in Table 4 denote the most significant process parameters for warpage. Melt temperature was found to be most significant process parameter followed by melt temperature and packing pressure. Injection time and cooling time had least effect on warpage of the bevel gear.

The optimum combination of process parameters is not included in the 27 FE analysis test runs given in Table 3. Therefore, a confirmation test is conducted to validate the results obtained by Taguchi method. The corresponding S/N ratio is 18.30 dB, which is higher than those obtained in orthogonal array design of experiment. It shows that the results obtained are better and warpage value is optimized.

3.2   *ANOVA*

The ANOVA is conducted on the data obtained from the FE analysis results using MIN-ITAB V14 software and the results are shown in Table 6. The confidence interval is set as 95% and significance level as 5% for the ANOVA analysis. If "P value" for any parameter is less

Table 4.   The response table of S/N ratios.

Level	A	B	C	D	E
Level 1	18.05	17.86	17.66	17.76	17.76
Level 2	17.76	17.75	17.80	17.76	17.77
Level 3	17.48	17.68	17.82	17.77	17.76
Difference	0.57	0.18	0.16	0.01	0.01

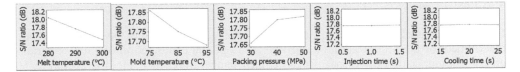

Figure 4.   Effect of the processing parameters on warpage.

Table 5.   The ANOVA table for part warpage.

Factor	SS	df	MS	F	P	PC (%)
Melt temperature	0.0003243	2	0.0001622	1569.20	0.000	83.07
Mold temperature	0.0000345	2	0.0000172	166.91	0.000	8.83
Packing pressure	0.0000298	2	0.0000149	144.33	0.000	7.63
Injection time	0.0000001	2	0.0000000	0.42	0.664	0.02
Cooling time	0.0000001	2	0.0000000	0.39	0.685	0.02
Error	0.0000017	16	0.0000001			0.43
Total	0.0003904	26				100.00

Table 6.   Comparison of warpage before and after optimization.

	A	B	C	D	E	Simulated warpage	ANN
Moldflow recommended process setting	290	90	40	1	20	0.1311	0.1311
Process setting after optimization	280	75	50	1.5	15	0.1216	0.1203
Change rate (%)						−7.24	−8.24

than 0.05, then the parameter has significant effect on the result. From the "P value" column of Table 5, it is clearly seen that melt temperature, mold temperature, and packing pressure have significant effect on the warpage values.

The percentage contribution of each parameter is calculated using the following equation:

$$(PC)_A = SS_A / SS_{Total} \tag{1}$$

where $(PC)_A$ = percentage contribution of factor A; SS_A = sum of squares of factor A; and SS_{Total} = sum of squares of all factors. Percentage contributions for other factors are calculated using the same method and are given in the last column of Table 5.

3.3   Neural network model

A neural network has one or more hidden layers placed between the input and output layers. The layers have processing units called neurons. These neurons are connected by variable weights. All the neurons in the preceding layer give total input to each neuron in the succeeding layer. In this study, neural network is used to predict the warpage values corresponding to optimal process setting determined by Taguchi method and Moldflow recommended process setting. MATLAB neural network toolbox is used for this study. A 5–10–10–1 feed-forward backpropogation neural network model is created and trained. The value of warpage by Moldflow recommended process setting and process setting after optimization are found out using this network. Comparison of the predicted and FE simulation values of warpage for the optimal setting and Moldflow recommended setting are given in Table 6.

4 CONCLUSION

Optimal settings of process parameters for the bevel gear made up of Nylon 66 A125 are determined for the optimization of warpage. The optimal combination of process parameters for minimization of warpage is 280°C melt temperature, 75°C mold temperature, 50 M Pa packing pressure, 1.5 s injection time and 15 s cooling time. Melt temperature is the most significant process parameter and contributes 83.07% to warpage. Mold temperature and packing pressure are the other significant process parameters while injection time and cooling time have least effect on warpage. The warpage value obtained in the Moldflow recommended process setting is reduced by 5.3% after optimization. A neural network is formulated using MATLAB neural network toolbox to predict the warpage values directly. The model predicts warpage values with good accuracy. The optimization methodology used in this study can be employed to minimize the time and cost associated with repetitive FE analyses.

REFERENCES

Chen, Chuang, Hsiao, Yang & Tsai., 2009, Simulation and experimental study in determining injection molding process parameters for thin-shell plastic parts via design of experiments analysis. Expert Systems with Applications 36: 10752–10759.

Chiang & Chang., 2006, Application of grey-fuzzy logic on the optimal process design of an injection-molded part with a thin shell feature. International Communications in Heat and Mass Transfer 33: 94–101.

Deng, Zhang & Lam., 2010, A hybrid of mode-pursuing sampling method and genetic algorithm for minimization of injection molding warpage. Materials and Design 31: 2118–2123.

Farshi, B., Gheshmi, S. & Miandoabchi, E., 2011, Optimization of injection molding process parameters using sequential simplex algorithm. Materials and Design 32: 414–423.

Hakimian, E. & Sulong, A.B., 2012, Analysis of warpage and shrinkage properties of injection-molded micro gears polymer composites using numerical simulations assisted by the Taguchi method. Materials and Design 42:62–71.

Kramschuster, A., Cavitt, R., Ermer, D., Chen Z. & Turng L., 2005, Quantitative study of shrinkage and warpage behaviour for microcellular and conventional injection molding. Polymer Engineering and Science: 1408–1418.

Ozcelik, B. & Erzurumlu, T., 2005, Determination of effecting dimensional parameters on warpage of thin shell plastic parts using integrated response surface method and genetic algorithm. International Communications in Heat and Mass Transfer 32: 1085–1094.

Taghizadeh, S., Özdemir, A. & Uluer, O. Warpage prediction in plastic injection molded part using artificial neural network. IJST. Transactions of Mechanical Engineering 37(M2): 149–160.

Wang, X., Zhao, G. & Wang, G., 2013, Research on the reduction of sink mark and warpage of the molded part in rapid heat cycle molding process. Materials and Design 47: 779–792.

Yin, F., Mao, H., Hua, L., Guo, W. & Shu, M., 2011, Back Propagation neural network modeling for warpage prediction and optimization of plastic products during injection molding. Materials and Design 32: 1844–1850.

Multi-disciplinary Sustainable Engineering: Current and Future Trends – Tekwani, Bhavsar & Modi (Eds)
© 2016 Taylor & Francis Group, London, ISBN 978-1-138-02845-6

Selection procedure of fin type and fin configuration for plate fin heat exchanger

B.A. Patel
B.H. Gardi College of Engineering and Technology, Rajkot, Gujarat, India

A.M. Lakdawala
Institute of Technology, Nirma University, Ahmedabad, Gujarat, India

Dileep Patil
Ingersoll Rand (India) Limited, Ahmedabad, Gujarat, India

ABSTRACT: The present study aims at selection of fin configuration for required heat duty. In this regard, a novel methodology is developed to select an appropriate fin by comparing different types and sizes of fin. This comparison is based on different constraints. These constraints are fixed frontal area, fixed heat exchanger volume and fixed pumping power. Different graphs are generated for plain fin, offset strip fin and louvered fin according to the constrains.

Keywords: plate fin heat exchanger; plain fin; strip fin; louvered fin; configuration

NOMENCLATURE

A	Heat exchanger area (m^2)
D	Hydraulic diameter (m)
f	Friction factor
j	Colburn factor
L	Length (m)
$\dot{m}$	Mass flow rate (kg/s)
NTU	Number of transfer units
P	Pumping power
Pr	Prandtl number
V	Volume (m^3)
ρ	Density of fluid (kg/m^3)
μ	Dynamic viscosity (Ns/m^2)

SUBSCRIPT

A	Fix area
fr	Frontal area
v	Fix volume
p	Fix pressure

SUPERSCRIPT

f	Fix parameter
p	Problem associate
s	Solution associate

1 INTRODUCTION

A proper selection of surface is one of the most important considerations in plate-fin heat exchanger design. There is no such thing as surface that is best for all applications. The particular application strongly influences the selection of surface to be used. Hence, this paper presents selection methodology for the type of fin, based on three types of constrains. These constrains are: fixed frontal area, fixed heat exchanger volume and fixed pumping power. This procedure will help to select the type of fin(configuration of fin) from Kays and London data. There are number of configurations for plain fin, strip fin and louvered fins provided by kays and London [Kays 1964]. However, at the first step of thermal designing of compact heat exchanger, a selection of proper fin type and configuration of fin on hot fluid side and cold fluid side is necessary. Moreover, the cold burn factor j and friction factor f for different fins are function of Reynolds numbers. These factors can be used to specify surface characteristics of different fin configurations. With the help of these parameters comparisons between the different fins on the basis of pumping power, heat transfer and size can be made.

Most common problems in the designing of heat exchanger are rating and sizing. Rating problem is concerned with determination of heat transfer rate and fluid outlet temperatures for prescribed fluid flow rates, while sizing procedure involves determination of heat exchanger dimensions, that is: selecting proper heat exchanger type and determining the size to meet requirements of specified hot and cold fluids inlet and outlet temperatures, flow rates and pressure drops [Shah 2003].

A family of methods for comparing compact heat transfer surface configurations are provided by T A Cowell [Cowell 1990]. A simple approach to surface selection based on the concept of Volume Performance Index (VPI) is presented in [Nunez 1989]. Surfaces, which result in the smallest volume will exhibit higher VPI [Nunez 1989]. Surfaces are compared on the basis of VPI and envelopes for best performance are produced.

2   SELECTION PROCEDURE OF FIN

Selection of fin of plate fin heat exchanger is required both sides (Hot side and Cold side). There is no method which gives direct selection of fin for selected application. For different fluid and operating condition each fin gives different effectiveness and pressure drop and also gives different size of heat exchanger. There are various methods for comparison of fin surfaces like Area goodness factor, Volume goodness factor and comparison with reference surface however, some difficulties remain with all of them. T A Cowell [Cowell 1990] showed the measurement for the relative values of required hydraulic diameter, frontal area, total volume and pumping power for different surface. Selection procedure based on three different constrains fix frontal area, fix pumping power and fix heat exchanger volume is shown below.

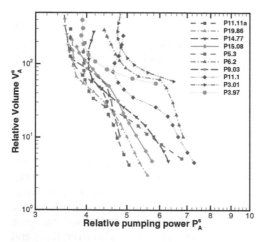

Figure 1. Relative volume vs Relative pumping power for plain fin.

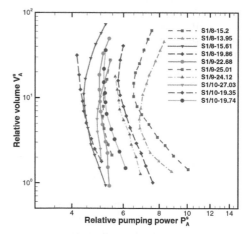

Figure 2. Relative volume vs Relative puming power for strip fin.

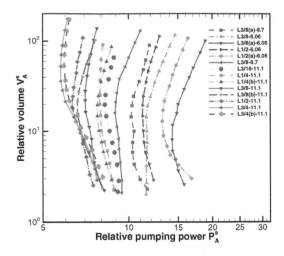

Figure 3.   Relative volume vs Relative pumping power for Louvered fin.

2.1   *Comparison of fixed frontal area*

Application of this condition allows the comparison of heat transfer surfaces without the need for scale to be specified. If A_f is fixed frontal area as per method given in [Cowell 1990]

$$D_A = \sigma \text{Re} \frac{\mu A^f}{\dot{m}} = D_A^s D_A^p$$

The subscript A implies the fixed frontal area condition, the superscript s implies the solution related component, and the p indicates the problem-related part

$$V_A = \frac{\sigma \text{Re}}{j} \cdot \frac{\mu A^{s^2} NTU^f \text{Pr}^{2/3}}{4\dot{m}} = V_A^s V_A^p$$

$$P_A = \frac{f}{\sigma^2 j} \cdot \frac{\dot{m}^3 NTU^f \text{Pr}^{2/3}}{2A^f \rho^2} = P_A^s P_A^p$$

2.2   *Comparison of fixed heat exchanger volume*

For fixed heat exchanger volume we can write equation of relative frontal area and relative pumping power as below,

$$D_v = \left(\sigma j \text{Re}^{1/2} \right) \cdot \left[\frac{4\dot{m} V^f}{NTU^f \mu \text{Pr}^{2/3}} \right]^{1/2} = D_v^s D_v^p$$

$$A_{f_v} = \left(\frac{j}{\sigma \text{Re}} \right)^{1/2} \cdot \left[\frac{4\dot{m} V^f}{NTU^f \mu \text{Pr}^{2/3}} \right]^{1/2} = A_{f_v}^s A_{f_v}^p$$

$$P_v = \left(\frac{f \text{Re}}{\sigma j^2} \right) \cdot \left[\frac{\mu \dot{m}^2 NTU^{f^2} \text{Pr}^{4/3}}{8 \rho^2 V^f} \right] = P_v^s P_v^p$$

2.3 Comparison of fixed pumping power

If comparisons are now made of the different geometries that are able to deliver the required heat transfer under the constraint that pumping power has a fixed value P'', then hydraulic diameter is given as follows,

$$D_p = \left(\frac{f\,\mathrm{Re}^2}{j}\right)^{1/2} \cdot \left[\frac{NTU^f \dot{m} \mu^2 \mathrm{Pr}^{2/3}}{2\rho^2 P^f}\right] = D_p^s D_p^p$$

$$A_p = \left(\frac{f}{j\sigma^2}\right) \cdot \left[\frac{NTU^f \dot{m}^3 \mathrm{Pr}^{2/3}}{2\rho^2 P^f}\right] = A_p^s A_p^p$$

$$V_p = \left(\frac{f\,\mathrm{Re}}{\sigma j^2}\right) \cdot \left[\frac{NTU^{f^2} \dot{m}^2 \mu \mathrm{Pr}^{4/3}}{8\rho^2 P^f}\right] = V_p^s V_p^p$$

Here, the comparison can be more easily interpreted if flow length is considered instead of the volume. Flow length is given by following equation,

$$L_p = \frac{V_p}{A_p} = \left(\frac{f\,\mathrm{Re}^2}{j^3}\right)^{1/2} \cdot \left[\frac{NTU^{f^3} \dot{m} \mu^2 \mathrm{Pr}^2}{32\rho^2 P^f}\right] = L_p^s L_p^p$$

Figures 7–9 are the plots of relative frontal area against Relative flow length for surfaces as plain fin, strip fin and louvered fin. A wide range of comparison can be rapidly made between different geometric solutions.

3 SELECTION PROCEDURE

The selection of fin configuration for the required heat duty and specific constrain can be made by using various graphs shown in Fig. 4 to Fig. 9. The comparison of the characteristics of different types of fins can be made by using the graphs of Pv vs Av as shown in Figs. 4–6 for plain fin, strip fin and louvered fin respectively. From these figures one can get relative frontal area for specified pumping power for all types of fins. Note that the fin having low frontal area is the best selection. Similarly Fig. 7, Fig. 8 and Fig. 9 can be used for the other constrains.

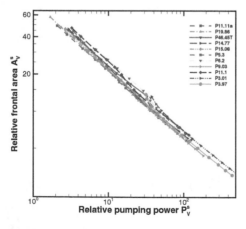

Figure 4. Relative pumping power vs Relative Frontal area for plain fin.

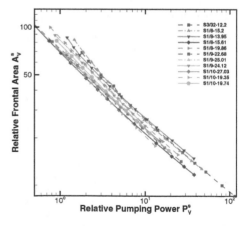

Figure 5. Relative pumping power vs Relative Frontal area for strip fin.

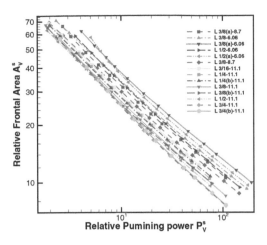

Figure 6. Relative pumping power vs Relative Frontal area for louvered fin.

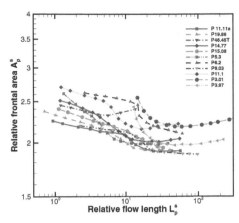

Figure 7. Relative frontal area vs Relative Flow length for plain fin.

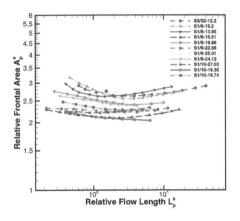

Figure 8. Relative frontal area vs Relative Flow length for strip fin.

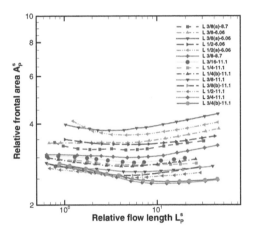

Figure 9. Relative frontal area vs Relative Flow length for louvered fin.

4 CONCLUSION

In the present work a methodology for selection of fin configuration is specified. The selection is based on various constrains. The methodology is found to be satisfactory tool for thermal design of compact heat exchanger.

REFERENCES

Cowell, T.A., 1990, A General Method for the Comparison of Compact Heat Transfer Surfaces, J. Heat Transfer, Vol 112(2), pp 288–294.

Kays W.M. and London, A.L.,1964, Compact heat exchanger, 3rd Edition.

Nunez M.P. Polley G.T., Reyes, E.T. Munoz, A.G., 1989 Surface selection and design of plate fin heat exchangers, App. Therm. Engg., Vol. 19, pp 917–931.

Shah R.K and Sekulic. D.P., 2003, Fundamentals of heat exchanger design. John Wiley and sons, Inc.

Application of fishbone diagram for modification of sheet metal tools: A case study

Kushal Pandya, Bimal Mawandiya, Bharat Modi & Shashikant Joshi
Department of Mechanical Engineering, Institute of Technology, Nirma University, Gujarat, India

ABSTRACT: In this case study, progressive tools used for manufacturing of the sheet metal components of Molded Case Circuit Breaker (MCCB) were studied. The progressive tools used are newly manufactured. By taking the trial of tooling one by one, problems related to tooling are identified. Corrective actions are taken to modify the tool and to eliminate all defects and assignable causes. Based on observations and by combining the causes of defects of all the tools, a detailed fishbone diagram is constructed. With the help of this study and modification of tooling, desired specifications are achieved. Sheet metal components produced by the rectified tool are then validated by calculating C_p & C_{pk} value for all the Critical to Quality (CTQ) dimensions as specified in drawing of component.

1   INTRODUCTION

A Molded Case Circuit Breaker (MCCB) is used to protect the electrical connections and electric circuits from damage caused by current overload and short circuit. In MCCB, operations like ON, OFF & Tripping is performed by the Tripping Mechanism, which consists of press components. Large numbers of Sheet metal tools are utilized to produce press components used in the Tripping Mechanism. For proper working of the tripping mechanism, press components must be as per the required specifications. Study of Sheet Metal Tooling is important, in order to achieve desired specifications.

1.1   *Press machine*

The press machine used in sheet metal forming operation is MCF 1000 (H Frame) 100T. This machine is only used for automatic blanking operation. It is a power press with 100 tonnage capacity. Its Frame is H type frame. Manufacturing line consists of a Decoiler, Straightener, Feeder and Press machine.

1.2   *Progressive die*

Progressive dies offer an effective way to convert raw material in the form of coil into a finished product with minimum material handling. When the material is fed from one station to another in the die, the material is progressively completed into finished part. The stock on progressive dies move from one station to another, with distinct operations being performed at every station. After the completion of each stroke, part material is fed by one progression. Usually, the material is perforated in initial stations, which are used by pilot for locating the strip (Vishwanath 2013).

2 LITERATURE REVIEW

The purpose of this literature review, is to develop a base of information in sheet metal tooling. An overview of the literature is presented for progressive tool and also for the problems faced during the working of tools

Podgornik et al. (2011) examined the probability of reducing lubrication and of replacing costly tungsten carbide material in blanking/piercing by introducing coatings in hard tool. Experimental results showed, that (AlCrN) coatings can be effectively used in blanking/piercing applications, on softer steel tools also. It reduces friction and wear, as well as lowers the costs of the tool.

Aurich et al. (2009) presented strategies for burr minimization. Simulation models of burr formation capable of indicating the interaction and dependencies of key process parameters. Specialized burr sensors are used for inspecting and detection of burr.

Nandedkar et al. experimental results showed, that for ratio of die radius to sheet thickness (R/t), minimum spring-back is observed up to a defined value of R/t ratio and it increases with increase in R/t ratio. It was also concluded that first the springback increases with increase in blank holder force, and then continuously decreases with increase in blank holder force.

3 METHODOLOGY

All Progressive tools are trialed on the press machine for problem identification. Before the trial, many factors are responsible for correct working of tool and they are listed in section 4. These factors have been studied and followed while running progressive tools. When tool is running, all the observations are noted down. Minor problems occurring during running of tool are solved simultaneously if possible. After completion of the trial, all the observations regarding tool, as well as the components are studied thoroughly. Based on the observations and detailed study of causes of defects, actions are taken. Factors affecting problems are identified and corrective actions are taken to rectify the tools. Rectifications of tool take considerable time as handling, disassembling and assembling of tool is difficult process. After rectification, tool is trialed again and checked for problems in running of tool. If the problems regarding running of tool or component dimensions still occur, then the tool needs to be rectified again as per observations in 2nd trial. If there is no problem in running of tool, then the component dimensions are checked as per design. If both, running of tool and the dimensions checked are satisfactory then more number of components are produced.

4 STANDARD OF PROCEDURE (SOP) FOR TOOL SETUP

- **Die placement** – The die should be set in a manner to facilitate the material feed parallel to the tool. Always place the center of the estimated tonnage in the center of the press. Die should always be in center along the length of bolster plate.
- **Die clamping** – Make sure that the clamps are in the correct position and they are not interfering with die parts. Make sure that the bolts and fasteners are fully tightened and secure.
- **Material properties and dimensions** – Before using coil, always check the hardness and grade as per requirement. Check coil width and thickness before entering in tool.
- **Pitch and progression** – The amount of feed pitch and progression should be set correctly in feeder. Overfeeding or underfeeding will result in a misfeed and can damage the die.
- **Feed line height** – Feed line height is the height at which the strip is fed into the die. Height of feeder should be at same height of die entry
- **Shut height** – Make sure that the shut height of the die is set correctly. Avoid hitting setup blocks. For setting shut height of new tool, move the ram of machine slowly to confirm the required shut height.
- **Coil Centering** – Coil should be in center, while placing in the straightener and feeder.

- **Stock condition** – Ensure that the coil is properly straightened and feeds smoothly through the die.
- **Die protection** – Ensure that all die protection equipments, such as misfeed sensors and proximity switches are installed and functioning properly.
- **Feed and pilot release** – Timing of the feed release should be exact, so that the pilots can easily locate the strip. Usually this is performed by moving the press down in inching mode and releasing the feed roller when the lead of the pilots has entered the pilot hole.
- **First hit condition** – Before taking the first stroke, make sure that the strip has entered in the die correctly. Strip should be passed through guide plate or guiding lifter in the die. Ensure that the strip is touched to the side clip stopper for correct pitch. Check the die thoroughly for any loose scrap present in die.
- **Lubrication** – Ensure that the coil is lubricated by Mineral turpentine oil (MTO) while running. Oil-mist unit should be in ON condition.

5 TOOL OBSERVATIONS AND ACTIONS TAKEN FOR RECTIFICATION

As explained in the methodology, tools were trialed on the press machine to find all defects in the tool. Same procedure was followed for all the tools. Some of the tools were even trialed for 2nd and 3rd time to produce correct components. Before the trial, all the items in SOP for tool setup should be performed. After trial run, the observations noted for all tools are listed below. As per observations and study of all the factors, actions taken are also listed below.

5.1 *Observations and the corresponding corrective actions taken are listed below in tabular form*

1. Observations identified during the trial and the corresponding corrective actions taken for rectification of tool 1 are listed as shown in Table 1. Component produced by progressive tool is moving contact.
2. Observations identified during the trial and the corresponding corrective actions taken for rectification of tool 2 are listed as shown in Table 2. Component produced by progressive tool is trip plate.
3. Observations identified during the trial and the corresponding corrective actions taken for rectification of tool 3 are listed as shown in Table 3. Component produced by progressive tool is latch link.

5.2 *Generic cause and effect diagram*

With the help of observations taken during working of all tools, detailed cause and effect diagram has been constructed. This cause and effect diagram shows all the possible causes

Table 1. Observations and corresponding corrective actions for tool no. 1.

Sr. No.	Observations	Remarks	Action taken
1	Components were not falling till 7 shots/stack	Angular relief given on the die is not effective.	Component Relief increased by 0.2 mm
2	Strip jammed.	Due to half piercing of one Hole.	Guide plate relieved by 0.4 mm after embossing station.
3	Strip jammed.	Strip bulging of 1 mm is taking place due to embossing and blanking collectively	Piercing slug relief changed from taper to step relief
4	7 Pilots broken	Due to strip jamming, misfeed occurred	Damaged pilots replaced and tool was ran with misfeed sensor

Table 2. Observations and corresponding corrective actions for tool no. 2.

Sr. No.	Observations	Remarks	Action taken
1	Misfeed sensor activates after 140 degree in every stroke.	Spring very sensitive	Misfeed pilot spring changed, hard spring used
2	Hole near bend line is elongated	Very near to bend line	Piercing punch diameter is increased by 0.05 mm
3	Micro slug at cut off station	–	New cut-off punch made
4	U-bend dimension deviated	Resizing station not effective	Shims of defined size added to resizing mechanism

Table 3. Observations and corresponding corrective actions for tool no. 3.

Sr. No.	Observations	Remarks	Action taken
1	No mis-feed sensor	No provision for mechanism of mis-feed sensor	Provision of mis-feed mechanism made
2	Piercing slug at first station coming up	Slug gets stuck to strip	New coil with correct hardness was used
3	U-bend dimension is deviated	Resizing in left side not performing	Re-striking punch changed from line contact to surface contact.
4	Cut-off punch broken	Due to high friction and heat	Punch was replaced by new PVD(AlCrN) coated punch

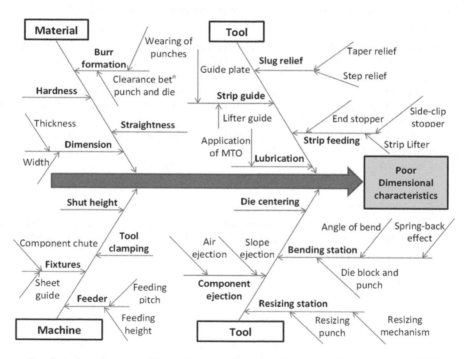

Figure 1. Genericcause and effect diagram of progressive tools for producing MCCB press component.

for defects and improper working of progressive tool used for manufacturing sheet metal components for MCCB. By referring this cause and effect diagram, problems related to progressive tool can be resolved easily. Detailed cause and effect diagram is shown in Figure 1.

6 VALIDATION OF RECTIFIED TOOLS

After rectification of the tool, it is trialed again to ensure that all the defects are eliminated. When it is ensured that all the defects are eliminated, some samples of components are checked for dimensional specification. If both, running of tool and dimensions checked are satisfactory, then more number of components are produced. Those components are then inspected for validation of rectified tool. Sample size of 30 numbers [Momtgomery 2009] are selected from the lot and inspected for CTQ (Critical to Quality) dimensions. The reading of CTQ dimensions for 30 samples are noted and C_p & C_{pk} value is calculated in Minitab Software. If the Cpk value for all CTQ dimensions comes above 1.33 (4 sigma) [Momtgomery 2009], then the components produced are under dimensional tolerance. Same procedure is followed for all the tools. Results of C_p & C_{pk} value are: Cpk value for hole with diameter 2.1 mm of tool 1 is **1.4**. C_{pk} value of U-bend dimension 11.2 mm of tool 2 is **1.41**. C_{pk} value of U-bend dimension 20 mm of tool 3 is **1.56**.

7 CONCLUSION

For solving problems related to sheet metal forming, detailed study of design, components and the working of progressive tool is essential. Factors affecting functioning of the tool, also plays important role in correct dimensional specifications. Fishbone diagram helps in identifying the significant causes of defects affecting the working of tool. Detailed Fishbone diagram is made by combining the defects occurred in all tools. 3 numbers of progressive tools were trialled and successfully modified to produce MCCB components with correct dimensional specifications. All the tools were validated by calculating C_p & C_{pk} value for CTQ (Critical To Quality) dimensions. C_{pk} value for CTQ dimensions are obtained above 1.33 (4 sigma) for all components. Same procedure can be followed to identify the possible defects and for the rectification of progressive tools used in the future to produce sheet metal components. By referring to the detailed fishbone diagram, significant defects of other progressive tool can be identified effortlessly.

REFERENCES

Aurich J.C., Dornfeld D., Arrazola P.J., Franke V., Leitz L., Min S., 2009. Burrs—Analysis, control and removal, *CIRP Annals—Manufacturing Technology 58, 519–542*.

Gawade S., & Nandedkar. V.M., Spring back in Sheet Metal Bending-A Review, *Journal of Mechanical and Civil Engineering (IOSR-JMCE) ISSN: 2278–1684, PP: 53–56*.

Podgornik B., Zajec B., Bay N., Vizintin J., 2011. Application of hard coatings for blanking and piercing tools, *Wear 270, 850–856*.

Momtgomery Douglas C., 2009. Introduction to statistical quality control, United States of America: John Wiley & Sons Inc.

Vishwanath M.C., Dr.Ramni, Sampath Kumar L, 2013. Design of progressive draw tool, *International Journal of Scientific and Research Publications, Volume 3, Issue 8, ISSN 2250-3153*.

Multi-disciplinary Sustainable Engineering: Current and Future Trends – Tekwani, Bhavsar & Modi (Eds)
© 2016 Taylor & Francis Group, London, ISBN 978-1-138-02845-6

A case study of productivity improvement of the grinding process in a rubber roller manufacturing industry

Darshan Maniyar
Lathia Industrial Supplies Company Pvt. Ltd., Naroda, Ahmedabad, Gujarat, India

Bimal Kumar Mawandiya, Bharatkumar Modi & Kaushik Patel
Department of Mechanical Engineering, Institute of Technology, Nirma University, Ahmedabad, Gujarat, India

ABSTRACT: In this paper, the grinding process of the rubber roller manufacturing is selected and the work study principles are applied for the productivity improvement. The activities of the worker are captured and arranged in the sequential order. The various flow process charts for man, machine, and materials are drawn, and time study is performed. Alternate grinding cycle is proposed and modified based on the results of the analysis of time study and process chart. With the modified process there is a productivity improvement of 50%.

Keywords: productivity improvement; rubber roller grinding; work study; method study; time study

1 INTRODUCTION

Productivity is one of the major factors that enables the manufacturing firm or organization to compete in the global market. Throughout the history of manufacturing, there has been a constant effort to improve the utilization of resources. Productivity studies aim at identifying how efficiently the resources in the system are used in producing desired production. In general terms, productivity can be defined as the ratio of output produced to input resources utilized in the production (Talib 2010).

To enhance the productivity, work study and time study are the most important principles (Barnes 1980). Time taken for process elements is one of the most important factors for improving productivity (Kanawaty 1992). Through time study, positive changes that occur in the organization can be observed (Pratiban).

An essential component for the textile and paper industries i.e., rubber roller that is shown in Figure 1, is the preliminary observations of the existing process of the rubber roller manufacturing that show ample scope for its improvement.

Figure 1. Rubber roller.

1.1 *Problem definition and research aim*

It is intended to carry out a research to enhance the productivity of rubber roller manufacturing industry through the elimination of non-value adding processes and/or through the implementation of enhanced manufacturing processes.

2 METHODOLOGY

1. Observe and describe the small process elements of each major process and its sub steps and perform time study i.e., time taken for each element.
2. Analyze the results of the time study in order to identify non-productive, excess time consuming, or unnecessary steps i.e., non-value adding and redundant activities.
3. Modify or eliminate the activities so as to reduce the process time and enhance the productivity.

3 GRINDING PROCESS

Grinding is a process of a rubber roller, in which material is removed and applied finishing on rubber covered on the outer periphery of the metal core using grinding wheel face on the center type cylindrical grinder machine shown in Figure 2.

3.1 *Process description*

1. Threaded rubber roller is brought on the machine by crane operator by a wrapped belt in the middle.
2. Head stock and tail stock are set to accommodate the roller between them as shown in Figure 2.
3. Roller center and machine bed center are aligned and roller is fixed between them accordingly.
4. Cloth wrapping is removed from roll and collected in the nearby tin.
5. Through-ness of roll and grinding wheel cutting length are adjusted through the stopper provided on the ways of the grinding machine.
6. The roller is ground with a grinding wheel in the face of grinding mode as shown in Figure 3. Rubber roller camber is set and the required depth of cut and feed are given.
7. As cut completes, powder is applied on both sides and the center and diameter is measured accordingly.
8. Hardness is verified at different ends.
9. Paper and cotton tape are wrapped on finished rubber roller by helper.
10. Roller is unloaded from the machine and kept at a designated area.
11. Machine is cleaned by the operator.

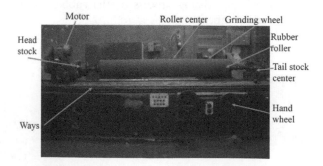

Figure 2. Roller held between head stock and tail stock on a grinding machine.

Figure 3. Face grinding of rubber roller on grinding machine.

266

Table 1. Process flow chart for man, machine, and material of a rubber roller grinding cycle.

Sr. No	Operator	Man Symbol						Machine Symbol						Material Symbol						Remarks
		○	⇨	D	□	▽	⊠	○	⇨	D	□	▽	⊠	○	⇨	D	□	▽	⊠	
1	In case of camber roll, goes behind the machine and provides camber by properly setting the knob provided.	●						●						●						
2	Turns on the Camber lever at the front of the machine.	●						●						●						
3	Brings in the grinding wheel and applies depth of cut by rotating the handwheel. At the same instant, provides feed to the machine table by moving lever in on position.	●						●								●					Feed in the grinding machine is hydraulically operated hence feed is automatic and varies with the material and size of roll and the depth of cut. The lathe bed moves horizontally as grinding wheel grinds the roll.
4	Cut running			●				●											●		
5	As cut completes, slows down machine by turning lever in off position.							●								●					Grinding cut is applied from left side of roll to right side of roll i.e. bed moves from right to left. The next cut continues from the place the previous cut ends.
6	Goes to fetch powder from machine table.			●				●						●							If powder gets used up, fetches new from near entrance. This process is done during the ongoing grinding cut.
7	Takes powder in paper and applies at both sides and center.	●								●									●		No powder is applied in case of ebonite rolls.
8	Goes to keep powder back at the machine table.			●						●						●					Stops the machine after application of powder.
9	Brings back the Micrometer for measuring diameter.			●						●						●					
10	Measures the diameter at sides and center.				●					●						●					If diameter is larger than required, applies another cut and follows previous grinding steps again.

The study of roller grinding process has been carried out by the application of method study and the process is divided into small process activities. The detailed process flow charts (man, machine, and material) consisting of a total of 132 activities have been drawn for the analysis purpose. However, due to the limitation of the number of pages for this research, only process flow charts (man, machine, and material) of the grinding cycle has been shown in Table 1.

4 ANALYSIS

The total activities for the grinding processes are 132 as summarized in Table 2. The activity of the grinding cycle is repeated eight to nine times with a depth of cut 0.05 mm each time to achieve the final dimension of the roller. Table 3 shows the various activities to be performed in one grinding cycle.

The total time to grind nine rollers for completing 132 activities is shown in Figure 4. The maximum and minimum times for grinding rubber rollers are 125 min and 57.23 min, respectively. Figure 4 shows the deviation in the process time as 66.77 min. Hence, the deviation in the process time needs to be minimized through improvement.

The time required for different activities like setup time, idle time, and working time are shown in Figure 5 through a pie-chart. The set-up time, ideal time, and the working times are 50.03 min (58%), 23.93 min (28%), and 11.82 min (14%), respectively. Hence, working and idle times of the machine need to be minimized through improvement.

From the Figure 6, it can be observed that the operation time for grinding varies from 52% to 71% of the total process time. Time study was performed for many rolls but the representative data for nine rollers are presented in the graphical form in Figure 6. Moreover, it has been observed that in the current practice, the number of grinding cycles is not based on the materials to be removed. In fact it is repeated four to eight times without any proper justification (Figure 7).

Table 2.	Summary of grinding process chart.					Table 3.	Activities for one cycle of grinding.			
Activity		Symbol	Man	Machine	Material	Activity		Symbol	Man	Machine	Material

Activity	Symbol	Man	Machine	Material
Operation	O	85	67	38
Inspection	D	12	0	12
Transportation	☐	11	30	3
Delay	⇨	21	42	68
Storage	▽	0	0	0
Operation and inspection	◉	3	0	11
Total number of activities for one grinding process		132	132	132

Activity	Symbol	Man	Machine	Material
Operation	O	9	6	3
Inspection	D	1	0	1
Transportation	☐	0	0	0
Delay	⇨	1	6	6
Storage	▽	0	0	0
Operation and inspection	◉	0	0	1
Total number of activities for one grinding cycle		11	11	11

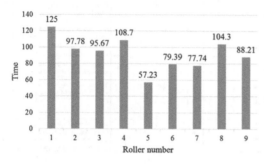

Figure 4.   Total time to grind the roller.

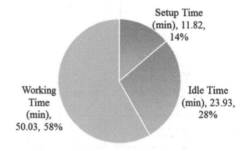

Figure 5.   Machine's average time analysis.

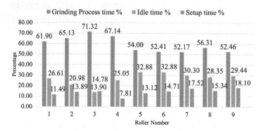

Figure 6.   Time analysis for specific roller.

Figure 7.   Materials removed and number of grinding cycles applied for specific roller.

5   RESULTS

5.1   *Implementation of grinding process*

It is decided that to reduce the grinding process time by reducing the number of cycles by increasing the depth of the cut. It has been proposed to finish grinding operation in three cycles namely rough cut, semi finish cut, and final cut. Rough cut makes the roller cylindrical where a depth of cut is selected up to 2 mm. Semi finish cut is taken as 1–2 mm as per the extra material left after the rough cut. Final cut brings the roller to the desired dimensions and 0.5–0.7 mm depth of the cut is taken. Several rollers over 10 days have been ground with the proposed grinding process. The quality of the rollers is found to be accepted. Hence,

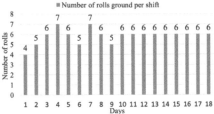

Figure 8. Roller grinding time span for first 10 days.

Figure 9. Grinding machine production of one shift for first 18 days.

the proposed process can be implemented. The proposed process reduces the grinding time. The grinding time for 60 rolls over the 10 days span has been shown in Figure 8.

Reasons for peak value for roller grinding time span for the first 10 days are presented in Figure 8 are as above:

a. Roller no. 8: Many cotton tape marks were there on the roll so cannot be decided to take the depth of the cut.
b. Roller no. 11: Camber was increased after third grinding cycle so one more grinding cycle was taken to achieve the required value of camber.
c. Roller no. 20 and 29: Had excessive rubber material, which was 10 mm more than the required size.

Grinding machine production of one shift for the first 18 days is shown in Figure 9. It can be observed from the figure that proposed process on the grinding machine increases the production gradually and becomes constant after 10 days. Therefore, the average number of rollers produced per shift is six, and the total number of activities is decreased to 77 from 132. While conducting experiments, no defects have been observed in rolls. Hence, it is proposed that an alternative grinding process is advantageous and the production per shift on grinding machine is improved by 50%.

6   CONCLUSION

Method study was carried out at the rubber roller grinding machine. The flow process chart (man, machine, and material) has been drawn and the rearrangement of activities are proposed. Time study was performed. Based on these, alternate grinding process had been proposed and the total grinding cycles were reduced from eight to nine grinding cycles to three grinding cycles i.e., rough cut, semi finish cut, and final cut. Production is increased from four rollers per shift to six rollers per shift. Considering 1.5 shift per day, improvement in the productivity is three rollers per day i.e., 50%. By implementing the improved alternates in the grinding process, it was estimated that the saving per day in rupees is 11,000, and the annual saving in rupees is around 30 lakhs per year.

REFERENCES

Abdul Talib Bon, Daiyanni Daim, April 1010, *Time Motion Study in Determination of Time Standard in Manpower Process, EnCon 2010 3rd Engineering Conference on Advancement in Mechanical and Manufacturing for Sustainable Environment, Malaysia.*
Barnes M. Ralph, 1980, *Motion and Time Study design and measurement of work, John Wiley, New York, USA.*
George Kanawaty, 1992, *Introduction to Work Study, International Labour Office, Geneva, fourth ed. pp. 17–132.*

Multi-disciplinary Sustainable Engineering: Current and Future Trends – Tekwani, Bhavsar & Modi (Eds)
© 2016 Taylor & Francis Group, London, ISBN 978-1-138-02845-6

Study on tensile and dry sliding wear behavior of epoxy based fiber reinforced composites

G.R. Arpitha
Malnad College of Engineering, Hassan, Karnataka, India
Visvesvaraya Technological University, Belagavi, Karnataka, India

B. Yogesha
Department of Mechanical Engineering, Malnad College of Engineering, Hassan, Karnataka, India

ABSTRACT: Composite materials have been the dominant emerging materials because of their low density, unlimited availability, good mechanical performance, and problem free disposal. Natural fibers offer a real alternative to the technical reinforcing fibers presently available. In this study, sisal/glass fiber reinforced epoxy composites are prepared and their tensile strength and dry sliding wear behavior of composites are evaluated. The results show that pure glass specimens have been at maximum tensile strength compared to pure sisal and other hybrid laminates. Taking hybrid specimen into account the specimen 60% sisal 40% glass shows maximum tensile strength than laminate 40% sisal 60% glass. The sisal fiber when hybridized with glass fiber provides a very good tensile strength. For dry slide wear, the frictional property of hybrid specimens is high with the highest weight loss indicating the average loss compared to dry slide wear of pure sisal and pure glass specimens which had lower weight loss.

1 INTRODUCTION

Materials are probably more deep rooted in our world than most of us realize, i.e. in transportation, housing, clothing, communication, food production, etc., virtually every segment of our lives are influenced by these materials daily. The development and advancement of societies are tied to the ability to produce and manipulate materials to fulfill their needs. The discipline of material science involves the investigation of relationship that exists between the structure and properties of the materials. The structure of the materials usually relates to the arrangement of its internal components. Solid materials have been classified into three basic classifications, viz., metals, ceramic, and polymers. This classification is primarily based on the chemical makeup and atomic structure, most materials fall into these categories. In addition to these three, other groups of important engineering materials are composite, semiconductor, and biomaterials. Many of our modern technologies require materials with unusual combinations of properties that cannot be met by the conventional metal alloys, ceramics, and polymeric materials. These can be combined in composite materials to produce unique characteristics. A composite material is defined as any substance which is made by physically combining two or more materials differing in composition or form to produce a multiphase material, which possesses superior properties that are not obtainable with any of the constituent materials acting alone. These constituents remain bonded together but retain their identity and properties. The composite material consists of matrix and reinforcement, matrix is a resin and reinforcement are a fiber the reinforcing material is usually stiffer, stronger than the matrix and there has to be a good adhesion between the components.

2 EXPERIMENTAL

2.1 *Materials*

In the present study, sisal fiber (Agave Sisalana) and glass fiber (woven mat form) are being investigated. The sisal fibers were obtained from Dharmapuri District, Tamilnadu, Chennai,

Table 1. Laminate description.

Laminate	Combinations Sisal (S) layers	%	Glass (G) layers	%	Series	Thickness (mm)	Laminate Weight (g)	Epoxy Weight (g)	Hardener Weight (g)
L1	4	100	–	–	SSSS	3.9	366	380	38
L2	–	–	5	100	GGGGG	3.8	305	200	20
L3	2	40	3	60	GSGSG	3.4	206	360	36
L4	4	65	2	35	SSGGSS	3.0	217	225	22.5
L5	3	60	2	40	SGSGS	3.2	239	225	22.5

India. The Glass-Fiber Reinforced Polymers (GFRPs) used for the fabrication is of a bidirectional mat having 360 gs M/s supplied by Suntech fiber Ltd. Bangalore. Commercially available epoxy (LY-556) and hardener (HY-951) was supplied by M/s zenith industrial suppliers, Bangalore.

2.2 Preparation of composite specimen

The fabrication of sisal/glass-epoxy hybrid composite materials is carried out through the hand lay-up technique. For the sample preparation, the first and foremost step is the arrangement of the tile mold which have smoothed and cleaned surfaces. Cover release sheet on an open mold. After that, the glass fiber and sisal fiber mats were cut into the dimension of 250×250. A measured amount of epoxy is taken for a different volume fraction of fiber and mixed with the hardener in the ratio of 10:1. The mixture was stirred properly for uniform mixing. Care had to be taken to avoid the formation of bubbles. A bottom tile piece was placed on the table and on top of that the Mylar sheet is placed. A well-mixed resin is poured on the Mylar sheet and it spreads all over the sheet. After that, a sisal mat is placed on the top of the resin. Again, a resin is poured on the top of the mat. A known amount of resin is poured every time. Care should be taken so that, the mats do not slide after placing. Again glass mat is placed on the top of the sisal mat and again resin is poured. The same procedure is repeated until required thickness of the laminate is obtained. After the fabrication, a released sheet is placed to cover matrix and care should be taken to avoid the formation of bubbles then another granite plate is placed on that to apply a load (40–50 Kg) on the matrix. The composite sheet takes 24 hours for curing in room temperature. After that, laminate is removed from the tile pieces and are then cut into the required dimensions depending on the standards. Five types of specimens are fabricated by adding different layers of sisal and glass fibers in different compositions. Table 1 shows different laminates with different combinations.

3 RESULT AND DISCUSSION

3.1 Tensile properties

Tests were carried out according to ASTM standards, for tensile test the sample was cut into the required dimension. The test was conducted on polymer testing UTM and the obtained results are tabulated below. The results show the tensile stress of sisal/glass and its hybrid composites. The laminate L4 which has 65% of sisal fiber and 35% of glass fiber has tensile stress of 297.97 N/ mm² which is less compared to laminates L2, L3, and L5 which has 100% glass fiber and 60% sisal fiber, 40% glass fiber and 40% sisal, 60% glass fiber respectively. The specimen L2 has maximum tensile stress i.e. 462 N/ mm² and L3 has a tensile stress of 309.4 N/mm² and L5 has a tensile stress of 330 N/mm². The specimen L1 is made up of 100% sisal fiber which has least tensile stress i.e. 174.2 N/mm². The specimen L1 has least ultimate tensile strength i.e. 174.254 N/mm² whereas specimen L2 has maximum UTS i.e. 462.002 but among hybrids L3 has least UTS and L5 have maximum UTS which is shown in the Table 5.1.1. The tensile strain on specimen L1, L2, L3, L4, and L5 is 0.0614, 0.0801, 0.07829, 0.07124, 0.07114 respectively. The specimen L2 has a maximum strain and specimen L1 has minimum strain. The tensile yield strength of L1, L2, L3, L4, and L5 are 147.54 N/mm², 259.26 N/mm², 211.7 N/mm², 208.18 N/mm² and 198.21 N/mm² respectively. The peak load of L1, L2, L3, L4, and L5 are 6327.007 N, 26334.003 N, 17635.398 N, 16984.181 N, 16875.424 N respectively. The tensile modulus of L1, L2, L3, L4, and L5 are 1639.12 N/mm², 5761.32 N/mm², 3951.88 N/mm², 208.18 N/mm², 4004.21 N/mm². Table 2 shows tensile test specimens results and Fig.2 shows Stress vs. Strain graphs of different laminates.

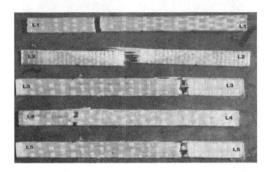

Figure 1.   Tensile testing specimens after testing.

Table 2.   Tensile test results.

Laminates	Stress (N/mm²)	Strain	Modulus (N/mm²)	Yield strength (N/mm²)	Peak load (N)	UTS (N/mm²)
L1	174.2	0.0614	1639.12	147.54	6327.007	174.254
L2	462	0.08019	5761.32	259.26	26334.003	462.002
L3	309.4	0.07829	3951.88	211.7	17635.398	309.398
L4	297.97	0.07124	208.18	208.18	16984.181	297.988
L5	330	0.07114	4004.21	198.21	16875.424	330.532

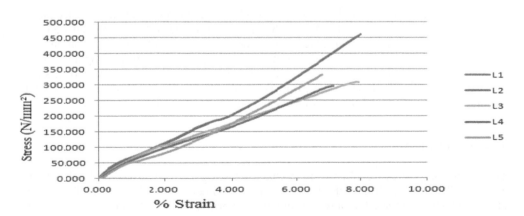

Figure 2.   Stress vs. Strain graphs of different laminates (Tensile test).

Table 3. Test parameters for varying load test.

Load	4 kg at 300 rpm
Test duration	30 minutes, Mass measured after every test
Wear track dia	80 mm
Test condition	Lubricant condition & at Ambient temperature.
Environment	Open to atmosphere, Humidity = 51.5% Rh, temperature = ambient room temperature.
Test was done by	Same operators complete testing for all specimens on the same test rig.

Figure 3. Dry sliding wear test specimens.

Table 4. Dry sliding wear test result.

Test No	Sample	Initial weight (g)	Final weight (g)	Weight loss (g)
1	Sisal	4.82048	4.81552	0.00496
		4.81049	4.80568	0.00481
		4.79220	4.78721	0.00499
2	Glass	4.53845	4.53558	0.00287
		4.54915	4.54629	0.00286
		4.53915	4.5364	0.00275
3	Combination (sisal/glass)	4.57726	4.57176	0.0055
		4.56627	4.56087	0.0054
		4.58670	4.5813	0.0054

3.2 *Dry sliding behavior of epoxy based fiber reinforced composites*

The sample 1 contains of sisal and its weight 4.82048 g initially, after the test its weight 4.81552 g with net weight loss of 0.00496 g. Sample 2 contains glass fiber of weight 4.53845 g will lose its component of weight 0.00287 g after the test. Similarly after the test, combination of sisal/glass (each of 50%) specimen of weight 4.57726 g after the test its weight 4.57176 g specimen lose its weight by 0.0055 g. From the results, combination of sisal/glass fiber specimen can wear most whereas pure glass fiber would wear least. Table 3 shows test parameters for varying test loads and Table 4 shows dry sliding wear test results of samples.

4 CONCLUSION

The hybrid composites are subjected to mechanical testing such as tensile and also dry sliding wear test. The test result indicates that pure glass specimen has the highest tensile strength of 462.002 Mpa compared to pure sisal and other hybrid laminates. Taking hybrid specimen into concentration the specimen L5 (60% sisal 40% glass) shows maximum tensile strength of 330.532 Mpa than laminate L3 (40% sisal 60% glass) which had more % of glass fiber yielded 309.398 Mpa. Thus it is concluded that sisal when hybridized with glass provides better tensile strength. Testing the specimens for dry slide wear, the frictional property of hybrid specimens is high with the highest weight loss indicating the average loss of .00543 g compared to dry slide wear of pure sisal and pure glass specimens which had lower weight loss.

REFERENCES

Geethamma VG, Thomas Mathew K, Lakshminarayanan R, Sabu Thomas, 1998, *Composite of short coir fibers and natural rubber Effect of chemical modification, loading and orientation of fiber*, Polymer; 6:1483–90.

Goulart S.A.S., Oliveira T.A., Teixeira A., Mileo P.C., Mulinari D.R.,2011, *Mechanical behaviour of polypropylene reinforced palm fibers composite*s, Procedia Engineering; 10:2034–2039.

Gowthami, A., K. Ramanaiah, A.V. Ratna Prasad, K. Hema Chandra Reddy, K. Mohana Rao, 2012. *Effect of silica on thermal and mechanical properties if sisal fiber reinforced polystercomposites,* Journal of material environment science, vol.4 (2), pp.199–204.

Hemalata Jena, Mihir Ku, Pandit, and Arun Ku, 2012, *Study the impact property of laminated Bamboo-Fiber composite filled with cenosphere*, International journal of environmental science and development,vol.3(5).pp.456–459.

Husseinsyah, S. and M. Mostapha, Zakaria, 2011, *the effect of filler content on properties of coconut shell filled polyster composites*, Malaysian polymer journal, vol.6. pp.87–9.

Joshi S V, Drzal L T, Mohanty A K, Arora S, 2004, *Are natural fiber composites environmentally superior to glass fiber-reinforced composites*, Compos Part A; 35:371–6.

Kai Yang, Mingyuan Gu, 2010. *Enhanced thermal conductivity of epoxy nanocomposites filled with hybrid filler system of triethylenetetramine-functionalized multi-walled carbon nanotube/silane-modifiednano-sized silicon carbide*. Composites: Part A 41 215–221.

Leandro José da Silva, Túlio Hallak Panzera, Vânia Regina Velloso, André Luis Christoforo, Fabrizio Scarpa, 2012, *Hybrid polymeric composites reinforced with sisal fibres and silica microparticles*. Composites: Part B 43 3436–3444.

Ramesha, M., K. Palanikumar, K. Hemachandra Reddy, 2013, *Comparative Evaluation on Properties of Hybrid Glass Fiber- Sisal/Jute Reinforced Epoxy Composites*. Procedia Engineering 51 745–750.

Ranganna, H., N. Karthikeyan, V. Nikhilmurthy, S. Raj Kumar, 2012, *Mechanical & Thermal Properties of Epoxy Based Hybrid Composites Reinforced with Sisal/Glass Fibres*. ISSN 2277–7156.

Sandhyarani Biswas, Alok Satapathy, Amar Patnaik *Effect of Ceramic Fillers on mechanical properties of Bamboo fiber Reinforced epoxy composites*, pp.1–6.

Se Young Kim, In Sub Han, Sang Kuk Woo, Kee Sung Lee, Do Kyung Kim, 2013, *Wear-mechanical properties of filler-added liquid silicon infiltration C/C–SiC composites*. Materials and Design 44 107–113.

Silva R.V. *Composito de resinapoliuretanaderivada de oleo de mamona e fibrasvegetais.* Tese (Doutorado)- Escola de, Engenharia de São Carlos, Universidade de São Paulo, Sao Carlos, 2003, p. 139.

Velmurugan, R., V. Manikandan, 2007, *Mechanical properties of palmyra/glass fiber hybrid composites*. Composites: Part A 38 2216–222.

Venkateshwaran, N., A. ElayaPerumal, A. Alavudeen, M. Thiruchitrambalam, 2011. *Mechanical and water absorption behaviour of banana/sisal reinforced hybrid composites 32 4017–4021.

Multi-disciplinary Sustainable Engineering: Current and Future Trends – Tekwani, Bhavsar & Modi (Eds)
© 2016 Taylor & Francis Group, London, ISBN 978-1-138-02845-6

Study and development of engineering standard for press working of stainless steel components used in switchgear

Ravikumar Jethi & Bharat Modi
Department of Mechanical Engineering, Institute of Technology, Nirma University, Ahmedabad, Gujarat, India

Kamlesh Chaturvedi
LCM, Larsen and Toubro Limited—Electrical and Automation, Vadodara, Gujarat, India

ABSTRACT: Nowadays, sheet metal forming is widely used in manufacturing processes in many companies because of its easy and mass production method. In switchgear (circuit breaker), many sheet metal components are used. They are produced by forming operations in mechanical or hydraulic press machines. Due to incorrect forming methods or process parameters, problems such as heavy burr, spring-back, wrong dimension, and deformation in components arise, which lead to part rejection. Due to these problems, there is waste of time and resources. The cause and effect analysis was carried out to tackle the problem of resetting and de-latching of the MCCB, and is presented in this paper. The component bracket was found to be dimensionally and surface finish wise inaccurate. This, in turn, would cause the functional problems in MCCB. One additional shaving operation was suggested to make the bracket dimensionally accurate. Triangular media were proposed for deburring operation to remove the burrs. The modifications adopted helped reduce rejection of parts and make the MCCB functionally correct.

Keywords: sheet metal forming; burr; spring-back

1 INTRODUCTION

A circuit breaker is a manually or automatically operated electrical switch designed to protect an electrical circuit from damage caused by an overload or short circuit. Its basic function is to detect a fault condition and to interrupt low current. It can be manually opened and closed, as well as automatically opened to protect the conductor or equipment from damage caused by excessive temperatures from over-current in the event of short circuit. In the latter event, the important consideration is the breaking and making time, which is customarily designed for few milliseconds to avoid a prolonged arcing and pre-arcing time that overheats and thus melts the moving and fixed contacts. The breaker is operated by a spring-operated mechanism that supplies the required mechanical energy to it. The drive is the heart of the circuit breaker. If it fails, the breaker cannot perform its intended function, and even if there is a delay in providing energy within few milliseconds, the breaker fails to perform its intended function. That is why the drive has to be precise to give adequate energy at all times through its life cycle.

1.1 *Need for standardization*

L&T EBG, Vadodara is a leading manufacturer of the MCCB and the ACB. It covers almost 63% marketing of the MCCB in India. For the MCCB, many sheet metal components are used, which are made in the press shop of L&T EBG. They make this component in thousands of numbers; however, due to some problem, the components are rejected due to damage, burr problem, and spring back effect. This paper is focused on developing the standardized procedure for the production of the component of the switchgear, namely the bracket.

2 STANDARDIZATION OF PROCEDURE FOR COMPONENTS USED IN THE SWITCHGEAR

There are many components used in the switchgear assembly. The switchgear may not function properly if the components used in the assembly are not dimensionally correct as per the design. Few switchgears have been found to be functionally faulty, and few components have been found to be dimensionally incorrect. Therefore, its causes were studied and remedies were suggested to rectify the same. The procedure for components, such as bracket, used in the switchgear was standardized and is presented in the following section.

2.1 Process standardization for the bracket

Bracket (CM 72008), as shown in Figure 1, is a component used in DN250 MCCB. The material used in this component is AISI TYPE GRADE 304 stainless steel of 1 mm thickness. This component is assembled with the drive shaft of the MCCB. Its mechanical properties are listed in Table 1.

2.1.1 Problems found in DN250 MCCB
MCCB DN250 encountered the problem of resetting and de-latching. The assembly of the drive shaft is shown in Figure 2. The probable causes were studied and are shown as the cause and effect diagram in Figure 3.

2.1.2 Dimensional and quality requirements for the bracket

- Center of the shaft to contact edge dimension, as shown in Figure 4, should be 10.02 to 10.10 mm;
- Surface roughness (Ra) at the contact edge should be less than 0.8 μm;
- Edge should be burr free; and
- Taper of the edge should be ≤ 0.2 mm.

Figure 1.   Bracket (CM 72008).

Table 1.   Mechanical properties of AISI 304 stainless steel.

Grade	304
Tensile strength (MPa)	520
Compression strength (MPa)	210
Hardness Rockwell B	90

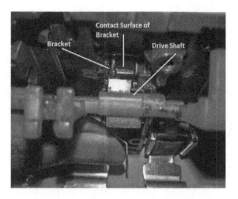

Figure 2.   Assembly of the drive shaft.

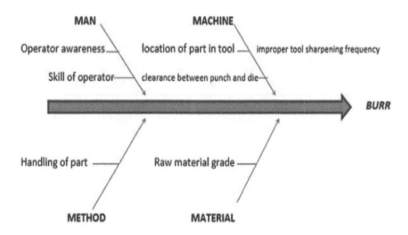

Figure 3.   Cause and effect diagram for burr.

2.1.3   *Manufacturing process for the bracket*

This component is produced in the progressive die. The first two operations, namely bending and shaving, are performed at the vendor's place. After the bending and shaving operation, the second shaving operation is performed at the company. The following steps are adopted to produce it within the prescribed dimensions:

1. Metal forming
2. Shaving operation
3. Deburring.

2.1.4   *Quality of the bracket*

2.1.4.1   *Dimensional accuracy of the bracket with the existing tool setup*

More than thousand components were produced, and the shaft center to edge distance was measured for every component. The typical measurement for a few components is presented in Table 2. The results are also presented as a process control chart. It was observed that the dimensions (X) are out of the accepted range and the process was also consistently out of control. Moreover, it indicates that if the component input dimension in the shaving operation is fixed by 0.1 mm, the final dimension would be under a specific dimension.

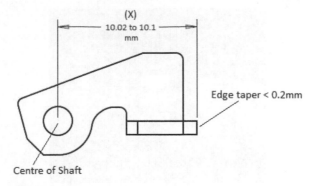

Figure 4. 2D view of the bracket.

Table 2. Data measured before setting the standardized ranges.

Component		
Sr. No	Before shaving	After shaving
1	10.27	10.11
2	10.32	10.15
3	10.25	10.1
4	10.33	10.14
5	10.3	10.12
6	10.27	10.12
7	10.32	10.15
8	10.26	10.09
9	10.31	10.13
10	10.36	10.16
11	10.28	10.11
12	10.34	10.15
13	10.25	10.1
14	10.32	10.15
15	10.27	10.13

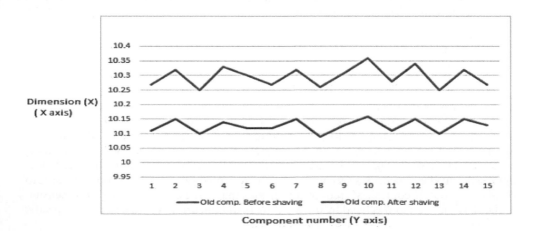

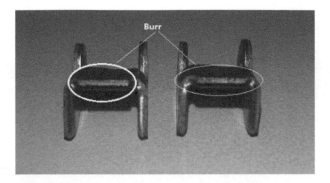

Figure 5.   Burrs on the bracket.

Figure 6.   Surface roughness of the bracket.

2.1.4.2   *Burrs on the edge*
The components were found with excessive burrs, as shown in Figure 5. The burrs are not acceptable as it causes the problem of de-latching.

2.1.4.3   *Surface finish at the edge*
Surface roughness was measured for each component and was found to be more than 0.8 μm. The surface appearance of the bracket is shown in Figure 6.

3   SUGGESTIONS FOR QUALITY IMPROVEMENT

3.1   *Including an extra shaving operation*

- The shaving operation produces a straight, square edge, generally to approximately 75% of the metal thickness. Two shaves make a better, straight edge (to approximately 90% of the metal thickness) than does a single-shaving operation [ASM Handbook 2006].
- According to the standardized procedure, we can obtain better results if we adopt a two time-shaving operation. So, the second shaving operation in the process was included.

3.2   *Setting the standardized allowance range before shaving*

- As per the standardized procedure, the allowance left for shaving should exceed the die clearance by approximately 0.15 mm [Tekaslan 2008].

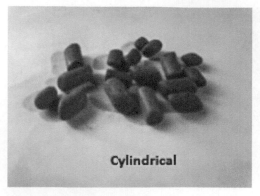

Figure 7. Deburring media.

- So, by adding 0.15 to the lower and the upper limit, the lower range would become 10.02+0.15 = 10.17 mm and the upper range 10.1+0.15 = 10.25 mm.
- Thus, we set the range of the input shaving material as 10.17 to 10.25 mm.

3.3 *Change deburring media from cylindrical to triangular*

- Triangular deburring media, as shown in Figure 7, provide shaving edges. Thus, a better deburring is expected.

3.4 *Process standardization*

Based on the above suggestions, the procedure is standardized as shown in the block diagram.

4 RESULTS AND DISCUSSION

4.1 *Dimensional accuracy of the bracket*

The components were produced as per the procedure explained in the standardized procedure. The components were checked for dimensional accuracy and quality. The results are summarized in Table 3.

The results are also presented as a process control chart. It was observed that the dimensions (X) were in the accepted range and the process was also consistent in control. The dimension of the bracket was in the range of 10.02 to 10.1 mm.

After applying the different standards on the component (CM 72008), we obtained the new result that satisfied the need of the component. The range of dimension of 10.02 to 10.10 mm was achieved by adding the second shaving operation. It was also useful to set the clearance between die and punch. Moreover, for burr-related problems, we changed the

Table 3.   Data after standardization.

Component		
Sr. No	Before shaving	After shaving
1	10.17	10.08
2	10.2	10.07
3	10.19	10.09
4	10.17	10.09
5	10.18	10.07
6	10.16	10.09
7	10.19	10.07
8	10.18	10.08
9	10.17	10.06
10	10.17	10.09
11	10.16	10.07
12	10.2	10.09
13	10.18	10.07
14	10.2	10.08
15	10.19	10.07

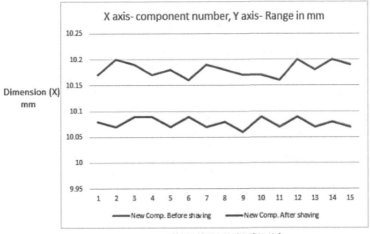

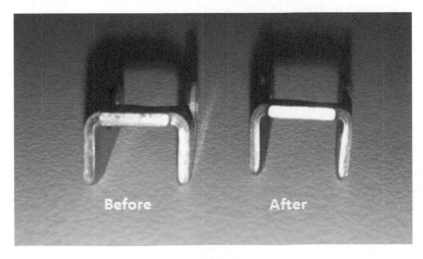

Figure 8.   Before and after image of the components.

media type from cylindrical to triangular shape so that the burr was removed from the edges and a good surface finish was achieved. Figure 8 shows the difference between the old and the new component before and after applying the standardized process.

5 CONCLUSION

- A cause and effect study was performed for the functionally incorrect MCCB.
- The bracket was found to be dimensionally inaccurate, causing the problem of de-latching and resetting in the MCCB.
- An additional shaving operation was proposed to make the bracket dimensionally accurate.
- From the process chart, it could be inferred that the new process was capable of producing the bracket as per specifications.
- Triangular deburring media were proposed to eliminate burrs, which were found to be promising.

REFERENCES

ASM Handbook, 2006, Volume 14B, Metalworking: Sheet Forming, ASM International.
Aurich, J.C., Dornfeld, D., Arrazola, P.J., Franke, V., Leitz, L. & Min, S. 2009, Burrs—Analysis, control and removal, CIRP Annals - Manufacturing Technology 58, 519–542.
Bing, G.J.A. & Wallbank, J. 2008, The effect of using a sprung stripper in sheet metal cutting, Journal of Materials Processing Technology 200,176–184.
Camacho, A.M., Torralvo, A.I., Bernal, C. & Sevilla, L. 2013, Investigations on Friction Factors in Metal Forming of Industrial Alloys, Procedia Engineering 63 (2013) 564–572.
Jochen Breitling, Bernd Pfeiffer, Tazlan Altan, Klaus Siegert, 1997, Process control in Blanking, Journal of Materials Processing Technology 71, 187–192.
Masahiro Sasadaa, Hitoshi Kobayashib, Isamu Aokia, 2006, Study on piercing mechanism of small holes, Journal of Materials Processing Technology 177, 649–652.
Ozgur Tekaslan, Nedim Gerger, Ulvi Seker, 2008, Determination of spring-back of stainless steel sheet metal in "V" bending dies, Materials and Design 29, 10431050 Book, Prakash Joshi, Press Tool Design.

Design and analysis of machine and mechanism

Error avoidance through design analysis of 5 axis vertical machining centre-components

Anjali N. Dave & P.A. Patel
Marwadi Education Foundation, Faculty of Technology, Rajkot, Gujarat, India

J.P. Mehta
Mech Power CNC, Rajkot, Gujarat, India

S.S. Khandare
BDCE, Sevagram, Wardha, Maharashtra, India

ABSTRACT: Metal Cutting Industry has done many researches to get products with high accuracy and low cost. To get accuracy of the product, errors should be eliminated or reduced. Many ways have been found for precise performance of machine tools like error avoidance, error compensation etc. This paper deals with the design and analysis of 5 axis Vertical Machining Centre column and head, in accordance to various relevant standards. Validation of structure is done through FEM analysis in stress, displacement and mode shape using creo parametric 2.0 simulation module of Pro/Engineer.

1 INTRODUCTION

The machine error analysis literature is vast and error definitions may vary considerably. In general if the produced part has a dimension other than the required dimension, it is considered as an error. The error may be, in a broader sense either a Machine error or a machining error. The errors include Geometric errors, joint kinematic errors, volumetric errors and thermal errors. The general approach to apply error avoidance is to build an accurate machine during its design and manufacturing stage, so that the error sources could be kept to a minimum extent (Dave, Mehta & khandare 2012, Dave, Mehta & Khandare 2013). Finite Element Analysis is used by machine designers to optimize structural elements for weight reduction, stress distribution and to limit deformation structural elements. Here, the Design analysis of the two main components i.e. column and head are done. This is to check, whether the components are designed for safe working condition, which further can reduce the error occurrence in part production.(Jian-feng 2008).

2 ANALYSIS OF STRUCTURE (MACHINE TOOL DESIGN HANDBOOK 2004, MEHTA 2006)

While analyzing the structure, to be on the safer side, most adverse conditions should be taken care of. So during considerations for drilling and milling operations, one should consider conditions, which generates maximum cutting force during operation.

2.1 *Cutting force calculations: Milling*

Cutting speed = $n = V_c * 1000 / Dc$; $S_m = S_z * Z * n$
Metal removal rate = $Q = b * t * s_m / 1000$
Power at the spindle = $N = U * K_h * K_{\lambda\_} * Q$
Tangential cutting force = $P = 6120 * N / V_c$

2.2 Cutting force calculations: Drilling

In the case of drilling, predominant effect is that of the thrust generated during operation (Subrahmanyam 2014).

Cutting speed = N = V*1000/πD

Power at spindle = N = 1.25*D2 *K* n* (0.056 + 1.5*S)/105

Thrust force = T_h = 1.16*K*D*(100*S) 0.85

Torque at spindle = T_s = 975* N / n

2.3 Finite element analysis of column and head

2.4 Load calculations and constraints for column

The cutting forces are not acting directly on the column, but get transmitted through turret, head, Z-slide & motor. To calculate its exact moment, a point on column is taken

Net load acting on column = W = $W_1+W_2+W_3+W_4+W_5$

Moment about X-axis = W l_y = $W_1 l_{1y} + W_2 l_{2y} + W_3 l_{3y} + W_4 l_{4y}$

The column will behave as a cantilever beam. The bottom face of a column is mounted on the base. So, moment about X-axis will try to bend column in Y-Z plane.

Moment about Y-axis = W l_x = $W_1 l_{1x} + W_2 l_{2x} + W_3 l_{3x +} W_4 l_{4x}$

This moment about Y-axes will try to bend column in X-Z plane.

Table 1. Common design strategy for structures.

Software used	Creo Parametric 2.0
Assumptions	Only Static loading Self weight of structure is ignored
Material to be used	Cast Iron ρ = Density 7.85*10–9 tonne/mm³ E = young's Modulus = 1.7 *105 MPa μ = Poisson's Ratio = 0.25
Types of element used	3D solid elements
Solution type	Standard design study, static analysis

Table 2. Analysis report for column.

Particulars	Details
Loads	1. Cutting force (Drilling thrust) 2. Weight of ATC 3. Weight of head, 4. Weight of ball screw 5. Weight of accumulator assembly
Constraints	Column is fixed on base, so face from where, column is assembled to base is taken fixed.

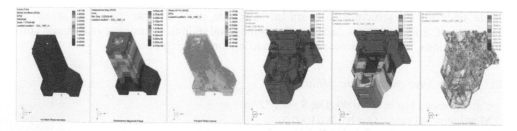

Figure 1. Stress and displacement-column and head.

Table 3.   FEA result for column before and after optimization.

	Analyzed results	
Parameter	Original	After optimization
Displacement	3.03*10–5 mm	3.27*10–5 mm
Maximum stress	4.87 N/mm^2	5.15 N/mm^2

Table 4.   Analysis report for head.

Particulars	Details
Loads	1. Dead weight of spindle
	2. Dead weight of encoder
	3. Dead weight of tool ejector assembly
	4. Dead weight of spindle motor assembly
	5. Thrust force during drilling
Constraints	Headstock is fixed on column, through LM blocks, so back face from where, Headstock is assembled to column is taken fixed

Table 5.   FEA result for head before and after optimization.

	Analyzed results	
Parameter	Original	After optimization
Displacement	0.04829 mm	0.06256 mm
Stress	8.349 N/mm^2	11.916 N/mm^2

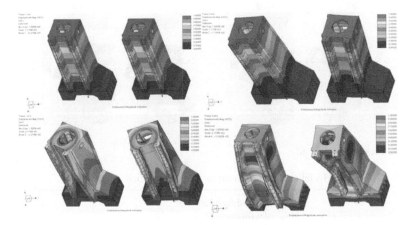

Figure 2.   Mode shapes of column.

2.6   *Remarks*

By reducing the dimensions of rib structure, the weight of structure can be reduced. Similar analysis is carried out on the optimized structure with same load and constraints. Optimized result is also within safe range. Maximum displacement value found in column is 0.003 μm and 0.06 μm. It is evident from the Graph that total deflection of the guide way at the top of column is not more than safe value in μm per meter in Y direction, which is well within the specified limits of tolerable displacement. Maximum von-mises stress value is 5.15 N/mm² and 11.92 N/mm², which is also satisfactory. Studying stress distribution result, it can be realized that except some localized stress concentration, stress distribution is uniform.

2.7 *Modal analysis of column*

Figure below shows various modes of vibration of the column. Here, only four modes and combination modes are observed. One can do analysis with more modes also.

As we know natural frequency of vibration happens without any external force. For the column the equation of vibration is (Singh 2004).

$$f = (1/2\pi) (K/m)^{1/2}$$

Table 6. Simulated modal frequency of column.

Mode No.	Natural frequency of vibration	
	Original	After optimization
Mode 1	82.7 Hz	83.6 Hz
Mode 2	115.6 Hz	114.66 Hz
Mode 3	227.4 Hz	221.26 Hz
Mode 4	319.6 Hz	289.59 Hz

2.8 *Modal analysis of head*

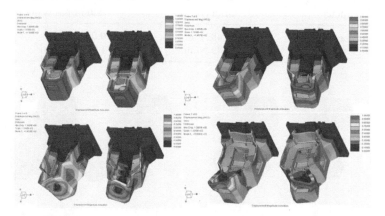

Figure 3. Mode shapes of head.

Table 7. Modal analysis-different modes of vibration.

Mode No.	Natural frequency of vibration(Hz)	
	Original	After optimization
Mode 1	150.662	137.68
Mode 2	190.754	183.44
Mode 3	345.379	328.25
Mode 4	580.089	573.53

2.9 Remarks of modal analysis

The result of basic mode shape of vibration for column and head is shown here. It is practically and comparatively very high than machine frequency range, which shows that column is stiff enough. The reason behind it is, properly and particularly spaced and arranged vertical and horizontal stiffeners. Horizontal stiffeners provide stability against torsion, while vertical stiffeners provide stability against bending.

The table reveals that the displacement and stress is changed and it is within the safe limit. Frequency has decreased in some mode of vibration.

3 ASSUMPTIONS

3.1 Assumptions for milling

Diameter of work piece (D_c): 125 mm;
Material to be cut: Medium carbon steel;
Cutting speed (V_C): 90 m/min;
Feed per minute (S_z): 0.2;
Depth of cut: 0.5 mm;
Tool nose radius: 0.5 mm;
Cutting Speed m/min. (V_c) = 90 m/min.

3.2 Assumptions for drilling

Diameter of drill (D): 32 mm;
Material to be cut: Medium Carbon Steel;
Cutting speed (v): 20 m/min;
Feed per revolution (s): 0.25 mm/rev;
K_h, K_r and U are constant factors for flank wear, rake angle and unit power (in Kw/cm^3/min) respectively.

4 NOMENCLATIURE

f = frequency of vibration, Hz
k = stiffness of link, kg/m
m = mass of structure, kg

5 CONCLUSION

The above analysis shows, that the static and dynamic analysis shows displacement, stress and modal frequency. Here, it is within the safe limit. After the optimization of rib size, the values are in safe limit. So the material can be saved and weight is reduced by 8%, which in turn reduces the material cost. The analysis can be extended to all main components like spindle, base, saddle, etc for various materials. Further dynamic, thermal load deflection, stress analysis can be done. The software analysis will save the cost and time. Further checking during assembly reduces the kinematic error.

ACKNOWLEDGEMENT

We thank Mac Power CNC Machines LTD, Rajkot, for providing knowledge in support of this research.

REFERENCES

Chandrupatla, Tirupati R., Belegundu, Ashok D. 2004. Introduction to finite elements in engineering.

CMTI, Machine tool design handbook. Tata Macgraw Hill, Edition.

Dave, Anjali N., Mehta, Mr. J.P., khandare, Dr. S.S. 2012. Geometric, Error Analysis & Measurement for Slant Bed CNC Turning Center, International conference of 'Benchmarks in Engineering Science and Technology', session–4.

Dave, Anjali N., Mehta, Mr. J.P., khandare, Dr. S.S. 2013. Accuracy Enhancement of 3 Axis Vertical milling Machine Centre, International and 16th National Conference on Machines and Mechanisms, IIT Roorkee, India,

Jian-feng Ma Yong-bin Yang Ling Zhao, Wu-yi Chen 2008. Structural bionic design and experimental verification of a machine tool column. School of mechanical engineering and Automation, Beihang University.

Kushnir, E., Hardinge Inc., Elmira 2005. Effect of machine tool structure dynamic on machine cutting performances, Proceedings of IMECE 2005 ASME, International mechanical engineering congress and exposition orlando, fl, usa november 5–11.

Macpower CNC automation catalogue.

Mehta, N.K. 2006. Machine tool design and numerical control. Tata Macgraw Hill, Edition.

Singh, V.P., 2004. Theory of Machines, Dhanpatrai Pub., Delhi, Edition.

Subrahmanyam, B.V., Rao, S., Krishna, V. Gopala, K., Rama, 2014. Static and Dynamic Analysis of Machine Tool Structures, IJRMET Vol. 4, Issue Spl – 1.

Multi-disciplinary Sustainable Engineering: Current and Future Trends – Tekwani, Bhavsar & Modi (Eds)
© 2016 Taylor & Francis Group, London, ISBN 978-1-138-02845-6

Investigation for multiple distributed defects in ball bearing using vibration signature analysis

Sham Kulkarni & S.B. Wadkar
Sinhgad College of Engineering, Vadgaon, Pune, Maharashtra, India

ABSTRACT: Vibration signature analysis of single row deep groove ball bearing with multiple surface roughness defects has been studied in this paper. As the localized bearing fault grows, becoming a distributed one generates a random signal with non-stationary components. In the present paper the effect of the surface roughness on the vibrational response of the outer race of ball bearing is analyzed. The effect of roughness, size, speed, and load on vibrational response has been investigated. It has been observed that the frequencies expected in the frequency spectrum of multiple defects are the same as the frequencies present in the frequency spectrum of a single roughness defect. However, peak amplitude of the vibration may vary because of the phase shift resulted from the time delay of the impact. Vibration signals are presented in the time and frequency domain. Also, the vibration spectrum produced by single and multiple roughness defects under pure radial load on the outer race of the bearing are compared. The experimental results are compared with ball pass frequency of the outer race. Frequency response obtained from experimental results is found to be identical with theoretical ball pass frequency of outer race.

Keywords: ball bearing; multiple distributed defect; surface roughness; vibration analysis

1   INTRODUCTION

In large area of machinery, deep groove ball bearings are essential parts of the rotating machinery that carry the load effectively. The distributed defects (such as surface roughness, waviness & off-size ball, etc.) appearing on the mating elements of the bearings are generated due to fatigue. A machine could be seriously risky if its faults occur in the bearings during service, which may incur safety problems and economical losses. Their movement & dynamics contributes to the overall vibration in a machine. Radially loaded rolling element bearings generate vibration even if they are geometrically perfect. This is because of the use of a finite number of rolling elements to carry the load. The vibrations spectra generated due to presence of distributed defects have been identified by various researchers in the time and frequency domain. Most of researchers have proposed experimental studies on vibration generation in bearing with a single local defect (Tondon 1999, Patil 2008). However, few studies have two local defect vibration analyses of rolling element bearings (McFaden 1985, Patel 2010). The multiple defects considering masses of rolling elements, shaft and housing have been proposed by Patel (Patel 2010). The vibrations generated by deep groove ball bearing having multiple local defects have been analyzed where time delay between two impacts have been derived and validated with experimental results (Patel 2014). Even when local defect grows, it becomes a distributed one, generating more complex signals with strong non-stationary contents. Time and frequency domain methods are used for monitoring the health of the bearings; however, correlation with the prediction of the amplitude of the spectral components with the extent of defect is necessary for its diagnostic purpose (Elia 2013).

The objective of this paper is to provide a better understanding of the vibration response generated due to the presence of multiple distributed defects on outer race of bearing performing an experimental investigation. The frequency spectra of the experimental results for a bearing having a single roughness defect and two roughness defects on outer race have been compared. Fourier transformation properties of the experimental signals have also been discussed in depth in this paper. The studies were carried out by several researchers related to vibrational analysis of local defects in bearing. Literature survey also indicates vibration studies considering single defect on races. There is scope to study the vibrational analysis of the multiple distributed defects on the bearing. They have worked on the bearing fault detection; however, there is the need to correlate the amplitude of spectral components to the extent of the defect. It is necessary to indicate here that multiple defects mean more than one defects on the same race.

1.1 Characteristic defect frequencies

The vibration generated by bearing appears at a different frequency range. Periodically occurring transient pulses are produced at frequencies determined by bearing geometry and speed. The frequencies of transient pulses depend on the characteristic frequencies of the bearing. The vibrations of the ball bearings with defect can be detected at various characteristic defect frequencies. For roughness fault, it is no longer impulsive, but rather has a randomly distributed phase, since the rolling elements are on different positions on the rough surface with discrete multiple points for every revolution. Theoretical defect frequencies are calculated by using the following formulae.

Cage fc is given by

$$fc = \frac{fs}{2}\left[1 - \frac{d}{D}\cos\alpha\right] \qquad\qquad (a)$$

Inner race defect frequency is given by $fi = z \times [fs - fc]$

$$fi = \frac{z \times fs}{2}\left[1 + \frac{d}{D}\cos\alpha\right] \qquad\qquad (b)$$

Outer race defect frequency is given by; $fo = z \times fc$

$$fo = \frac{z \times fs}{2}\left[1 - \frac{d}{D}\cos\alpha\right] \qquad\qquad (c)$$

Ball defect frequency is given by; $fbd = 2 \times fb$

$$fbd = \frac{D \times fs}{2 \times d}\left[1 - \frac{d^2}{D^2}\cos^2\alpha\right] \qquad\qquad (d)$$

Where d = rolling element diameter; D = pitch diameter & z = numbers of the balls.

The current paper illustrates how the rough surface of the distributed defects can be generated and incorporated in the bearing function, reviews and discusses the nature of the bearing signals from the perspective of cyclostationarity.

1.2 Bearing defect

The artificial defects were introduced separately on the outer race of the single row deep groove ball bearing using electric discharge machining. Experiments were performed on separate test bearings having variable defect sizes and defect depth constants. Dynamic analysis was carried out for different speeds and varying radial load.

Initially, the measurements of the vibration amplitude of healthy bearings were carried out for reference. Later, the experiments were performed on defective test bearings having a roughness of 4.57 micron (6 amps current). Figure 1 shows bearings with single and multiple

Table 1. Theoretical calculation of the frequencies at different speeds.

Sr. No.	Speed in RPM	Rotational Frequency (Hz)				
		fs	fc	fo	fbd	fi
1	600	10	3.48	24.37	29.89	45.63
2	900	15	5.22	36.55	44.83	68.44
3	1200	20	6.96	48.74	59.77	91.25

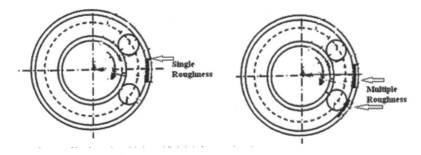

Figure 1. Single and multiple artificial defects on bearing outer race.

Table 2. Bearing properties.

Property	Value
Bearing outside diameter, Do	85 mm
Bearing bore diameter, di	30 mm
Bearing width, B	23 mm
Ball diameter, d	17.463 mm
Number of balls, z	7
Maximum dynamic load capacity	46339 N
Maximum static load capacity	25186 N
Pitch Diameter D	57.5 mm

defects on outer race having artificial roughness of 6 amp current. For the same distributed defect, amplitude of vibrations was confirmed after measuring repeatedly for 3 trials for an approximate time period of 15 sec.

The effort had been carried out in measuring the amplitudes of vibrations on outer race defected bearings having two defects by changing the position of the defect in the load zone located at 15–15, 30–30, and 45–45 degree. In the entire experimental work healthy support bearings were used. Signals were acquired in the frequency domain. Experimental work has been carried out by keeping outer race stationary. Signals were acquired in the frequency domain and interpreted in the proper format.

2 EXPERIMENTATION

2.1 *Experimental set up*

The setup consists of a shaft supported on healthy bearings and driven by a variable speed motor. The test bearing (single row deep groove ball bearing) was placed between two healthy bearings on the shaft. The drive to test rig was provided with a DC motor through coupling as shown in Figure 2. Different speeds were obtained using the control panel of the DC motor.

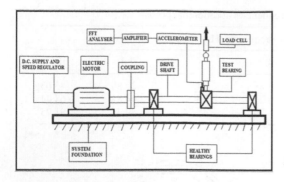 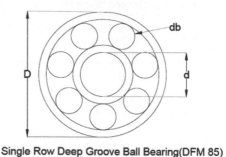

Single Row Deep Groove Ball Bearing(DFM 85)

Figure 2.   Experimental setup and geometry of ball bearing.

A tensile loading arrangement was used to apply radial load on the bearing. A piezo-electric accelerometer with a magnetic sensitivity of 10 mV per-m/s² was used to measure the vibrations. It was mounted closer to the housing of the test bearing. The accelerometer was connected to the charge amplifier (FFT Analyser), the output of which was connected to a computer. The relevant hardware and the software required to acquire the data was stored and displayed the time domain signal that were installed in the computer used for this work.

3   RESULTS AND DISCUSSION

For the comparison of the frequency spectra of the two surface roughness defects, where the second defect was created at 15 degree, 30 degree, and 45 degree, and frequency magnitude was measured. The magnitude of Fourier transform has also been obtained. The deformation magnitude increases when the defect and ball contact occurs in the loaded zone, and it decreases in the non-loaded zone.

3.1   *Comparison of the vibrational response of the healthy bearing with the defective bearings having multiple defects on the outer race*

Figures 3–6 shows the vibrational response [Acceleration (m/s2) Vs. Frequency (Hz)] of the healthy bearing and 6 amp defect bearing having 15–15, 30–30 and 45–45 degree defect locations respectively. All the responses were taken at identical conditions i.e. 10 kg load and 900 rpm speed. From these four graphs, it is very easy to understand the trend of the change of the vibrational response with change in defect locations in case of the bearing having two defects.

Peak amplitude of vibrations were observed at BPFO (from equation (c)) to be less than one in case of the healthy bearing as compared to those vibrations obtained from 15–15, 30–30 and 45–45 degree defect locations of 6 amp defect bearing.

The impulses are obtained after every 0.02897seconds which is corresponding to the reciprocal of the outer ring defect frequency (fo) 34.52 Hz at 900 rpm as shown in Figure 4(b). The strong peak at 34.52 Hz (fo) is clearly shown in Figures 3–6. Some harmonic peaks were also observed at fo+fs, fo+2fs, fo–fs, 2fo, 3fo etc., where fs is the shaft rotation frequency 15 Hz. With the increase in the angle of defect, amplitude of the vibration goes on increasing.

3.2   *Comparison of vibration response of the defective bearing having multiple 6 amp defect; at different loads, speeds and defect locations*

From the Figures 7–9 it was observed that for multiple defects of 6 amps at various locations:

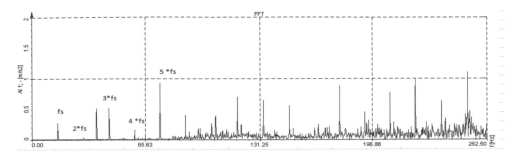

Figure 3. Vibration response of the healthy bearing.

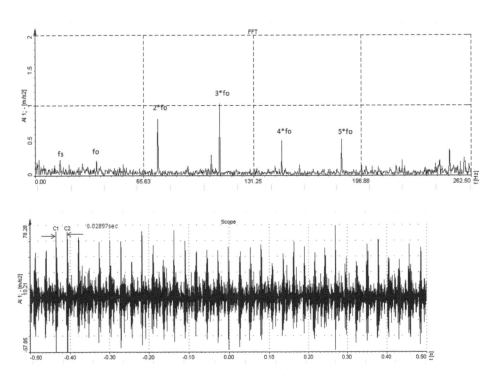

Figure 4. Vibration response of the bearing with 6 amp defect (a) frequency (b) time at 15–15 degree.

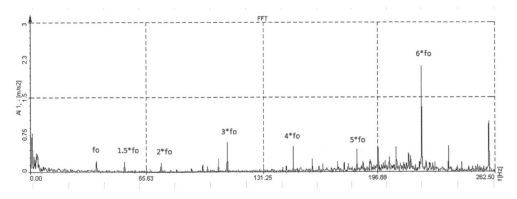

Figure 5. Vibration response of bearing with 6 amp defect at 30–30 degree location.

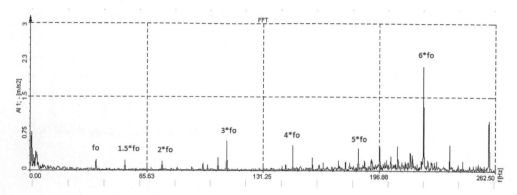

Figure 6. Vibration response of bearing with 6 amp defect at 45–45 degree location.

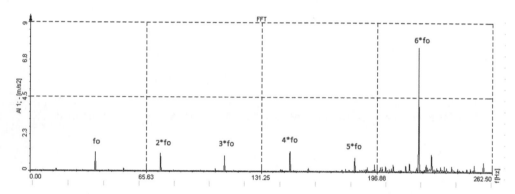

Figure 7. Variation of the peak acceleration value with the defect location for 600 rpm.

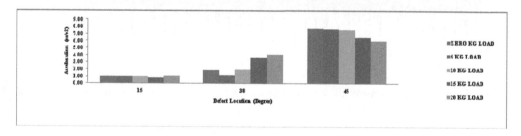

Figure 8. Variation of the peak acceleration value with the defect location for 900 rpm.

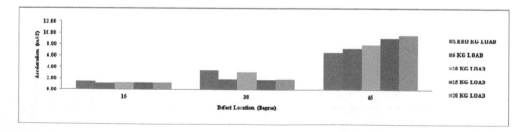

Figure 9. Variation of peak acceleration value with defect location for 1200 rpm.

a. Effect of Load: -At 15–15 degree defect location, as load increases from 0 kg to 20 kg, amplitude of the vibration increases for all speeds. At 30–30 and 45–45 degree defect locations as load increases from 0 kg to 20 kg, the amplitude of vibration increases for all speeds except at 1200 rpm.
b. Effect of Defect Location: As the defect angle was increased for all speeds, the vibration amplitude increased from 15–15 to 45–45 degree multiple defect location.
c. Effect of Speed: As the speed was increased from 300 to 1500 rpm, vibration amplitude of multiple defects went on increasing for all the defect locations.

3.3 *Comparison of the vibrational response of the defective bearing having multiple 6 amp defects; at different speeds, at 10 and 20 kg loads and 45 degree defect location*

From Figures 10a and 10b, it was observed that for 10 and 20 kg loads as the speed increased from 300 to 1500 rpm, there was an increase in peak amplitudes of vibration, respectively. Same trend was observed for 0 kg, 5 kg and 10 kg loads.

3.4 *Comparison of the vibrational response of the defective bearing having single and multiple 6 amp defect; at different loads, speeds, and 45 degree defect location*

From the above graphs, it was clearly observed that as the number of defects was increased from one to two, there was a sudden increase in vibration amplitude. There were a few exceptions as seen in Figure 10 (b) at 900 rpm (15 and 20 kg load), where vibration amplitude was reduced. These were examined in case of multiple defects because the ball remains in contact with the defect surface for more time which may reduce the amplitude of the vibration.

From the results and discussion, it is clear that the frequencies expected in the frequency spectrum of the two defects are the same as the frequencies present in the frequency spectrum of the single roughness. However, the amplitude of the vibration may increase or decrease because of the phase shift resulted from the time delay of the impulse.

4 CONCLUSION

1. The roughness of the contact surfaces influences the excitation process, which is therefore, most accurately represented by frequency domain models using bearing element geometry of the contact.

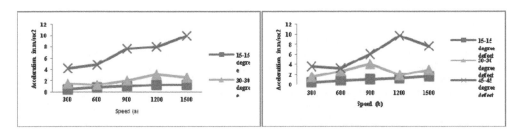

Figure 10. Variation of peak acceleration value with (a) 10 kg load (b) 20 kg load.

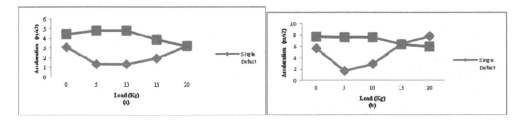

Figure 11. Variation of the peak acceleration value with load for (a) 600 rpm (b) 900 rpm.

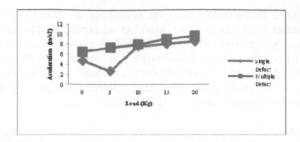

Figure 12.   Variation of peak acceleration value with load for 1200 rpm.

2. For the outer race roughness, the spectrum had components at outer race defect frequency and its harmonics. Other orders of roughness generate sidebands at multiples of cage frequency about these peaks.
3. In case of multiple outer race defect peak amplitude of the vibration for various parameters such as roughness, size, and the position of the roughness defect were observed at the characteristic roughness defect frequency. It was also emphasized that the amplitude levels of the defect frequency are unpredictable with the increase in the load at the test bearing.
4. With the increase in speed, amplitude of vibration in both single and multiple defects, roughness increases.
5. The Experimental model predicts the amplitudes of the spectral components due to multiple outer race roughness to be higher as compared to those that are due to single outer race roughness.

REFERENCES

Abhay, U. 2013. Vibration signature analysis of defective deep groove ball bearings using numerical and experimental approach. International Journal of Scientific & Engineering Research, volume 4, Issue 6.
Sunnersjo, C.S. 1985. Rolling element Vibrations-The effects of geometrical imperfections and wear" Journal of sound and vibration. 98(4), 455–474.
Koulocheris, D. 2014. Wear and multiple fault diagnosis on rolling bearings using vibration signal analysis. International Journal of Engineering Science Invention ISSN 2319–6734, volume 3, Issue 4, PP 11–19.
Elia, G.D. 2013. Combined blind separation and cyclostationary techniques for monitoring distributed wear in gear box rolling bearings. Engineering department in Ferrara, University of Ferrara Italy.
Harris. 2001. Rolling bearing analysis, fourth edition, A Wiley—publication.
Igarashi, T., Kato, J. 1985. Studies on the vibration and sound of defective rolling bearings third report: vibration of ball bearing with multiple defects. JSME Bulletin, p. 28492–499.
McFadden, P.D. Smith. J.D., 1985. The vibration produced by multiple point defects in a rolling element bearing". Journal of Sound and Vibration. P 98. 263–273.4.
Patil, M.S, Mathew, J., Rajendrakumar, P.K. 2008. Bearing signature analysis as a medium for fault detection: a review. ASME Transaction on Tribology. 32 p. 130014001–014007.
Patel, V.N, Tandon, N., Pandey, R.K. 2010. A dynamic model for vibration studies of deep groove ball bearings considering single and multiple defects in races. ASME Transaction on Tribology. 98 p.132041101–1-041101–10.
Sadettin, O., Nizami, A., 2006. Vibration Monitoring for defect diagnosis of Rolling Element Bearings as a predictive maintenance tool: Compressive case studies. Elseveir Publication NDT & E International, PP 293–298.
Rao, S.S. Mechanical Vibration, Fourth Edition Pearson Publication.
Tandon, N., Choudhury, A. 1999. A review of vibration and acoustic measurement methods for the detection of defects in rolling element bearings. Tribology International. 32 p. 469–480.
Patel, V.N., 2014. Vibrations Generated by Rolling Element Bearings having Multiple Local Defects on Races, 2nd International Conference on Innovations in Automation and Mechatronics Engineering, ICIAME, Elsevier publication Procedia Technology 14, pp. 312–319.

Multi-disciplinary Sustainable Engineering: Current and Future Trends – Tekwani, Bhavsar & Modi (Eds)
© *2016 Taylor & Francis Group, London, ISBN 978-1-138-02845-6*

Stress analysis of channel shell to nozzle junction of high pressure feedwater heater

Vatsal N. Desai, Darshita J. Shah & Dhaval B. Shah
Department of Mechanical Engineering, Institute of Technology, Nirma University, Ahmedabad, Gujarat, India

ABSTRACT: A feedwater heater is a heat exchanger designed to preheat boiler feedwater by means of condensing steam extracted from a steam turbine. High pressure feedwater heater is employed just before the boiler so that, the water fed to the boiler is preheated and at higher pressure. This increases the boiler efficiency as thermal shock is reduced. The aim of the present work is to analyse the junction of Channel shell, which contains the feedwater to nozzle junction. The nozzles are considered as the feedwater inlet and outlet. The analysis methodology opted is according to ASME Section VIII Div.2.

Keywords: stress analysis; pressure feedwater heater; channel shell; nozzle junction

NOMENCLATURE

P	Design Pressure
R	Inside Radius of Channel shell
E	Joint Efficiency
S	Allowable Stress Value of material
tr	Required thickness of Shell
t	Nominal thickness of Shell
CA	Corrosion Allowance

1   INTRODUCTION

All modern large steam power plants use a process of regenerative feedwater heating to increase the overall cycle efficiency of plant and to minimize induced thermal stresses in the boiler. In large power plants, feedwater heating generally takes place in multiple stages. Feedwater heaters are mostly shell and tube type heat exchangers.

Nozzles serve as entryway and exit way for shell side and tube side fluid. The size and orientation of the nozzles depend on the amount of fluid flow and flow velocity required.

ASME Section VIII Div.1 has rules for design of Shell, Dished end etc. i.e. pressure parts. But the only loading considered is pressure. But when the nozzle is connected with a flange, which is in turn bolted or welded to some other part, then there will be forces and moments induced. Such type of case is not addressed in Div.1.

Hence, for the cases where there is external loading except pressure and the possibility of failure is high then ASME Section VIII Div. 2 can be referred, which provides guidelines for analysis of such parts where external loading is applicable and chances of failure are high.

2   MECHANICAL DESIGN OF CHANNEL SHELL

2.1   *Design inputs*

P   =   18.34   MPa
R   =   724   mm
S   =   138   MPa
E   =   1.0   –
CA   =   3.2   mm

2.2   *Channel shell thickness calculation as per UG 27*

2.2.1   *For circumferential stress*

$$t_r = \frac{P * R}{S * E - 0.6 * P} + CA = 55.98 \text{ mm}$$

2.2.2   *For longitudinal stress*

$$t_r = \frac{P * R}{2 * S * E + 0.4 * P} + CA = 25.51 \text{ mm}$$

2.2.3   *Minimum thickness according to UG-16(b)(4)*

$$t_r = 2.5 + CA = 5.7 \text{ mm}$$

The maximum value of thickness required is 55.98. Hence the plate of 58 mm size is chosen.

$$t = 58 \text{ mm}$$

3   ANALYSIS METHODOLOGY

For analysis of Channel Shell to nozzle junction, according to ASME Section VIII Div.2, procedure followed is:

Step 1   Elastic Stress Analysis of the Assembly is carried out. The material to be considered should be linearly elastic.

Step 2   From the point at which maximum stress is generated, a Stress Classification Line (SCL) normal to the point is drawn.

Step 3   Stresses are linearized along the SCL and stresses for different combinations such as Membrane, Membrane plus Bending and Peak are obtained.

Step 4   For each case of stress there are well defined limits in the Code. Hence, the values obtained are verified against the limits.

4   STATIC STRUCTURAL ANALYSIS

4.1   *Inputs for finite element analyses*

A.   Geometry:
    Shell I.D.: 724 mm Shell
    Thickness: 58 mm
    Length of Shell: 980 mm
    Nozzle I.D.: 80 mm
    Nozzle thickness: 60 mm
    Nozzle Height: 200 mm

B. Material Data:
   The material properties for structural steel as given in Table 1.

C. Mechanical Boundary Conditions:
   1. Internal Pressure: 18.34 MPa.
   2. Fixed at both thickness faces.
   3. Nozzle loads on both nozzles are same and stated in the Table 2.

4.2 *Finite element model*

Adhering to the given geometry inputs, a CAD model was prepared and imported in ANSYS Work-bench. Figure 1 shows the CAD model, while Figure 2 shows the meshed model. For meshing, hexahedron elements were used. Figure 3 shows the loading and boundary conditions applied on the model.

Table 1.   Material properties of structural steel.

Property	Value	Unit
Density	7850	Kg/m³
Young's modulus	2.00E + 5	MPa
Poisson's ratio	0.3	–
Bulk modulus	1.67E + 5	MPa
Shear modulus	7.69E + 5	MPa

Table 2.   Nozzle loads.

Nozzle	Fx	Fy	Fz	Mx	My	Mz
N1, N2	7340	5500	7340	4400	2930	3820

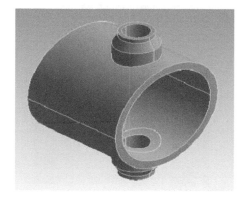

Figure 1.   CAD model.

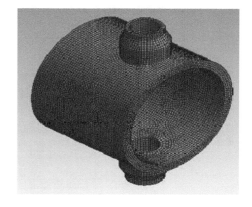

Figure 2.   Meshed model.

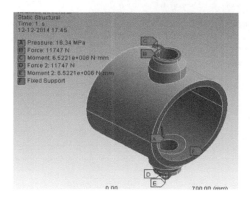

Figure 3.   Loading conditions.

303

4.3 *Equivalent stress obtained*

Figure 4 shows the stress distribution obtained after various mechanical loadings. The maximum value of stress 350.96 MPa, is generated at the inner junction of shell and nozzle.

5   LINEARIZATION OF STRESS

Figure 5 shows the cross-sectional view of the model such that the point of maximum stress is visible. Figure 6 shows Stress Classification Line (SCL) is drawn normal to the point of maximum stress. Figure 7 shows the linearized distribution of stress along the SCL.

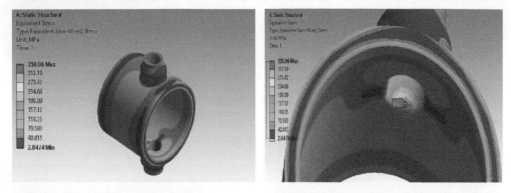

Figure 4.   Equivalent stress distribution at inner junction of shell and nozzle.

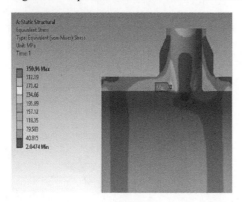

Figure 5.   Sectional view.   Figure 6.   SCL.

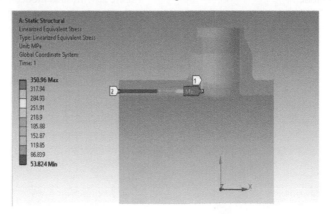

Figure 7.   Linearized stress distribution.

6   RESULTS AND DISCUSSION

From the results of analysis, it can be observed that the maximum stress occurs at the junction of Pressure Vessel and the nozzle. High stress concentration is developed at this location due to abrupt change in the geometry and the consequent change in stress flow. Table 3 compares the values of maximum stress obtained for different load cases with their respective allowable values. From Table 4, it is evident that although the value of stress at junction is high, the component is still safe as the stress values are below allowable limits.

Table 3.   Values of stress along different points along the SCL.

0	Length [mm]	Membrane [MPa]	Bending [MPa]	Membrane+Bending [MPa]	Peak [MPa]	Total [MPa]
1	0	97.72	113.4	203.44	154.95	350.96
2	7.8638	97.72	108.68	198.86	74.925	269.15
3	15.728	97.72	103.95	194.3	41.368	232.23
4	23.591	97.72	99.227	189.74	20.656	203.76
5	31.455	97.72	94.502	185.19	18.041	182.17
6	39.319	97.72	89.776	180.65	19.594	171.91
7	47.182	97.72	85.051	176.12	23.406	161.61
8	55.046	97.72	80.326	171.59	27.667	152.17
9	62.91	97.72	75.601	167.08	27.441	146.78
10	70.774	97.72	70.876	162.59	26.756	141.61
11	78.638	97.72	66.151	158.1	26.119	136.62
12	86.501	97.72	61.426	153.63	24.837	132.54
13	94.365	97.72	56.701	149.18	23.304	128.91
14	102.23	97.72	51.976	144.74	21.336	125.91
15	110.09	97.72	47.251	140.32	19.303	123
16	117.96	97.72	42.526	135.92	17.421	120.07
17	125.82	97.72	37.801	131.55	15.659	117.34
18	133.68	97.72	33.076	127.19	14.345	114.45
19	141.55	97.72	28.35	122.87	13.289	111.7
20	149.41	97.72	23.625	118.58	12.378	109.03
21	157.28	97.72	18.9	114.32	11.878	106.37
22	165.14	97.72	14.175	110.1	11.875	103.69
23	173	97.72	9.4502	105.92	12.366	100.74
24	180.87	97.72	4.7251	101.8	13.187	97.82
25	188.73	97.72	1.67E-14	97.72	14.14	94.69
26	196.59	97.72	4.7251	93.706	14.925	91.26
27	204.46	97.72	9.4502	89.761	15.619	88.084
28	212.32	97.72	14.175	85.894	16.315	84.934
29	220.19	97.72	18.9	82.118	16.839	81.524
30	228.05	97.72	23.625	78.445	17.196	77.842

Table 4. Comparison of stresses.

Linearized stress	Stress value (in MPa)	Allowable limit (in MPa)
Membrane	97.72	138
Bending	113.44	171.6
Membrane + Bending	203	207
Total	350	414

REFERENCES

ASME Boiler and Pressure Vessel Code. 2013. Section VIII—Division I. Pressure Vessels, UG, UHX.

Bednar, H.H., 1990. Pressure vessel design handbook. Krieger Publishing Company.

Brownell, L.E. and Young, E.H., 1959. Process equipment design: vessel design. John Wiley & Sons.

Chen, J.S., Lu, J.J. and Yu, M.H. 2014. The effect of analysis model on the stress intensity calculation for the nozzle attached to pressure vessel under internal pressure loading. *International Journal of Pressure Vessels and Piping*, pp. 117–118.

Heat Exchange Institute. 2006. Standards of Closed Feedwater Heaters.

Kuppan, T. 2000. Heat exchanger design handbook, volume 126. Marcel Dekker New York.

Moss, D.R. 2004. Pressure vessel design manual. Elsevier.

Reinhardt, W.D. 2001. Yield criteria for the elastic-plastic design of tubesheets with triangular penetration pattern. *Journal of pressure vessel technology*, pp. 118–123.

Shah, R.K. and Sekulic, D.P. 2003. Fundamentals of heat exchanger design. John Wiley & Sons.

Soler, A.I. and Singh, K.P. 1987. An elastic-plastic analysis of the integral tubesheet in U-tube heat exchangers towards an ASME code oriented approach. *International journal of pressure vessels and piping*, 27(5):367–384.

Sriharsha, K., Venkata, R.M. and Mallikarjun, M.V. 2012. Strength analysis of tube to tube sheet joint in shell and tube heat exchanger. *International Journal of Science, Engineering and Technology Research*, 1(4): pp-43.

Shaque, M.A. and Khan. 2010. Stress distributions in a horizontal pressure vessel and the saddle supports. *International Journal of Pressure Vessels and Piping*, 87.

Tooth, A.S., John, S.T., Cheung, Heong, W.N., Lin, S.O. and Nadarajah, C. 1998. An alternative way to support horizontal pressure vessels subject to thermal loading. *International journal of pressure vessels and piping*, 75(8):617–623.

Tubular Exchanger Manufacturers Association. TEMA, 2009. Standards of Tubular Exchanger Manufacturers Association.

Ukadgaonker, V.G., Kale, P.A., Agnihotri, N.A. and Shanmuga R. 1996. Review of analysis of tube sheets. *International journal of pressure vessels and piping*, 67(3): 279–297.

Multi-disciplinary Sustainable Engineering: Current and Future Trends – Tekwani, Bhavsar & Modi (Eds)
© 2016 Taylor & Francis Group, London, ISBN 978-1-138-02845-6

Design and FE analysis of components of bi-drum boiler

Ayaz Girach, Shashikant J. Joshi & Dhaval B. Shah
Department of Mechanical Engineering, Institute of Technology, Nirma University, Ahmedabad, Gujarat, India

ABSTRACT: In early days, the Sterling boilers contained four drums, one at the bottom and three at the top, connected by bank tubes. As the boiler technology advanced in capacity and pressure, they switched over to bi-drum and single drum boiler. The present day bi-drum boilers are designed for high pressure, flow, and temperature. Bi-drum boilers are used both for power generation and process steam generators. An attempt is made to design **Bi-Drum** boiler based on *Indian Boiler Regulation (IBR)*. In this paper, stress analysis of the main pressure parts are also covered to ensure that stress value at any point is not exceeding its allowable limit.

Keywords: Bi-drum boiler; steam drum; mud drum; dished head; Indian boiler regulation

1 INTRODUCTION

Boilers have been used for over a century and a half in many countries of Europe and America and also in Japan, South Africa, Russia, and Australia. In the recent years, with rapid industrialization of the developing world, vast power-generating capacities have been added, particularly in countries such as China and India (Rayaprolu, 2009).

However, according to the Indian Boiler Act, 1923, a boiler is a closed pressure vessel with capacity exceeding 2235 liters used for generating steam under pressure. It includes all the mountings fitted to such vessels which remain wholly or partly under pressure when steam is shut off (Chattopadhyay, 2001).

The basic design steps in boiler design:

1. Calculate boiler heat duty
2. Calculate burner heat duty and fuel firing rate
3. Fix the furnace geometry
4. Calculate furnace volume required
5. Calculate required furnace exit gas temperature
6. Calculate required effective projected radiant surface
7. Calculate convection bank surface
8. Select steam drum size
9. Workout boiler mountings and fittings

Carry out stress analysis calculation for the configured boiler.

2 PRELIMANARY DESIGN DATA

Preliminary design data and specified materials are given by customer as per their requirement.

1. Working Pressure 11.25 kg/cm^2
2. Evaporation Capacity 12000 kg/hr

From the above data Maximum Metal Working Temperature (MMWT) is calculated as per IBR regulation 338/271. It states that saturation temperature at the corresponding temperature plus 28°C.

- Saturation temperature corresponding to 11.25 Kg/cm^2 = 188°C.
- Maximum Metal Working Temperature at 11.25 Kg/cm^2 = 188 + 28 = 216°C

This temperature is used for calculating allowable stress for material while designing.

3 DESIGN OF STEAM DRUM

Initial size of steam drum is taken as Ø1220 × 16 mm. Material for construction is ASME SA 515/516 Gr. 70 or Equivalent. Allowable stress for the material at MMWT is 1407 kg/cm^2.

As per IBR regulation 270, strength calculation for steam drum is carried out by using formula (1)

$$WP = \frac{2f\,E(t-0.762)}{D+t-0.762} \tag{1}$$

where, t: Minimum thickness required in mm
WP: Working Pressure in kg/cm^2
D: Internal diameter of drum in mm
f: Allowable stress for material ASME SA 515/516 Gr. 70 or Equivalent as per ASME section-II part-D at metal working temperature.
E: Least of the ligament efficiency

Minimum ligament efficiency is calculated as 0.46. Using the formula (1), the calculated thickness is 11.46 mm which is less than provided thickness of 16 mm. Figure 1 shows the solid model of steam drum and Figure 2 shows the solid model of steam drum assembly.

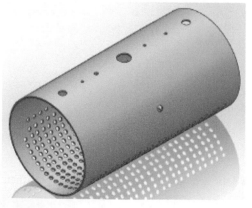

Figure 1. Solid model of steam drum.

Figure 2. Solid model of steam drum assembly.

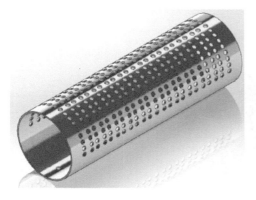

Figure 3.   Solid model of mud drum.   Figure 4.   Solid model of mud drum assembly.

4   DESIGN OF MUD DRUM

Initial size of mud drum is taken as Ø760 × 16 mm. Material for construction is ASME SA 515/516 Gr. 70 or Equivalent. Allowable stress for the material at MMWT is 1407 kg/cm^2.

As per IBR regulation 270, strength calculation for steam drum is carried out by using formula (2)

$$WP = \frac{2f\ E(t-0.762)}{D+t-0.762} \qquad (2)$$

where, t: Minimum thickness required in mm
   WP: Working Pressure in kg/cm^2
   D: Internal diameter of drum in mm
   f: Allowable stress for material ASME SA 515/516 Gr. 70 or Equivalent as per ASME section-II part-D at metal working temperature.
   E: Least of the ligament efficiency

Minimum ligament efficiency is calculated as 0.46. Using the formula (2), the calculated thickness is 7.43 mm which is less than provided thickness of 16 mm. Figure 3 shows the solid model of mud drum and Figure 4 shows the solid model of mud drum assembly.

5   DESIGN OF DISHED HEAD

Material for construction is ASME SA 515/516 Gr. 70 or Equivalent. Allowable stress for the material at MMWT is 1407 kg/cm^2. As per IBR regulation 276, strength calculation for steam drum is carried out by using formula (3).

$$T = \frac{PD}{2fE - P} + 0.762 \qquad (3)$$

where, P: Maximum working pressure in kg/cm^2 = 11.25
   D: Internal diameter of the Dished Head in mm = 1220
   f: Allowable stress for material ASME SA 515/516 Gr. 70 or Equivalent as per ASME section-II part-D at metal working temperature
   E: 1

Using the formula (3), the calculated thickness is 5.66 mm which is less than provided thickness of 16 mm. The solid model for the same is shown in Figure 5.

Figure 5.   Solid model of 2:1 Semi-ellipsoidal dished end.

6   FINITE ELEMENT ANALYSIS

Stress analysis of steam drum, mud drum and dished head is carried out in solid works software which includes the data for maximum principal stress. Methodology for carrying stress analysis in solid works is as follow.

1. Solid Modeling
2. Boundary conditions
    i. Fixtures
    ii. External Loads
3. Meshing
4. Solution
5. Result

6.1   *FEA analysis of steam drum*

Both end face is restricted to have movement in all the direction i.e. zero degree of freedom. It is indicated by the 'Green Arrow'. The other boundary condition is load which is in form of internal pressure which act on the inside surface of steam drum is indicated by 'Red Arrow' as shown in Figure 6. As shown in Figure 7 meshing is carried out and detail of meshing property and parameters are shown in Figure 8. Figure 9 shows the maximum principal stress of steam drum which is 1271.6 kg/cm^2.

6.2   *FEA analysis of steam drum*

Both end face is restricted to have movement in all the direction i.e. zero degree of freedom. It is indicated by the green arrow. The other boundary condition is load which is in form of internal pressure which act on the inside surface of steam drum is indicated by red arrow as shown in Figure 10. As shown in Figure 11 meshing is carried out and detail of meshing property and parameters are shown in Figure 12. Figure 13 shows the maximum principal stress of mud drum which is 1369.6 kg/cm^2.

6.3   *FEA analysis of dished head*

The end face is restricted to have movement in all the direction i.e. zero degree of freedom. It is indicated by the green arrow. The other boundary condition is load which is in form of internal pressure which act on the inside concave surface of dished head ad is indicated by red arrow. Figure 15 shows the maximum principal stress of dished head which is 1225 kg/cm^2.

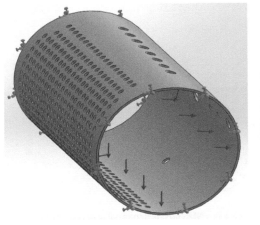

Figure 6.   Boundary conditions of steam drum.

Figure 7.   Meshing of steam drum.

Mesh Details	
Study name	Steam_Drum (-Default-)
Mesh type	Solid Mesh
Mesher Used	Standard mesh
Automatic Transition	Off
Include Mesh Auto Loops	Off
Jacobian points	16 points
Element size	43.1325 mm
Tolerance	2.15663 mm
Mesh quality	High
Total nodes	194013
Total elements	101666
Maximum Aspect Ratio	179.08
Percentage of elements with Aspect Ratio < 3	94.5
Percentage of elements with Aspect Ratio > 10	0.00492
% of distorted elements (Jacobian)	0

Figure 8.   Meshing of steam drum.

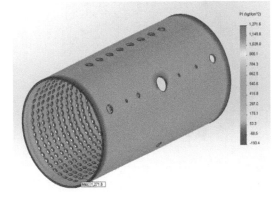

Figure 9.   Maximum principal stress of steam drums.

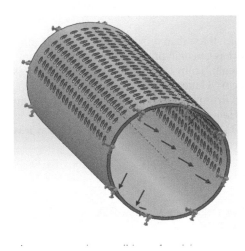

Figure 10.   Boundary conditions of mud drum.

Figure 11.   Meshing of mud drum.

Mesh Details	
Study name	Mud_drum_Static (-Default-)
Mesh type	Solid Mesh
Mesher Used	Standard mesh
Automatic Transition	Off
Include Mesh Auto Loops	Off
Jacobian points	16 points
Element size	33.8408 mm
Tolerance	1.69204 mm
Mesh quality	High
Total nodes	192196
Total elements	102080
Maximum Aspect Ratio	142.94
Percentage of elements with Aspect Ratio < 3	97.7
Percentage of elements with Aspect Ratio > 10	0.00098
% of distorted elements (Jacobian)	0

Figure 12. Meshing of mud drum.

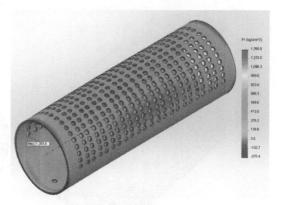

Figure 13. Maximum principal stress of mud drum.

Figure 14. Boundary conditions of dished head.

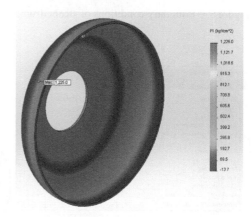

Figure 15. Maximum principal stress of dished head.

7 CONCLUSIONS

Design of steam drum, mud drum and dished head has been carried out and the stress analysis of the same has been presented in this paper. Following points have been concluded from the present design and analysis.

- Calculated thickness of steam drum is 11.46 mm which is less than provided thickness of 16 mm.
- Calculated thickness of mud drum is 7.43 mm which is less than provided thickness of 16 mm.
- Steam Drum—Maximum stress generated in steam drum is 1271.6 kg/cm² which is less than the maximum allowable stress i.e. 1407 kg/cm². Hence our design is safe.
- Mud Drum—Maximum stress generated in mud drum is 1369.6 kg/cm² which is less than the maximum allowable stress i.e. 1407 kg/cm². Hence our design is safe.
- Dished Head—Maximum stress generated in Dished Head is 1225 kg/cm² which is less than the maximum allowable stress i.e. 1407 kg/cm². Hence our design is safe.

REFERENCES

Chattopadhyay P. 2001. Boiler operation engineering, New Delhi: Tata McGraw Hill.
Indian Boiler Regulation, 2010.
Rayaprolu K. 2009. Boilers for power and process, New York: CRC Press.
Solidworks—2013 solid modeling software help.
Solidworks—2013 analysis software help.

Design of storage rack for rubber coated rolls

Chirag Patel, Shashikant J. Joshi & Dhaval B. Shah
*Department of Mechanical Engineering, Institute of Technology, Nirma University,
Ahmedabad, Gujarat, India*

ABSTRACT: The roll is an important component of the machines used in the industries like textile, paper and steel etc. The present process being adopted involve the storing of rolls randomly one on other because there is no specific structure for storage of roll. This leads to: damage of the rolls, time for loading and unloading of selected rolls is very high, congestion on the shop floor, increase in non-conformation, damage to calendared rubber sheet. Design calculation is made for all the components of storage rack. The geometric modelling is done. It is estimated that the total cost of manufacturing storage rack for rubber coated rolls is 2 lac. Space saved by storage rack 37.50% of earlier. The monetary benefit due to space saving by storage rack for rubber coated rolls is Rs. 7.2 lac.

Keywords: storage rack; rubber coated roll; contact stress

1   INTRODUCTION

The roll is an important component of the machines used in the industries like textile, paper and steel etc. The surface finish and hardness of the roll is very important because it is used to pass out certain value thickness of paper, cloth etc. It has to handle very carefully because it is coated by rubber with accurate surface finish. The image of the rubber covered roll is shown in Figure 1.

The present process being adopted involve the storing of rolls randomly one on other because there is no specific structure for storage of roll as shown in Figure 2.

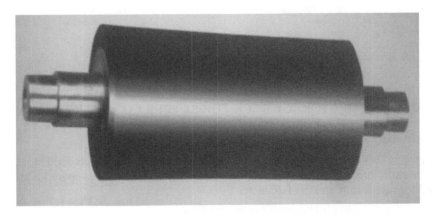

Figure 1.   Image of the rubber covered roll (*Lathia Industrial Manual*).

Figure 2.   Storing of rubber coated rolls (*Lathia Industrial Manual*).

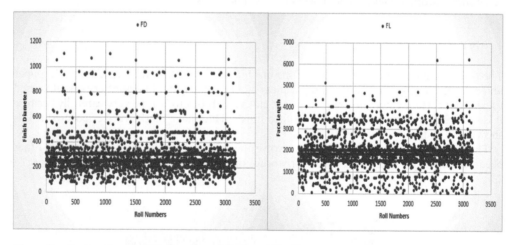

Figure 3.   Scatter chart for face length and diameter of roll.

2   MODELLING OF STORAGE RACK FOR RUBBER COATED ROLLS

2.1   *Preliminary study*

The size of rubber coated rolls is not fixed. So in order to design widely used storage rack, last six months production data are studied and scatter charts are prepared as shown in Figure 3 for face length and final diameter.

2.2   *Conceptual design of the storage rack*

The conceptual design for the storage rack is carried out using solid modeling software. The line diagram and 3D CAD model for the same as shown in Figure 4.

314

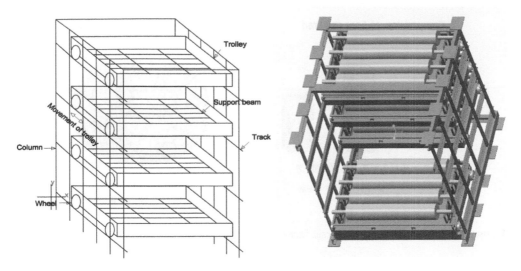

Figure 4. Conceptual design of storage rack for rubber coated rolls.

Table 1. Design of main components of storage rack.

Sr. No.	Component	Criterion	Result/Comment
1	Support beam	Fixed at both ends and loaded at middle by 5000 N	$40 \times 40 \times 6$ angle beam is selected
2	Trolley frame	Fixed at both ends and uniformly loaded by 4.90 N/mm	MC150 section is selected
3	Connecting rod	Cantilever beam loaded at tip by 17550 N	Solid rod with dia. 30 mm is selected
4	Wheel bearing pair	Radial load on each bearing 8775 N	Bearing pair 6206–6207 is selected
5	Track	Fixed at both ends and loaded at middle by 17550 N	MC150 beam is selected
6	Column	Loaded by 70200 N with 46.97 mm eccentricity	MC75 column is selected

3 DESIGN OF STORAGE RACK FOR RUBBER COATED ROLLS

3.1 *Design of main components*

Table 1 shows design of support beam, trolley frame, connecting rod, wheel bearing, track and column. The support beam, trolley frame, connecting rod and track are designed according to PSG design data book (design data PSG) and are selected from I.S. Specifications for dimensions for hot rolled steel beam, column, channel and angle sections (Third Revision) 808: 1989 (I.S. Specifications (third revision) 808:1989). Wheel bearing pairs are designed according to procedure given in Design of machine element by V B Bhandari (Bhandari page 573–577) and are selected from SKF bearing catalogue (SKF bearing catalogue). Column is designed according to design of eccentric loaded column in Shigley's Mechanical Engineering Design (budynas page 96–100, page 177–182).

3.2 *Contact stress between wheel and track*

Here the metal wheel will be rolled over the metal track so contact stress would be generated, and the contact type is cylinder to flat surface.

Contact stress is given by (Norton page 225–237),

$$\sigma_c = 0.418 \times \left[\frac{P'E}{R} \right]^{\frac{1}{2}}$$

(1)

315

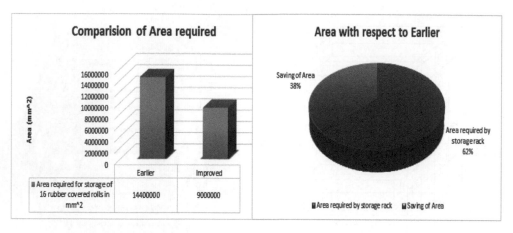

Figure 5. Floor space saving by storage rack.

where $P' = \dfrac{p}{a} = \dfrac{17550}{50} = 351\ N\,/\,mm^2$

$$\sigma_c = 0.418 \times \left[\frac{351 \times 2 \times 10^5}{90} \right]^{\frac{1}{2}} = 412.17\ N\,/\,mm^2$$

Now allowable contact stress,

$$[\sigma_c] = C_R \times BHN = 25 \times 150 = 3750 N\,/\,mm^2$$

As $[\sigma_c] > \sigma_c$, so wheel and track are safe.

3.3 *Floor space saving by storage rack*

In Figure 5, bar chart shows the comparison of the area required for storage of 16 rubber coated rolls with and without storage rack. Pie chart shows that 38% of earlier floor space can be saved by the use of storage rack. Monetary benefit from that floor space saving is 7.2 lac.

4 CONCLUSIONS

Following points have been concluded from the present design of storage rack for rubber coated rolls.

- The present process being adopted involve the storing of rolls randomly one on other.
- This research paper shows the design of storage rack for rubber coated rolls.
- Designed structure utilizes vertical space which leads to saving of floor space 38% of the earlier space requirement for storage of 16 rubber coated rolls.
- The monitory value of the space saving is around 7.2 lac.

REFERENCES

Bhandari V. B., "Design of machine elements", (Page 573–577).
Booklet of Lathia Industrial Supplies Pvt. Ltd.
Budynas Nisbett," Shigley's Mechanical Engineering Design", (Page 96–100) (Page 177–182).
Faculty of Mechanical Engineering, Design Data PSG College of Technology, DPV Printers (1995).
I.S. Specifications for dimensions for hot rolled steel beam, column, channel and angle sections (Third Revision) 808: 1989.
Norton Robert L., "Machine Design", (Page 225–237).
SKF bearing catalogue.

Multi-disciplinary Sustainable Engineering: Current and Future Trends – Tekwani, Bhavsar & Modi (Eds)
© 2016 Taylor & Francis Group, London, ISBN 978-1-138-02845-6

Design of caustic recovery tank for bottle washer machine

Jaimin Pandya
Department of Mechanical Engineering, Atmiya Institute of Technology and Science, Gujarat, India

Reena Trivedi
Department of Mechanical Engineering, Institute of Technology, Nirma University, Gujarat, India

Gokulesh Patel
KHS Machinery Pvt Ltd., Gujarat, India

ABSTRACT: In the beverage industries, water is an essential raw material with a consumption of approximately 2.5 liters per liter of beverage. Water used by a bottle washing machine is about half of the total of water used in the form of caustic soda solution. Caustic soda solution can be reused after removing dirt from it through filtering and sedimentation processes. Thus, water can be saved as well as the cost of caustic soda solution be reduced. In the presented work, an above ground steel sedimentation tank is designed as a caustic recovery system. The design is based on various codes and standards. It includes the design of the tank shell, roof and bottom. It also includes the design of leg supports, thermal insulation, and attachments such as stiffening ring, lifting lug, manholes, stairs, and the design of piping attachments such as air vents and tank cleaning nozzle.

1 INTRODUCTION

In the beverage industries, water used by a bottle washing machine is about 1.25 liters per liter of beverage, in the form of caustic soda solution. So, the recovery of caustic soda solution is a very important factor in the water resources management in industrial activities and development. It can be reused after removing dirt from it through filtering and sedimentation process. In the presented work, an above ground steel sedimentation tank is designed as a caustic recovery system, because inspection and maintenance are very difficult in case of underground tanks and the mobility of system is also very important. The tank design should be based on various codes and standards, like Indian standard, API standards and Eurocodes. API 650 and IS 803 have a similar approach for design and Eurocode is based on the Load Resistance Design or Limit State Design method, which factorizes load and resistance properties based on reliability theory. So, Eurocode requires more and accurate information. Eurocode is more detailed; it covers a large range of structures. API, on the other hand, uses the Allowable Stress Design method based on the unfactored values and does not tell anything about the probability of failure. The methods given in API 650 are only for tank design and far more straight forward than the Eurocode. So, API 650 makes the design work easier.

2 BASICS OF SEDIMENTATION TANK

In most of the cases, a sedimentation tank is nothing more than a large tank, often with a special solid material removal system. Types of dirt present in the caustic zone of a bottle washer may include colors, fibers, glue, label, beverage residues, dust, straws, sand, lemon slices, cigarette butts, etc. In most of the cases, the caustic zone of a bottle washer has the following conditions:

Table 1. Technical parameter of bottle washer caustic zone.

Characteristics	Bottle washing machine
Volume	40 m³ (40,000 liter)
Temperature	60°C
NaOH concentration	1.5% to 3%
PH value	8–10
Total Suspended Solids mg/l	120
Time required for sedimentation process	8–12 hours

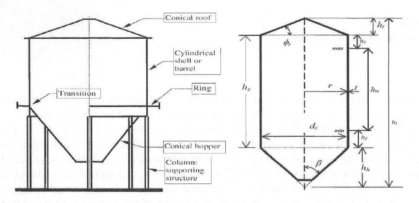

Figure 1. Geometry of caustic recovery tank.

In the geometric design of the tank, a conical bottom is required to remove dirt. Half apex angle of conical bottom should be in standard value so that manufacturing cost can be reduced and should be less than or equal to 45°. The diameter of the tank will affect the transportation of the tank. A tank with a very high diameter requires special permissions and transportation cover, so it is taken as 2.6 meter for this case and the height is taken as 8 meter. For high H/D ratio, steel required will be more. Roof angle can be taken as 15° which is most conventional for a tank roof.

3 DESIGN OF CAUSTIC RECOVERY TANK

3.1 Design of tank wall, roof and bottom

Material for the tank should be according to API 650 section 4; for tank wall steel A516 grade 70 was selected. It had an allowable stress of 160 MPa for design and 180 MPa for testing.

For the fabrication of 8m cylindrical shell, API 650 and IS 803 suggest four numbers of courses. A516 Gr70 steel plates of 2 m width were used for the said designing. According to API 650 and IS 803, for tank having nominal tank diameter less than or equal to 15 m should have minimum cylindrical shell thickness of 5 mm in corroded condition. So, with corrosion allowance of 3 mm, cylindrical shell thickness was 8 mm. API 650 and IS 803 suggest following formula to decide shell thickness:

$$t = \frac{4.9 \times D \times G \times (H - 0.3)}{E \times S} + Corrosion\ Allowance \tag{1}$$

where E is weld efficiency and S is allowable strength according to API 650.

For the roof of the tank, a conical roof is mostly used. For a tank with a small diameter there is no need of a supported roof and so a self-supporting cone roof can be used. It can

318

be formed such that the surface of a right cone is supported only at its periphery. In API 650 and IS 803 a roof thickness of less than 5 mm is not allowed. Including corrosion allowance, the roof wall thickness should be 8 mm. IS 803 suggests the following formula to decide roof thickness:

$$t = \frac{D}{5 \times \sin \varphi}$$

(2)

where, φ = roof angle.

For the conical bottom, design is not covered in API 650 and IS 803. But IS 803 and API 650 consider a minimum thickness of 6 mm in corroded condition, with a corrosion allowance of 3 mm and for economical fabrication the conical bottom can be considered of standard thickness i.e. 10 mm.

3.2 Calculation of thermal insulation for tank

Insulation of a tank is required to prevent heat loss, because the temperature of caustic soda solution is about 60°C in a bottle washer and Rockwool insulation products are the economical choice for insulating material, and for cladding stainless steel or aluminized steel is used. Total heat loss can be calculated by considering an equivalent hollow composite cylinder for the tank as follows:

$$Q_i = \frac{2\pi L_i \left(T_{NaOH} - T_{Air}\right)}{\dfrac{1}{h_{NaOH} \times r_1} + \dfrac{\ln \frac{r_2}{r_1}}{K_{SA516}} + \dfrac{\ln \frac{r_3}{r_2}}{K_{incsulation}} + \dfrac{\ln \frac{r_4}{r_3}}{K_{cladding}} + \dfrac{1}{h_{Air} \times r_4}}$$

(3)

Total heat transfer in given time will be equal to heat loss by hot caustic soda solution.

$$Q \times t = m \times C \times (T_1 - T_2)$$

(4)

To avoid heat transfer by convection, the insulation thickness can be decided such that outside cladding metal surface will remain the same as the atmospheric temperature for the assumed time period. So, the following result can be obtained by the explained theory and the optimum thickness can be found out, which is 40 mm in this case.

3.3 Design of stiffening rings

A tank needs to be provided with stiffening rings to maintain roundness when the tank is subjected to external loads like wind load. Here, L shape angle is selected for the stiffening ring. According to API 650, the required minimum section modulus (in cm^3) of the stiffening ring can be determined by the following equation:

$$Z = \frac{D^2 H}{17} \times \left(\frac{V}{190}\right)^2$$

(5)

where V is design wind speed in km/h, Diameter and Height are in meter.

It can be calculated from Indian standard IS 875: part 3 from the basic wind speed (V_b). Basic wind speed (V_b) for any site can be obtained from Figure 1 of IS 875: part 3 and should be modified to include the following effects to get the design wind velocity at any height (V_z) for the chosen structure. It can be mathematically expressed as follows:

$$V_z = k_1 \times k_2 \times k_3 \times V_b$$

(6)

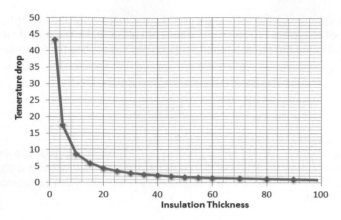

Figure 2.   Variation of temperature drop with thermal insulation.

where, K_1 = risk coefficient = 1.08 for assumed basic wind speed and structure, K_2 = terrain, height and structure size factor = 1.09 for terrain category 1 and class A structure with respect to height of the structure, K_3 = topography factor = 1.00

So Design wind speed V_z = 233 km/h or 65 m/s.

Design wind pressure $P_z = 0.6 \times (V_z)^2 = 2535$ Pa.

According to API 650, the minimum size of the angle for use in a built-up stiffening ring should be $65 \times 65 \times 6$ mm. Four stiffening rings can be used as matter of safety with an equal distance of 1600 mm. Fillet weld joints of 12 mm between rings and tank shell are assumed to be sufficient.

3.4   *Design of leg support*

Caustic recovery tanks require access under the tank for sludge outlet, so leg support can be selected. For large tanks, a round steel pipe is particularly suitable, because it has the same area moment of inertia in all directions and good torsional resistance. The pipe must be painted for corrosion resistance. The seal welds on both ends should be adequate for protection against atmospheric dampness and corrosion. Centroidal axes of pipes were used such that the legs coincide with the centerline of the tank shell; thus, eliminating the eccentric loading in the column. Loads on support legs were vertical due to weight and horizontal due to wind forces. Generally, additional moments from piping or other equipment were not being considered. It was assumed that tops of the support legs were to be welded to a rigid vessel wall which is actually flexible.

No of legs, N = 6 and Projected area for wind load, A = 29.3 m^3.

Mass of the vessel in empty condition, M_e = 7000 kg; weight W_e = 68700 N

Mass of the vessel in operating condition, M_o = 55000 kg; weight W_o = 539825 N

Force coefficient for circular structure considering H/D ratio, C_f = 0.8

Wind force Pw = $C_f \times A \times P_z = 0.8 \times 29.3 \times 2535 = 59420$ N.

Moment at base due to wind loading = $M_b = P \times (H + L) = 386230$ Nm

Maximum total compressive axial load will be at downwind side always in filled condition.

$$C_o = \frac{W_o}{N} + \frac{4 \times M_b}{N \times D} = 189033\,N \tag{7}$$

Maximum total tensile axial load will be at upwind side always in empty condition.

$$T_e = -\frac{W_e}{N} + \frac{4 \times M_b}{N \times D} = 87592.5\,N \tag{8}$$

320

For the lateral load F per column F = P/6 = 9.9 kN.

For pipe leg support DN 300 pipe with 12 mm thickness is selected. So, Axial compressive stress will be of 16.08 MPa and bending stress will be of 13.44 MPa.

For combined column stress in compression, the most widely used procedure is straight-line interaction method, used in the AISC handbook and for this case the design is safe under combined stress.

$$\frac{f_a}{F_a} + \frac{f_b}{F_b} = 0.16 \le 1 \qquad (9)$$

The design of support should be safe under buckling and for a matter of safety a braced leg support must also be used. For bracing, a horizontal pipe can be used of DN 150 (OD 165.1 mm) and 8 mm thickness. A thickness of 10 to 20 mm for reinforcement pad would be sufficient as a transition ring. Leg to shell transition ring connecting weld should have a minimum weld size of 8 mm. Base plate attachment weld should have a minimum weld size of 10 mm. A base plate of 30 mm thickness can be selected. Generally, 8 to 12 anchor bolts of M 30 are used in these structures.

3.5 *Design of lifting lug*

According to standard SANS 10131, lifting lug geometry of the lifting lug can be as follows:

Lifting lugs is used for loading and unloading tanks in empty condition and to place tanks on foundations. According to API 650, there should be a minimum of two lugs on each tank. Generally lifting lugs are fabricated from A36 Steel. The thickness of lifting lug can be found out by considering it as a simple cantilever beam. For this case the thickness of lug can be 40 mm and the doubling plate can be 20 mm thick. Lug to plate fillet weld 12 mm and plate to tank fillet weld 10 mm in 20 mm thick plate.

3.6 *Design of piping attachments*

Following pipes should be attached to the tank:

1. Inlet of standard pipe size DN 100 and overflow line of standard pipe size DN 150 at upper part of cylindrical shell, outlet of standard pipe size DN 125 at bottom part of cylindrical shell and sludge outlet of standard pipe size DN 300 at bottom of the tank.
2. Air vent on the roof, each tank should have provisions for both normal and emergency venting. Two pipes of diameter 4 inch (DN 100) can be used as air vents.
3. Cleaning pipe for nozzles on the roof. For caustic recovery tank having diameter of 2.6 m and height of 8 m, cleaning nozzle should have minimum opening size of DN 150, operating pressure of 2 bar and flow rate of 50 l/min.
4. Attachment to push solid wastes out of DN 80 or DN 100, if solid waste are frozen.

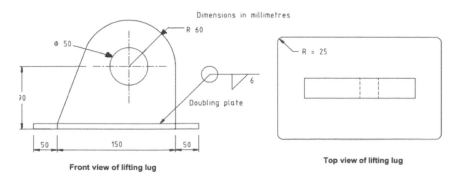

Figure 3.   Lifting lug geometry according to SANS 10131.

321

Each of the piping attachments should have proper support with tank shell. Design of pipe, piping supports, reinforcement and other parameter of design should be according to section 5.7 of API 650 based on fabrication and erection criterion.

3.7 *Design of attachments for maintenance*

A caustic soda recovery tank would require two manholes for maintenance, on the roof and at the bottom part of the cylindrical shell. Shell manhole will be under hydrostatic loading, so the design of shell manhole is different than of roof manhole. Both roof and shell manholes should be designed according to section 5.8 of API 650. Both inside and outside stairs should be designed according Australian standard AS 1657.

4 CONCLUSION

Above ground a steel sedimentation tank can be used for the caustic recovery of a bottle washer. In this paper, a standard design procedure of a caustic recovery tank for a bottle washer has been developed and explained through various codes and standards with relevant examples.

REFERENCES

Almukhtar R and Ageena N, 2012, Water Recycling/Reuse in Factories Case study Soft Drink Factory, Engineering & Technology Journal, Vol.30, No.1.
American Petroleum Institute, 2013, API 650: Welded Steel Tanks for Oil Storage.
Bureau of Indian Standards, 2006, IS 803: Code of Practice for Design, Fabrication and Erection of Vertical Mild Steel Cylindrical Welded Oil Storage Tanks.
Bureau of Indian Standards, 2006, IS 875, Code of Practice for Design Loads.
Henry M. Bednar, 1986, Pressure Vessel Design Handbook, second edition.
Moss Dennis, 2003, Pressure vessel design manual: illustrated procedures for solving major pressure vessel design problems, third edition.
Standards South Africa, 2004, SANS 10131 Above-ground storage tanks for petroleum products.
Yunus Cenjal, 2002, Heat Transfer: A practical approach, second edition.

Multi-disciplinary Sustainable Engineering: Current and Future Trends – Tekwani, Bhavsar & Modi (Eds)
© 2016 Taylor & Francis Group, London, ISBN 978-1-138-02845-6

Design and development of deployable mechanism for spacecraft reflector

Valay M. Patel
Department of Mechanical Engineering, Institute of Technology, Nirma University, Ahmedabad, Gujarat, India

A.R. Srinivas & Piyush Shukla
Space Applications Centre (ISRO), Ahmedabad, Gujarat, India

Dhaval V. Patel
Department of Mechanical Engineering, Institute of Technology, Nirma University, Ahmedabad, Gujarat, India

ABSTRACT: Reflectors are widely used in space communication as receive or transmit antenna; as mirrors in remote sensing payloads. To maximize the gain energy and ground resolution, reflector aperture diameter needs to be as large as possible. Design and development of large reflector is highly complex and the challenge is that they cannot be launched in deployed configuration due to limited payload fairing in launch vehicle. Space industries are looking at developing deployable reflector in space. Larger volume reflector up to the size of 30 m diameter are the challenge to be realized because of the mechanism involved in the deployment.

This paper addresses one of deployable antenna for a communication payload using a scissor mechanism. The present study covers design, development and realization of an antenna of 3.0 meter in diameter and demonstrate it in a prototype of 0.2 meter diameter. The antenna prototype has been realized using desktop manufacturing and the folding and deploying concepts have been demonstrated.

This paper brings out the detailed development aspect of the design & realization of prototype antenna.

Keywords: reflectors; deployable; stow; antenna; scissor mechanism; triad link; planar mechanism; 3D printing; GUI, algorithm

1 INTRODUCTION

With advancements in satellite communication technology and deeper societal impact of space technology, there has been a steep rise in demand of high gain space antenna systems. High gain antennas (Youchi 2011) are required for high speed & loss less wireless communication and/or data transfers between satellites and ground stations. Conventionally, large sized antennas are placed on communication satellites and antenna sizes face limitation due to limited volume of launch vehicle fairing. In order to reduce cost of launching, multiple small size satellite having small antenna are launched to space; this limits the efficient communication between ground station and satellite.

Such limitation of space & requirement of large size reflectors in space can be catered by use of collapsible & deployable antenna structures (Research Areas in Space, 2013). Such antenna can be kept in stowed condition during launch & can be deployed once in designated orbit. The design of deployment mechanisms and their development is a multidisciplinary

Table 1. Comparisons of gain and mass of populated deployable antenna.

Type	Developer	Maximum Gain (dBi)	Mass (kg)
Deployable UHF/VHF	Innovative Solutions in Space	Not Mention	0.10
Deployable High Gain Antenna	BDS Phantomworks	18	1.0
Deployable High Gain Antenna	SERC	15	1.0
Deployable Helical Antenna	Astro Aerospace	13	0.3

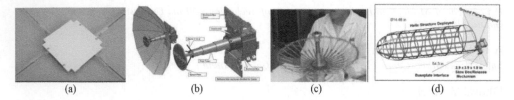

(a)   (b)   (c)   (d)

Figure 1. Image of deployable antenna (a) Innovative Solutions in Space [ISIS_Ants_Brochure_v.7.11] (b) BDS Phantomworks [Charles S] (c) SERC [Small Spacecraft Technology State of the Art] (d) Astro Aerospace [D. Ochoa, K. Hummer and M. Ciffone].

task which involves knowledge of mechanism design, structural design, electronic drive & instrumentation and computer interface.

Deployable UHF/VHF antenna (ISIS_Ants_Brochure_v.7.11) was firstly developed by Innovative Solutions in space, Netherland for small satellites CubeSat. This actuator operated antenna is consuming maximum 2W power from the satellites to deploy 4 monopole aerials. BDS Phantomworks (Charles 2012) and Space Engineering Research Centre (Small Spacecraft Technology State of the Art, 2014) (SERC), USA were independently built up deployable high gain parabolic antenna for small satellites application. First deployable helical antenna was prepared by Astro Aerospace (D. Ochoa, K. Hummer and M. Ciffone). This self-deployable antenna has 300:1 volume reduction ratio therefore it could be fitted in 0.5U CubeSat. Maximum gain and overall mass are the most important parameter for deployable antenna are shown in Table 1.

Small satellites have limitations over the availability of power, hence low powered actuator or stored elastic strain energy should be used for deployment of antenna. Previously, various antenna topologies like crossed log periodic, conical log spiral, helical pantograph, and eight-helix [G. Olson, S. Pellegrino, J. Costantine and J. Banik] etc. were explored which are operated through elastic strain energy. These kind of antennas have some limitations like; it can be deployed only once, antennas release mechanism and risk of full deployment.

2 MECHANISM DESIGN

One of the primary objectives of deployable antenna design is to minimize stowed volume as much as possible and conform to RF designed shape in deployed configuration. The design of deployment mechanism should be such that it complies with both classical mechanics and antenna theory. Triad element structure and scissor mechanism (linear pantograph) concepts were utilized for development of cylindrical pantograph deployable structure for antenna.

Linear pantograph is the simplest type of motion transfer mechanism in which linear elements are interconnected with a hinge joint. Varying the angle between any two elements, changes the angle between all hinges and overall length of the mechanism.

Triad elements are specially designed for hinge joint location i.e. centre hinge joint doesn't lie in line with forcing and working end rather it is placed at offset from the line as shown in Figure 2(a). So a series of such triad links can be used to form perimeter of a closed circle. The offset in the hinge point allows in plane radial expansion and contraction [You, Z., Pellegrino, S.,1997]. This motion has been exploited in the present work for stowing and deployment of antenna.

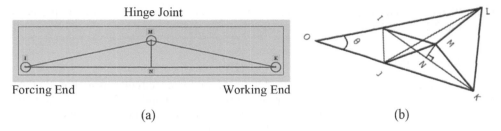

Hinge Joint

Forcing End

Working End

(a)

(b)

Figure 2. (a) Joint in triad element (b) Triad element.

To expand radially, the element subtends a fixed angle (θ) and expands by side as shown in Figure 2(b). The angle θ is found out by:

$$\theta = \frac{360°}{n} \qquad (1)$$

where n = number of element's pair

The deployed circle radius (R) is the input parameter of the mechanism. Thus the perpendicular distance between centre hinge and side hinge is achieved by:

$$\overline{KN} = R * \tan\frac{\theta}{2} \qquad (2)$$

To form a closed loop mechanism the length MN and KN of the element must satisfy the following relation,

$$\overline{MN} = \overline{KN} * \tan\frac{\theta}{2} \qquad (3)$$

These triad elements are connected with each other and form close loop mechanism. This mechanism provides an advantage of single point drive for deployment and stowing.

2.1 Design of triad link

For expandable circle pantograph mechanism the link should have an offset at centre hinge point as per the calculated angle. The links are connected with each other by hinge joint so the relative distance between hinges of the single link is responsible for deployment. The hinge joints of an element are calculated from the triad element theory discussed earlier. This triad type elements should be created as long as the triangle is formed.

Two input parameters are required for a link design of a triad planar structure. Deployed diameter of the antenna and number of pairs of element are these two variables parameters. For sample calculation, consider 15 pair of triad scissor elements are complete pantograph circle with the 600 mm deployed diameter.

Necessitated input data: n = 15, R = 300 mm.

$$\theta = \frac{360°}{n} = \frac{360°}{15}$$
$$\theta = 24°$$

Equation (2) & (3) can be used to calculate offset $\overline{KN}$ and $\overline{MN}$. Value of $\overline{KN}$ = 63.767 mm and $\overline{MN}$ = 13.582 mm is obtained. As shown in Figure 2(b) ΔKMN and ΔIMN are similar triangle so the proper location of all three hinges (K, M, I) is acquired. Deployed diameter and number of link are the two key variables that affect the area reduction ratio in the mechanism. Planar mechanism design and effect of two variables on area reduction are presented in section 2.2, 2.3 & 2.4.

2.2 *Design of planar mechanism*

As discussed above the triad links arranged in series can form a circular assembly. 3D CAD Modal design [Mastering Autodesk Inventor 2015] of a typical link is shown in Figure 3(b). Thickness of the triad link is 2.5 mm. The design makes use of fasteners to produce an interface hinge joint having very low co-efficient of friction.

Figure 3(c) shows 3D CAD simulation of complete assembly having 20 number of triad links. The DOF of the assembly [J. Uicker, G. Pennock, J. Shigley] can be calculated from Kutzbach criterion i.e. 3(n-1)-2j, where n is the total no of links and j is the number of simple hinge connections. The DOF for the present assembly is 1.

Number of links and diameter of circle are two key parameters in designing mechanisms. Section 2.3 and 2.4 represent a comparative study and effect of change in links and diameter.

2.3 *Design with constant inner deployed diameter*

In this design the required inner deployed diameter is kept constant and number of links is varied. The mechanism has been designed by varying number of links ranging from 10–40 and the percentage of area reduction versus number of links are shown in Figure 4.

The graphs conclude that the stowed and deployed diameter of planar mechanism is inversely proportional to the number of links. Reduction ratio of area increases as number of links are increased.

2.4 *Design with constant number of links*

The second design study was done by keeping number of links constant and changing the deployed diameter of the planar mechanism. Various circular patterns were formulated as

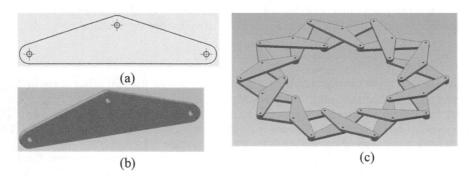

(a)

(b)

(c)

Figure 3. (a) Triad element triangle (b) CAD modal of triad link (c) CAD assembly of planar mechanism.

Table 2. Design parameter comparison.

Design Parameter	Deploy Diameter = 150 mm				
Number of Links	*10*	*16*	*20*	*30*	*40*
Stow Dia. (mm)	220	132	106	75	55
Deploy Dia. (mm)	230	180	167	157	155
Area Reduction (%)	8.51	46.22	59.71	77.18	87.41

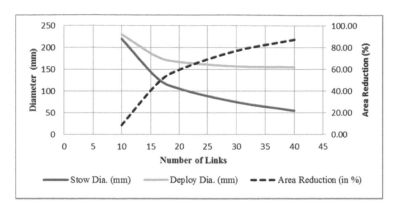

Figure 4.   Graph of number of link vs diameter (mm) and area reduction (%).

Table 3.   Design parameter comparison.

Design Parameter	Constant Number of links = 20			
Deployed Inner Diameter	*100*	*150*	*200*	*250*
Stow Dia. (mm)	74	107	139	172
Deploy Outer Dia. (mm)	112	167	218	271
Area Reduction (%)	56.35	58.95	59.34	59.72

per the deployed diameter value. Table 3 presents variations in various design parameters by varying deployed inner diameter.

Stowed and deployed outer diameters are linearly proportional to deployed inner diameter of planar mechanism pattern. Percentage reduction in area increases as the deployed inner diameter increases.

Keeping in view the above parametric studies and the design goals specified for antenna deployment (~80% reduction in area), a design configuration with 20 links was selected for prototype development.

2.5   Design of mechanism assembly

In order to grow the mechanism out of expansions plane, scissor mechanism has been used which connects two individual planes of expansion using triad links. The legs of scissor mechanism are connected with two different triad elements. Scissor mechanism and circular planar mechanism are kept perpendicular with respect to their functioning plane. They may not be directly connected with each other therefore 'L' shaped connecting links are used for interfacing the two mechanisms. A CAD simulation of the concept is shown in Figure 6. The entire assembly is driven through a single drive location as shown in Figure 6.

2.6   Drive assembly

All links of the cylindrical pantograph mechanism are connected with revolute joints. Advantage of this type of revolute joint is that motion can be easily transformed from input to the required location with lesser frictional losses. Mechanism has only one mobility therefore single actuator was utilized for deployed and stowed purpose.

DC motor drive is used for repeated deployment and stowed functioning. Entire mechanism revolves with respect to one ground link. Ground link is fixed with platform and driver link (Figure 7(a)) is designed to have coupling with DC drive output shaft. Both were replaced

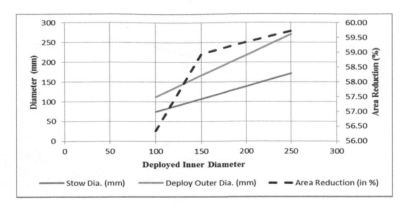

Figure 5. Graph of deployed inner diameter (mm) vs diameter (mm) and area reduction (%).

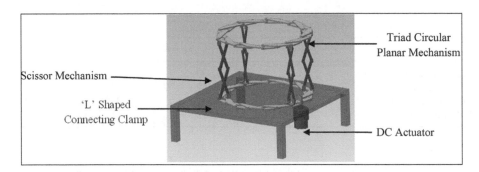

Figure 6. CAD Modal of deployable antenna.

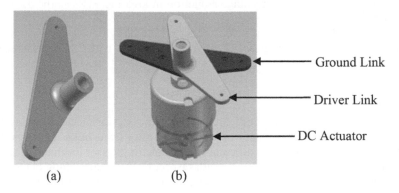

(a)   (b)

Figure 7. CAD modal of (a) Driver link (b) Actuator arrangement.

with one pair of the bottom triad element in mechanism. As shown in Figure 7(b) shaft of the motor was directly coupled with ground and driver link.

3 REALIZATION AND ASSEMBLY

A working prototype for lab demonstration was developed using the designed mechanism. Acrylonitrile Butadiene Styrene (ABS) material has been used for manufacturing links for development of lab model using Fusion Deposition based 3D printing [NETOPIA] technology.

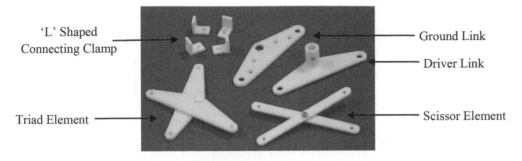

Figure 8.   Realized assembly parts.

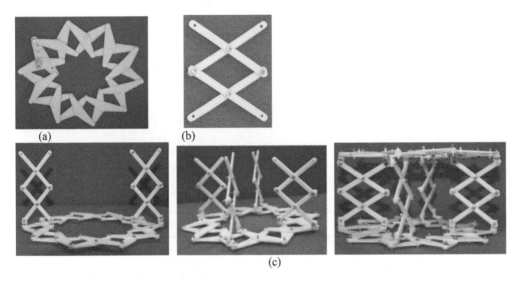

Figure 9.   (a) Assembly of planar mechanism (b) Assembly of scissor elements (c) Assembly sequence of mechanism.

The design can be realized using other conventional material and manufacturing processes as per the strength/stiffness requirements of the space mission. Figure 8 shows various assembly parts manufactured using ABS material. Assembly operations like boring, match hole drilling etc. has been performed on the parts. A total of 40 triad links including drive link and 16 links for scissors mechanism were developed and assembled using interference fit hinge pins.

The assembly of mechanism began by arranging the triad links in circular pattern as shown in Figure 9(a) and assembling the scissors mechanism as shown in Figure 9(b). The two assemblies were then joined together using L-brackets as shown in Figure 9(c). The use of hinge pins provide very low friction revolute joints.

4   DRIVE AND CONTROL SYSTEMS

Prototyped antenna deployment mechanism is to be driven by a DC motor. A computer based GUI was developed using MatLab [B. Dash and V. Vasudevan] for direction control and driving of DC motor. The computer program is fed into a microcontroller card [Michael McRoberts] shown in Figure 10(c). Typical circuit diagram consisting of driver IC, control card and DC motor is shown in Figure 10(b). Table 4 presents configuration detail of various electrical components

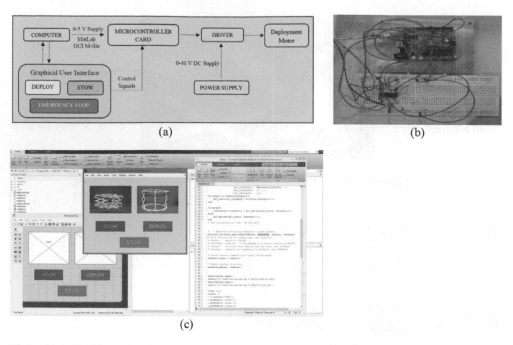

Figure 10. (a) Schematic of control algorithm (b) Realized control algorithm (c) GUI developed in MatLab.

Table 4. Specification of electronics component.

No.	Parts	Specifications
1	DC Motor	➤ Power supply – 12 volts ➤ Current – 3 A ➤ Rotational Speed – 10 rpm ➤ Max Torque Transmission – 4.12 Nm
2	Microcontroller Card	➤ Arduino Uno R3 ➤ AT mega 328 microcontroller ➤ Input voltage – 7–12V ➤ 14 Digital I/O Pins
3	Integrated Circuit	➤ Power Supply 50 v ➤ Logic Supply Voltage 7 V ➤ Motor direction control
4	Jumper wire & Bread Board	➤ Standard 7″ M/M and M/F ➤ Overall size – 170 X 65 mm ➤ Pin-projection hole – 65 X 10

used for driving the DC motor. A schematic of control algorithm is shown in Figure 10(a) and the realized test set-up is shown in Figure 11(c).

5 TESTING AND MEASUREMENT

The test setup as discussed in section 4 has been used for multiple stowing and deployment operations. Figure 11(a) shows the mechanism in completely deployed condition and Figure 11(b) shows the assembly in completely stowed condition. The dimensional results

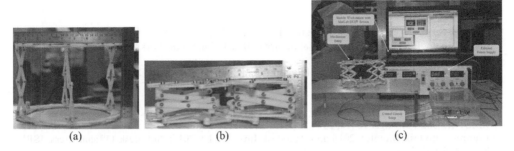

| (a) | (b) | (c) |

Figure 11.   (a) Completely deployed condition (b) Completely stowed condition (c) Realized test set-up.

Table 5.   Cases of repeatability.

Cycle	Deployed Dia. (mm)	Deployed Height (mm)	Stowed Dia. (mm)	Stowed Height (mm)	Reduction in Volume (%)
1	220.0	150.0	143	50.5	85.78
2	219.0	151.2	144.0	52.5	84.99
3	222.5	149	142.8	48.0	86.73
4	218.5	151	143.6	49.6	85.81

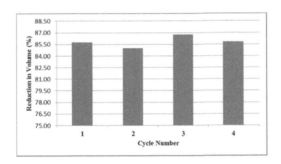

Figure 12.   Cycle vs reduction in volume.

obtained during multiple testing of the mechanism is presented in Table 5 and the test data has been used to calculate% volume reduction in each case.

Total volume reduction in deployed and stowed condition is 85.92% which was calculated from the experimental evaluation of the mechanism.

As observed from the test measurement presented above, the prototype has a repeatability within 2.03%.

6   CONCLUSION

A design of foldable antenna structure for space application has been simulated, realized and tested by demonstration of working prototype. The prototype has been driven using an electric DC drive and computer based GUI for direction control and the mechanism was successfully tested multiple times. The test results have shown a volume reduction of ~86% against targeted value of >80% with a repeatability of 2.03%. The design can be extended as it could be used for space base deployable antenna system in future.

REFERENCES

Charles S. "Scott" MacGillivraym, Miniature Deployable High Gain Antenna for CubeSats, The Boeing Company Boeing Defense, Space & Security (BDS) / Phantom Works, Huntington Beach, CA. April 22nd, 2011.

Dash B. and Vasudevan V. 'GUI/Simulink Based Interactive Interface for a DC Motor with PI Controller' International Journal of Scientific & Engineering Research Volume 2, Issue 12, December-2011 1 ISSN 2229–5518

ISIS_Ants_Brochure_v.7.11"Deployable UHF and VHF antennas" by Innovative Solutions In Space (ISIS) Molengraaffsingel 12–14, 2629 JD, Delft, The Netherlands.

Mastering Autodesk Inventor 2015 and Autodesk Inventor LT 2015: Autodesk Official Press, ISBN: 978-1-118-86213-1

Michael McRoberts, 'Beginning Arduino' Second Edition, Technology in Action.

NETOPIA-Forum for the Digital Society, '3D Printing: Technology and Beyond' November 2013

Ochoa D., Hummer K. and Ciffone M." Deployable Helical Antenna for Nano-Satellites", 28th Annual AIAA/USU Conference on Small Satellites.

Olson G. Pellegrino S. Banik J. and Costantine J. "Deployable Helical Antennas for CubeSats", American Institute of Aeronautics and Astronautics.

Olson G., Pellegrino S., Costantine J. and Banik J. "Structural Architectures for a Deployable Wideband UHF Antenna", American Institute of Aeronautics and Astronautics.

Research Areas in Space—A Compendium For Preparing Project Proposals by Universities/Institutes—RESPOND Program by Indian Space Research Organization, Bangalore August 2013

Uicker J., Pennock G., Shigley J. Theory of Machines and Mechanisms, oxford.

"Small Spacecraft Technology State of the Art", Mission Design Division Staff Ames Research Center, Moffett Field, California, NASA/TP–2014–216648/REV1, July2014.

Yoichi Koishi, 2011"Future satellite communication and enabling technologies", APRSAF-18

You, Z., Pellegrino, S., "Foldable bar structures," *International Journal of Solids and Structures*, Vol. 34, No. 15, P. 1825–1847, May 1997.

Multi-disciplinary Sustainable Engineering: Current and Future Trends – Tekwani, Bhavsar & Modi (Eds)
© 2016 Taylor & Francis Group, London, ISBN 978-1-138-02845-6

Random vibration fatigue life estimation of lower suspension arm of a Sedan car

Pinank A. Patel & Vivek G. Patel
Marwadi Education Foundation, Faculty of Engineering, Rajkot, Gujarat, India

S.S. Khandare
B.D. College of Engineering, Sevagram, Wardha, Maharashtra, India

ABSTRACT: The present work explains the random vibration based fatigue analysis of lower suspension arm of a sedan car. In the present work, we have used the three interval method based on Gaussian distribution to determine the random vibration fatigue life of the component. Initially, the load due to road surface variation is collected using data logger. The data logger along with accelerometer gave the time domain acceleration data. These measured values of acceleration are then used and using Fast Fourier Transform (FFT) analysis the frequency domain Power Spectral Density (PSD) achieved. Finite element method modal analysis is performed to get the natural frequencies and corresponding vibration modes of the structure under study (LSA). Using random vibration analysis PSD acceleration is applied to the structure. PSD acceleration analysis gives us the stress value for the given loading, finally three interval methods and Miner's cumulative fatigue damage rule are used, and fatigue life of lower suspension arm is verified.

1 INTRODUCTION

In practice, the real fatigue loadings are random in nature in respect of frequency and force magnitude. Power Spectrum Density (PSD) explains how the power of the time signal gets distributed among frequencies (Raheman 2009). In frequency domain method, there are peak distribution method and amplitude distribution method; whereas in time-domain method trigonometric series method, the inverse Fourier transform method, the Parameter model method, etc. are used. The complex vibration fatigue life under random load is hard to estimate (Andrew 1999). Narrowband and Broadband random load are studied based on the PSD function of vibration components for fatigue life estimation by (Slavic 2013). Then Petrucci (2004) proposed the Sample method of structure random vibration fatigue life estimation; through this method we can use spectral density in the frequency domain to describe the case of Broadband random vibration load. Steinberg proposed a three interval method based on Gaussian distribution. Among these methods, the three interval method is used in this research work as it provides better and accurate results, and it is easy to use.

The suspension system is used to absorb shocks and jerks from the road or rough terrain. The suspension allows a wheel to move independently to the body. Comfort can be achieved by isolating the riders from road disturbances like bumps or potholes. Control is achieved by keeping car body from rolling and pitching, and maintaining good contact between road and tire (Patel 2013). Fatigue analysis of lower suspension arm can be carried out in both time domain and frequency domain. Rain flow counting is used to calculate the number of cycles at each stress level based on time history in Time domain analysis (Bahkali 2013). Frequency domain is more useful in case of random load histories as compared to Time domain analysis. The Dirlik approach is widely used for frequency domain fatigue analysis. For fatigue analysis in frequency domain other than Dirlik method, Tovo–Benasciutti and Zhao–Baker methods are also used by researchers.

Figure 1. Road data collection by using data logger.

2 FINITE ELEMENT BASED LIFE ESTIMATION OF LSA UNDER RANDOM VIBRATIONS

The finite element analysis provides a CAE tool for calculation of stress of the complex structural parts. For the stress analysis, the Random vibration fatigue analysis method is used. The method makes use of the finite element and Steinberg's three interval method to compute the life of the lower suspension arm of a sedan car.

2.1 Road data collection

Initially, the load due to road surface variation is collected using data logger. The data logger along with accelerometer gave the time domain acceleration data. These measured values of acceleration are then used and using fast Fourier transform analysis the frequency domain PSD is achieved.

2.2 Finite element analysis

The CAD model of lower suspension arm of a sedan car developed using CAD software and then imported to ANSYS. The mechanical properties of AL 7075-T6 (aluminum alloy) considered from ASM material data handBook volume 19: fatigue and fracture added to the Finite Element Method (FEM) software (ANSYS). The suspension arm can rotate about Y axis when displaced due to external load therefore in boundary condition the remote displacement with rotation about Y-axis to be free is applied.

2.3 Modal analysis

Modal analysis will help to generate a mathematical model of the structure. With the help of FEM methods, response to dynamic forces can be calculated which helps in designing of the structure. For design of structure under dynamic loading conditions, the important parameters are natural frequencies and mode shapes. Results of modal analysis provides the vibration parameters like natural frequencies and mode shapes of a structural component.

According to above Power spectrum density curve we can represent the power spectrum density function as follows:

$G(f) = 0.001f + 0.1341$ $7.59 \leq f < 124.89$
$G(f) = 0.0007f + 0.1735$ $124.89 \leq f < 292.18$
$G(f) = 0.378$ $292.18 \leq f < 376.83$
$G(f) = 0.0002f + 0.3003$ $376.83 \leq f < 548.43$
$G(f) = -0.00096f + 0.9673$ $548.43 \leq f < 773.29$

Table 1. Mechanical properties of AL7075T6 (Peter 1996).

Property	
Density	2795 Kg/m³
Young's modulus E	7.11E + 10 Pa
Poison's ratio	0.3
Strength coefficient σ_f`	886 MPa
Strength exponent B	−0.076
Ductility coefficient ε_f`	0.446
Ductility exponent C	−0.759
Cyclic strength coefficient k`	913 Mpa
Cyclic strain hardening exponent n`	0.088

Table 2. Natural frequencies of lower suspension arm.

Mode number	Frequency	Power spectrum density
1	7.59	0.141
2	124.89	0.259
3	292.18	0.378
4	376.83	0.378
5	548.43	0.421
6	773.29	0.194

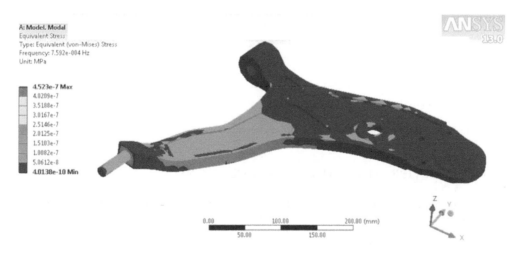

Figure 2. Vibration mode of first order natural frequency.

2.4 PSD response analysis

The input condition, for PSD response analysis, is acceleration due to the road surface and a load spectrum. With PSD response and the results of the modal analysis, we can calculate the random vibration response and value of stress due to this loading. The value of maximum stress achieved here is the most dangerous point and is considered as the input condition of the vibration fatigue life analysis.

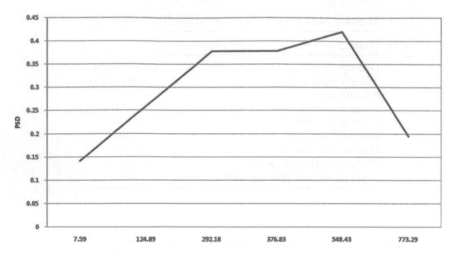

Figure 3.   PSD function curve.

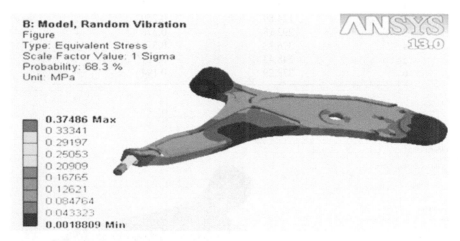

Figure 4.   Stress diagram of LSA under random vibrations.

3   RANDOM VIBRATION FATIGUE CALCULATIONS

The minor's cumulative fatigue damage theory is used for calculating vibration fatigue life; The Steinberg's three interval method is used to compute lives of the structural parts. According to Steinberg's three interval method, we can divide the stresses into three stress intervals as $-1\sigma \sim 1\sigma$, $-2\sigma \sim 2\sigma$, $-3\sigma \sim 3\sigma$. Steinberg's method assumes that if the stress is out of these three intervals, it will not cause any damage. The stress that is greater than 3σ occurs only within 0.27% of the time and makes no difference.

Assuming the random vibration time is T for which the component has to operate without any sign of failure, n-order moment of inertia is:

$$Mn = \int_{-\infty}^{+\infty} f^n G(f) df \qquad (1)$$

The moment of inertia is the area under the curve of the PSD function.
G(f) is the PSD function and represents the frequency.
Vibration average frequency f_0 can be calculated by the following formula,

$$f_0 = \frac{1}{2\pi}\sqrt{\frac{m_2}{m_0}} \qquad (2)$$

After finding average frequency, we calculate each value according to:

$$n_{1\sigma} = 0.683 \ T \ f_0 \quad n_{2\sigma} = 0.271 \ T \ f_0, \ \text{and} \ n_{3\sigma} = 0.0433 \ T \ f_0.$$

Using standard manual, we get the corresponding $N_1\sigma$, $N_2\sigma$, and $N_3\sigma$.
Now using Miner's rule calculate cumulative fatigue damage,

$$D = \frac{n_{1\sigma}}{N_{1\sigma}} + \frac{n_{2\sigma}}{N_{2\sigma}} + \frac{n_{3\sigma}}{N_{3\sigma}} \qquad (3)$$

In the above equation $ni\sigma$ (i = 1, 2, and 3) is the actual number of cycles when the stress amplitude equal to σi (i = 1, 2, and 3) or less than that is applied on LSA. In the above equation, $Ni\sigma$ (i = 1, 2, and 3) are the relevant life in number of cycles. If the final total fatigue damage D is less than 1, it means that there is no structural damage. So if this condition is satisfied (D < 1) then we can say that the design is able to satisfy the fatigue needs. If D ≥ 1 then, we need to redesign to meet the demand of fatigue. According to random vibration stress analysis, the maximum stress acting on the LSA is 0.37486. Using three interval methods:

$$M_0 = \int_{-\infty}^{+\infty} f^n G(f) df$$
$$= 245.216$$

$$M_2 = \int_{-\infty}^{+\infty} f^2 G(f) df$$
$$= 50112530$$

$$f_0 = \frac{1}{2_\pi}\sqrt{\frac{m_2}{m_0}} = 71.948 \ Hz$$

$$n1\sigma = 0.683 * T * f_0 = 0.683 * 9 * 10^6 * 71.948 = 442.26 * \ 10^6$$

$$n2\sigma = 0.271 * T * f_0 = 0.271 * 9 * 10^6 * 71.948 = 175.48 * 10^6$$

$$n3\sigma = 0.0433 * T * f_0 = 0.0433 * 9 * 10^6 * 71.948 = 28.038 * 10^6$$

According to Standard manual for AL7075T6: when 1σ = 0.37486 Mpa, $N1\sigma$ = ∞. When 2σ = 0.74972 Mpa, $N2\sigma$ = ∞. When 3σ = 1.12458 Mpa, $N3\sigma$ = ∞.
Total damage is:

$$D = \frac{n_{1\sigma}}{N_{1\sigma}} + \frac{n_{2\sigma}}{N_{2\sigma}} + \frac{n_{3\sigma}}{N_{3\sigma}} = \ll 1$$

So we can conclude that the lower suspension arm is safe under the given loading conditions.

4   CONCLUSION

This paper explains the random vibration fatigue effects on lower suspension arm of a sedan car. Frequency response based fatigue analysis method is used for lower suspension arm of a sedan car. Results of the FEA show that the life predicted from the vibration fatigue analysis is consistently higher, and the component is safe under random vibration loading. Furthermore, the vibration fatigue analysis can improve the understanding of system behaviors in terms of frequency characteristics of the structures, loads, and their couplings.

REFERENCES

Andrew, H.,1999. A Frequency domain approaches for fatigue life estimation from Finite Element Analysis, International Conference on Damage Assessment of Structures.

Bahkali, A. 2013. NVH and random vibration fatigue analysis of a landing gear's leg for an unmanned aerial vehicle using LS-DYNA. *European LS-DYNA Conference.*

Patel, P. 2013. Comparative study of SAE 1045 (carbon steel) and aluminum alloy 7075-t6 for lower suspension arm of a sedan car. *International journal of advanced research in engineering and Technology, 2,*119–124.

Peter, A. 1996. Fatigue and Fracture. ASM Handbook (564–565).

Petrucci, G. 2004. Fatigue life prediction under wideband Random loading, Fatigue and Fracture of Engineering Materials and Structures

Raheman, M.M. 2009. Finite element based vibration fatigue analysis for a new free piston engine component. *Arabian journal of science and engineering, 34,* 231–246.

Slavic, J. 2013. Frequency domain methods for a vibration fatigue life estimation-application to real data. *International Journal of fatigue,* 47, 8–17.

Si-Xue Qu, 2013. Analysis of random vibration life of mechanical parts of an actuating cylinder based on the finite element. *International Conference on Quality, Reliability, Risk, Maintenance, and Safety Engineering,* 917–920.

Multi-disciplinary Sustainable Engineering: Current and Future Trends – Tekwani, Bhavsar & Modi (Eds)
© 2016 Taylor & Francis Group, London, ISBN 978-1-138-02845-6

Fixation mechanism for dielectric material in electromechanical packages for space payload

D.S. Chaudhari, D.A. Vartak, K.B. Vyas & R.M. Parmar
Space Applications Centre (ISRO), Ahmedabad, Gujarat, India

ABSTRACT: An electromechanical package is designed to operate over designated microwave frequencies in space payloads. Substrate of thermostat microwave material – TMM10i material needs to be fixed directly on package with screws and it can be attached to carrier plate. Analysis of the fixation mechanism is challenging and important as stress developed in this brittle TMM10i substrate at various stages like assembly, testing, etc may result in cracking of substrate. It may eventually fail the electromechanical package and ultimately result into a total failure of payload and the mission. This paper describes the fixation mechanism of TMM10i substrate by fasteners and by attaching substrate to carrier plate. Efforts are made to analyze torque value for fixing substrate & sufficient stress margin against its fractural strength. FE analysis is carried out to find suitable material for carrier plate & stress developed at various stages during assembly and testing. It also derived the safety margin against failure of substrate.

1 INTRODUCTION

Electromechanical package provides electrical interfaces in the components in the spacecraft system to ensure high reliability. The package consists of substrate mounted on carrier plates, stepped cover and electronic devices, which are also mounted on substrates and tray surface. TMM10i needs to use as a substrate for its advantage of high frequency carrier, which is designed for microstrip application. Fixing mechanism of TMM10i depends on the thickness of substrate and its applications. Moreover, electromechanical package has to withstand stringent conditions during launch and the space environment. Any failure in substrate results in breaking of microwave integrated circuits.

Housing of microwave electronic components is carried out in a well-designed enclosures knows as electromechanical packages (Fig. 1). These are designed for microwave integrated circuits and printed circuit boards. The electromechanical packages contain small electronic components which are mounted on substrates.

Figure 1.   Electromechanical package.

Presently alumina substrate is used with kovar carrier plate as its coefficient of thermal expansion is suitable for avoiding any thermal stresses. TMM10i is an advanced circuit material for RF and microwave circuit with wide range of dielectric constant. However, the electrical andmechanical properties of TMM10i laminates offer combines the benefits for both ceramic and traditional PTFE microwave circuit laminates which can aptly replace the alumina. TMM10i substrates with thickness 1.27 mm (0.05") & 20 mil (0.02") are required to be used in the package for microstrip application.

Mechanism for the fixation of TMM10i substrate to package is developed by two ways:

- Substrate with 1.27 mm thickness – It is directly mounted by fasteners on the package.
- Substrate with 20 mil thickness – It is attached with carrier plate made of suitable space qualified materials like kovar, aluminum alloy, and copper.

2 FIXATION MECHANISM BY FASTENERS

TMM10i substrate with thickness 1.27 mm (Fig. 2) is mounted directly on an aluminum alloy package using hexagonal socket head screws with torque. This fixation mechanism offers several advantages like ease in rework process, usage of desired size of substrate, and weight reduction by avoiding the carrier plate. The fixation methodology depends on various factors like size of screws used, warpage of substrate, electronics components on substrates, brittleness of substrate etc. However, the factor of substrate-brittleness makes this methodology critically significant. Also, it is important that the substrates mounted on the package should be properly tightened. If it is over tightened, it may add extra stress on the thread and lead to the cracking of substrate. And if it is under-tightened, then due to vibration during transportation or other operations, there is a high possibility of loosening of screws, which may come out from the assembly. To avoid this known problem area, it is absolutely necessary to tighten every substrate assembly with an appropriate torque value.

2.1 *Evaluation of torque and its effect*

The torque value depends on type of material being assembled, size of screws & safe load applied.

$$Torque\ T = K \times D \times W \tag{1}$$

where T = torque in Nm; K = friction factor; D = screw diameter in meter; W = load applied in N.

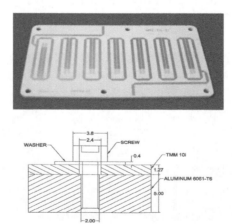

Figure 2.   TMM10i substrates 1.27 mm (thick) & assembly of with aluminium box.

2.1.1 *Importance of friction factor K*

Friction factor "K" plays a very important role in torque application. "K" depends on the material, size, surface friction, and threading of the screw. Its value for the screw bearing area and thread area of screw is different. TMM10i is in contact with under head of SS304 screw, whereas helicoil of SS304 is in contact with threads of screw.

Where, "p" is the lead (pitch) of the thread, r_t is the radius of the bolt, r_c is the radius of the Head, μ_t is the coefficient of friction (COF) between threads & helicoil, μ_c is the coefficient of friction between head & TMM10i and α is one half of the thread angle (usually 30°) (Fig. 3).

K = Friction factor (Nut factor)
$K = K_1 + K_2 + K_3$
Where, K_1 = Geometric factor
K_2 = Thread friction related factor
K_3 = Under head friction related factor

Using the formula to find out $K1$, $K2$, & $K3$

$$K_1 = \frac{p}{2\pi D} = 0.0318$$

$p = 0.4$ mm for $M2$ screw, $D = 2$ mm diameter for $M2$

$$K_2 = \frac{\mu_t r_t}{D \cos \alpha} = 0.083$$

$r_t = 0.87$ mm, α = half of the thread angle (usually 30°), $\mu_t = 0.18$. Coefficient of friction for SS304 thread

$$k_3 = \frac{\mu_c r_c}{D} = 0.168$$

$r_c = 1.05$ mm (Diameter of plain washer 4.5 mm, head diameter of bolt 3.8 & diameter of hole of substrate TMM10i 2.4 mm, contact radius: screw head to plain washer $r_c = 0.7$ & washer to TMM10i $r_c = 1.05$), COF on SS304 washer & TMM10i (μ_c) is 0.20.

$$\text{Friction factor K} = 0.0318 + 0.083 + 0.168 = 0.283 \tag{2}$$

2.1.2 *Safe applied load (W)*

While assembling the TMM10i substrate in the aluminium alloy package, over tightened of screws may damage threads of package due to its softness. Safe applied load (W) avoids the shearing failure of threads. Yield strength of aluminium alloy is considered 240 MPa.

$$W = A_b \times \sigma_y$$

A_b = bolt stress area, σ_y = yield strength of aluminium

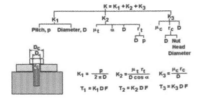 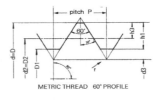

Figure 3.   Details of friction factor K & nomenclature of metric thread.

$$A_b = \frac{\pi}{4} \; d^2 = 2.38 \; mm^2$$

A_b = bolt stress area, σ_y = yield strength of aluminium, $d = D - 0.648$ p, where D = major diameter of bolt = 2 mm, p = pitch = 0.4 mm.

Considering, 75% of yield strength of aluminium (factor of safety) = 180 MPa.

$$Safe \; applied \; load \; W = 428.4 \; N \tag{3}$$

2.1.3 Effect of preload of bolt on TMM10i material

Stress experienced by TMM10i material due to preload of the bolt.

Stress on TMM10i $[\sigma]$ = Preload of bolt/Stress area

$$\sigma = \frac{W}{\frac{\pi}{4}[D1^2 - d1^2]} = 37.67 \; MPa$$

$D1$ = diameter of plain washer, $d1$ = diameter of hole on TMM10i

Stress produced on substrate due to preload is less than its flexural strength nearly by 2.5 times which does not lead any failure of substrate.

2.1.4 Calculation of torque (preload)

From above, torque $T = K \times D \times W$

Now, putting values from (2), (3) & considering D = 1.7 mm (effective diameter),

$$T = 0.206 \; Nm \approx 0.21 \; Nm \tag{4}$$

2.2 Effect of load on TMM10i substrate during dynamic test

TMM10i substrate with 0.21 Nm preload has to sustain the load experienced during vibration without structural degradation. Load experienced by the substrate is derived from the response against input qualification levels 11.8 g & 17.5 g in parallel to mounting & perpendicular to mounting plane.

$$Mass \; of \; substrate \; (TMM10i) = 11.91 \; g.$$

Load experienced by TMM10i substrate during vibration is very less compared with the applied load by preload. There is sufficient margin against the load experienced during dynamic testing. It also ensures assembly integrity of substrate with box.

2.3 Finite element analysis

FE analysis in Hyperworks is carried out to simulate stress induced (Fig. 4) during vibration test. Boundary condition: constrain at screw location (fixed) & load at CG 109 g.

Table 1. Gpeak and load calculation from vibration plots.

Axis	Amplification in random response g/g	Response at 11.8 & 17.5 g in x, y, & z axis g	Gpeak actual g	Load on substrate Kg
Y	1.33	15.69	47.07	0.560
X	2	23.6	70.8	0.843
Z	2.08	36.4	109.2	1.300

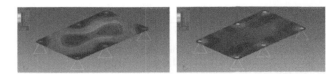

Figure 4.   Displacement and stress pattern at 109 g load.

Table 2.   Displacement and maximum stress at 109 g.

Input g	Displacement mm	Max principle stress MPa	Safe strength MPa
109	0.015	6.99	94

Figure 5.   Substrate TMM10i and fixture—test box.

Substrate TMM10i

Carrier Plate

Figure 6.   TMM10i substrates 20 mil and carrier plate configuration.

Stress induced in TMM10i substrate at 109 g is 7 MPa which has sufficient margin against its flexural strength.

2.4   *Experimental analysis*

Experimental study is carried out to find maximum torque value at which TMM10i substrate breaks. TMM10i substrate is mounted on special type fixture of aluminium material (test box) made for experimental study (Fig. 5). Helicoil was inserted into the tapped holes of fixture to simulate the actual condition. Torque was increased step by step and after each tightening visual inspection was carried out. Torque 0.10 Nm was applied to all six screws initially and gradually it was increased upto 0.50 Nm.

The substrate broke at 0.50 Nm, which indicated a sufficient margin against the derived torque 0.21 Nm, without any fracture.

3   FIXATION MECHANISM BY ADHESIVE BONDING

TMM10i substrate (20 mil thick) can be directly attached to carrier plate. Metallic carrier plate is used to support fragile TMM10i substrate and is soldered by preform (Au80Sn20) (Fig. 6). Space qualified materials like kovar, aluminum alloy 6061T6 and copper can be used as a carrier plate. During attachment, Coefficients of Thermal Expansion (CTE) of two bonding materials play an important role in generating thermo-mechanical strains. Any failure in TMM10i substrate results into failure of the package. Analysis is carried out to determine the space qualified metallic material suitable for carrier plate to avoid the production of stress in substrate at various stages like attachment and testing.

Table 3. Physical properties of materials.

	Modulus of elasticity (MPa)	Density (g/cc)	Poisson ratio μ	Co – efficient of thermal expansion (ppm/C)	Yield stress/ flexural strength (MPa)
Kovar	1.4E + 5	8.3	0.30	5.02	276
Preform (Au80Sn20)	6.8E + 4	15	0.405	16	275
Substrate (TMM10i)	16547	2.77	0.22	20	90
Copper	117E + 3	8.9	0.340	18	100
Aluminum alloy 6061T6	7.0E + 4	2.7	0.3	24	280

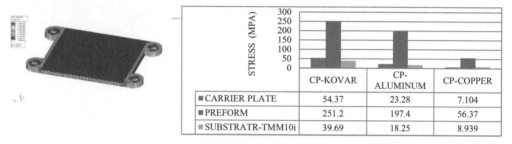

	CP-KOVAR	CP-ALUMINUM	CP-COPPER
■ CARRIER PLATE	54.37	23.28	7.104
■ PREFORM	251.2	197.4	56.37
■ SUBSTRATR-TMM10i	39.69	18.25	8.939

Figure 7. Stress produced during attachment of substrate with carrier plate.

3.1 Carrier plate assembly configuration

Substrate is attached with solder preform on the carrier plate at 300°C. Subsequently, it is slowly cooled to solidify preform and make a perfect joint between substrate and carrier plate.

3.2 Material properties

As per this table, materials for carrier plate like aluminum alloy—Al6061T6 & copper have CTE closer to TMM10i substrate.

3.3 Thermo structural analysis

Analysis is carried out on carrier plate assembly with different materials like kovar, aluminum alloy, and copper along with TMM10i as a substrate material.

- Deflection & stress developed during the attachment of substrate with CP are derived.
- All lugs are under free condition & temperature load given to all nodes.

Case 1: Stress during attachment of substrate
  Carrier plate is heated on heating pad at 300°C in free condition. The preform (Au80Sn20) material is placed on the carrier plate (Melting point: preform: 280°C). Substrate is placed on preform. Analysis is carried out by simulating actual condition during the substrate attachment procedure.
  Boundary condition for analysis:

- Analysis is carried out by taking temperature on the surface of substrate 280°C.
- Carrier plate with 300°C temp. at base & 280°C temp. at preform (melting point).
- Ambient temp. is considered 25°C.
- Carrier plate is in free condition.

	CP - Kovar	CP-Aluminum	CP- Copper
■ SUBSTRATE TMM10i	19.56	7.945	10.35
■ CARRIER PLATE	305.8	52.08	127.3
■ PREFORM	237.6	136.6	143.6

Figure 8. Displacement of CP assembly along with aluminium plate and stress on components of CP assembly during thermal cycling at thermal load 85°C.

Table 4. Stress induced in substrate during attachment with carrier plate of kovar, aluminum, and copper and stress margin.

Carrier plate material	Stress developed on substrate TMM10i during attachment MPa	Flexural strength of TMM10i substrate MPa	Stress margin	Weight of CP assembly grams
Kovar	39.69		2.26	7.56
Aluminium alloy	18.25	90	4.93	3.57
Copper	8.939		10.06	8.00

Case 2: Stress during thermal cycling

Analysis is carried out to find the stress developed on CP components due to thermal load during thermal cycling test at 85° on carrier plate assembly along with aluminum plate. CP with kovar, aluminum, and copper material is taken for the analysis.

Boundary condition for analysis:

- Overall temperature 85°C.
- Ambient temperature 25°C.
- CP is constrained with four lugs on aluminum plate (100 × 100 × 5 mm).

4 OBSERVATION

Fixing mechanism by fasteners:

- TMM10i substrate clamped directly on aluminium plate with torque 0.21 Nm for M2 screws. Stress produced 37.67 MPa (due to torque 0.21 Nm) in bolt, which does not lead any failure in TMM10i substrate.
- Load experienced by TMM10i substrate during vibration is much lesser than clamping load. Moreover, FEA shows that stress induced in TMM10i substrate at 109 g is 7 MPa which has the sufficient margin against its flexural strength. It ensures integrity of substrate with box.
- It is observed from experimental analysis, that screw head shears at torque falls in range of 0.50–0.55 Nm, when used with washer. But screws without any washer, break the substrate at 0.50 Nm due to penetration of head.

Fixing mechanism by adhesive onding:

- FE analysis shows the stresses produced in substrates during TMM10i attachment with aluminum alloy and copper CP are 18.25 MPa & 8.93 MPa which is less than that of kovar CP (39.69 MPa).
- Stress produced on substrate during thermal cycling with aluminum alloy CP is 7.94 MPa, whereas with kovar and copper 19.56 & 10.35 MPa.

345

5 CONCLUSION

TMM10i substrate with 1.27 mm thickness can be mounted directly on aluminium plate with torque 0.21 Nm for M2 screws. TMM10i substrate withstands against preload without degradation and loosening during dynamic testing.

Aluminum alloy 6061T6 can be used to attach TMM10i substrate. It is the best suitable space qualified material than kovar and copper. Stress produced in substrate during attachment and thermal cycling test is less than flexural strength of TMM10i substrate. There is reduction in stresses by 45.98% & 40.16% in TMM10i substrate with aluminum alloy during attachment and thermal cycling that of kovar. Moreover aluminum alloy provides advantage in weight reduction by 44.62% and corrosion resistance against copper.

The fixation mechanism for substrate in electromechanical package described above ensures the integrity of assembly during the assembling and testing without any failure. This helps in avoiding heavy penalty in mission schedule, cost, and technology up gradation.

ACKNOWLEDGMENTS

I would like to thank all the authors Mr. Arup Hait, Sci/Eng SG, Mr. Ulkesh Desai, Sci/Eng SF, Space Applications Centre-ISRO for their guidance and valuable suggestions. Also I would like to acknowledge the authors Dr. BS Munjal, Head-SSD & Mr. K.S. Parikh, Deputy Director SNAA, of Space Applications Centre-ISRO for their technical support which I received during the course of preparation of this paper.

REFERENCES

ASM Handbook. 1997, Volume 2, Non-ferrous Metals & Alloys, American Society of Metals.
Altair HyperMesh Manual. 2012., Version 12.0.
Barrett Richard T. 1990, Fastener Design Manual, 15–16 Nasa Reference Publication 1228.
Bickford John H. 2007, Introduction to Design & behaviour of bolted joint, forth edition CRC press-Davis JR, Concise Meal Hand book.
ISRO Pax-300 - workmanship guideline.
Michael Pecht. 1991, Handbook of Electronics Package Design, Marcel Dekker Inc., N.Y.
MIL-STD-883 Test Methods for Microwave Circuits ISAC Bangalore, Environmental Test Level Specification.
Rogger Corporation Advance Circuit Materials. TMM Thermoset Microwave Materials
Vartak Dhaval & Prof. Manglik VK. 2014, Experimental analysis of substrate failure of microwave electronics packages for communication payloads at International journal of emerging technologies and applications in engineering, technology and sciences.
Vartak Dhaval & Prof. Manglik VK. March 2014, Failure Analysis of Substrate of Carrier Plate of Electromechanical Package for Space Payload in IJERT, volume 3, issue 3.

Multi-disciplinary Sustainable Engineering: Current and Future Trends – Tekwani, Bhavsar & Modi (Eds)
© 2016 Taylor & Francis Group, London, ISBN 978-1-138-02845-6

Performance enhancement of mixed flow submersible pump using CFD

N.C. Patel
CFD Analysis Centre, Jyoti Ltd., Vadodara, India

S.V. Jain & R.N. Patel
Department of Mechanical Engineering, Institute of Technology, Nirma University, Ahmedabad, Gujarat, India

ABSTRACT: In the present study, three-stage mixed flow submersible pump is analyzed using CFD software ANSYS-CFX. The impeller and the bowl of first stage are analyzed by using the rotating and stationary frame of reference, while the interface is taken as frozen rotor. To consider turbulence effects, four different turbulence models are used viz. k-ε, SST, BSL, and SSG. As boundary conditions, mass flow rate is specified at impeller inlet while relative pressure is defined at bowl outlet. At design condition, the pump efficiency was found as 73.54% and 57.71% respectively, with CFD and experimental results provided by manufacturer. The reason for higher efficiency with CFD analysis could be negligence of mechanical losses, volumetric losses and surface friction effects. To improve the efficiency, impeller was modified by reducing the number of blades. The efficiency and head of modified impeller were increased by 0.95% and 2.65% and power input is decreased by 3.5%.

1 INTRODUCTION

A submersible pump is a device mainly used for oil and water extraction, which comprises of an impeller rotating at a high speed and hermetically sealed motor within a stationary casing. The major benefit of this type of pump is that it prevents pump cavitation, which is one of the major problems with a high elevation difference between the fluid surface and pump. It can be classified into two types: (1) Mixed flow, (2) Radial flow.

In order to enhance the performance of the pump, CFD techniques are becoming popular in pump industry. For the knowledge of the flow inside the pump, results of CFD analysis becomes useful than the experimental test results. With the help of advanced CFD software, the flow phenomena like flow separation, inlet pre-swirl, and outlet recirculation can be predicted. By this it saves cost and time of the product development process.

Manivannan (2010) did a detailed CFD analysis to predict the flow pattern inside the impeller; the results show that the percentage increase in the head, power rating, and efficiency are 13.66%, 12.16%, and 18.18% respectively by changing the inlet and outlet vane angles. Chaudhari et al. (2013) carried out the analysis of mixed flow pump impeller to improve the head. Muggli et al. (2002) analyzed the behavior of the flow in a vertical semi-axial mixed flow pump by numerical flow simulations of the entire stage, and the results have been compared to test data. Patel and Ramakrishnan presented CFD analysis of mixed flow pump at duty point and also at part load for the prediction of the flow in the single stage by steady state simulation. Miner (2001) presented the results of a study using a coarse grid to analyze the flow in the impeller of a mixed flow pump using FLOTRAN, which concludes that using small models, CFD can be used effectively in the design process. Si et al. (2013) carried out the multi-objective optimization in low-specific-speed pumps, which indicate that the inlet setting angle of the guide vane influences efficiency significantly and that the outlet angle of blades has an effect on the head and shaft power.

In this paper, CFD analysis of three stage mixed flow submersible pump (discharge: 0.03 m³/s, head: 8 m per stage) is carried out with the existing and modified designs to improve its performance.

2 CFD ANALYSIS OF MIXED FLOW SUBMERSIBLE PUMP

2.1 Governing equations of CFD

The generalized mass conservation equation is given below:

$$\frac{\partial \rho}{\partial t} + div(\rho v) = 0 \tag{1}$$

The generalized momentum conservation equation (Navier-Stokes Equation) is given below:

$$\frac{\partial(\rho u)}{\partial t} + div(\rho uu) = -\frac{\partial p}{\partial x} + div(\mu \, gradu) + SMx \tag{2}$$

where ρ = density of water (kg/m³), t = time (s), v = velocity m/s, u = velocity in x-direction (m/s), p = pressure (N/m²), μ = dynamic viscosity (Ns/m²), S = source term.

2.2 Modeling and meshing

Figure 1 shows the 2-D drawings of impeller and bowl from which 3-D models of impeller and bowl are created.

Computational model of impeller and bowl used for analysis, are created using the CAD software SolidWorks. The meshing of the computational models is done using the tetrahedral mesh with the help of ANSYS-CFX mesh option. Figure 2 shows the computational and meshed models of impeller and bowl.

2.3 Boundary conditions and solution techniques

As boundary conditions, mass flow rate is specified at impeller inlet and relative static pressure and direction are defined at bowl outlet. Water is used as a fluid to carry out all the simulations. The impeller domain was analyzed in rotating frame of reference with rotating speed of 2880 rpm, whereas bowl domain was analyzed in stationary frame of reference. The interface between the impeller and bowl was considered as the frozen rotor. For the pressure-velocity coupling SIMPLE was used. To consider the effect of turbulence, four turbulence models are used viz. k-ε, SST, BSL and SSG. The solution was run in double precision mode of CFX-Solver and convergence criteria were set to 0.001.

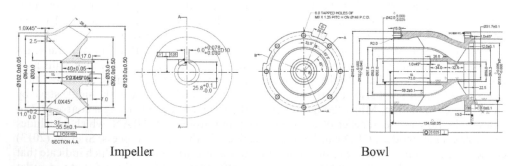

Impeller   Bowl

Figure 1.   2-D drawings of impeller and bowl.

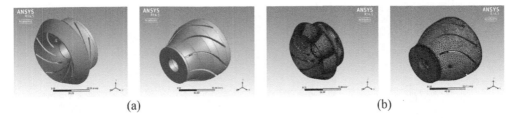

(a)   (b)

Figure 2.   (a) Computational models of impeller and bowl, (b) Meshed models of impeller and bowl.

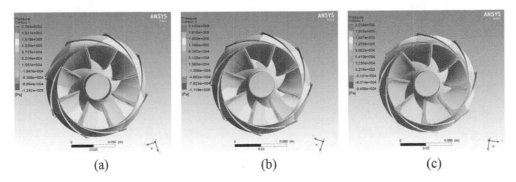

(a)   (b)   (c)

Figure 3.   Pressure contours in impeller at (a) 0.7 Q_{bep} (b) Q_{bep} (c) 1.3 Q_{bep}.

2.4   *Results and discussion*

Initially, to study the effects of grid size variation, analysis was done by increasing the grid elements from the range of 11–30 lacs. After 20 lacs grid elements, the efficiency variation was found within 1%. Hence, further analysis was done with 20 lacs elements. Among various turbulence models, the results obtained with k-ε model were found closest to the experimental results. The pressure contours obtained from analysis for impeller at different flow rates are shown in Figure 3. These contours help in determining the regions of lower pressure which may be prone to cavitation. Due to the energy transfer from rotating impeller, the static pressure increases from impeller inlet to outlet. At the leading edge of blades the value of static pressure is at minimum, and static pressure increases along the flow.

The velocity vectors for part load, bep and over rated condition of flow in the impeller are shown in Figure 4. It shows the direction of flow. Smooth flow can be seen at rated and over rated condition, and at part load condition recirculation zone was found which may lead to decrease in efficiency.

The streamlines for various load conditions of flow in impeller are shown in Figure 5. For Q_{bep} and 1.3 Q_{bep} the flow pattern is uniform, while for 0.7 Q_{bep} vortex formation takes place which increases the losses, and hence efficiency decreases.

The results are plotted in terms of non-dimensional parameters e.g. head number (ψ), discharge number (ϕ), power number (π). The validation of CFD results with model testing results, provided by manufacturer are shown in Figure 6 (Patel, 2015). The maximum efficiency of pump was found as 57.71% and 73.54% with model testing and CFD respectively. The reason for higher efficiency in CFD could be negligence of mechanical losses, volumetric losses and surface friction along with modelling errors. However, the trends for different curves are similar.

$$\psi = \frac{g.H}{n^2.D^2} \tag{3}$$

$$\phi = \frac{Q}{n.D^3} \tag{4}$$

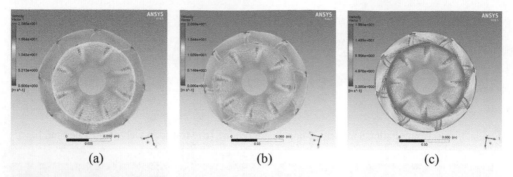

Figure 4. Velocity vectors in impeller at (a) 0.7 Q_{bep} (b) Q_{bep} (c) 1.3 Q_{bep}.

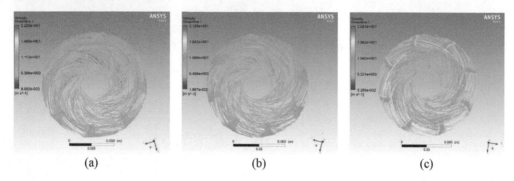

Figure 5. Streamlines in impeller at (a) 0.7 Q_{bep} (b) Q_{bep} (c) 1.3 Q_{bep}.

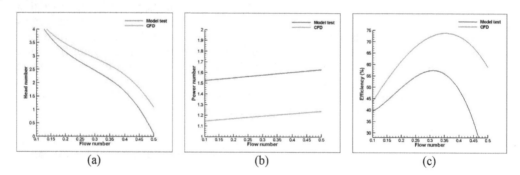

Figure 6. (a) Flow number vs head number (b) Flow number vs power number (c) Flow number vs efficiency.

$$\pi = \frac{P}{\rho.n^3.D^5} \tag{5}$$

where g = acceleration due to gravity (m/s²), H = head (m), n = rotational speed (rps), D = impeller diameter (m), Q = discharge (m³/s), P = input power (W).

3 CFD ANALYSIS OF MODIFIED PUMP

Lobanoff (1985) reported that for better efficiency of pump the number of blades should be six or seven. In the existing pump, the number of blades in impeller and bowl were eight and six respectively. As bowl is having six blades, in the impeller six blades are not recommended.

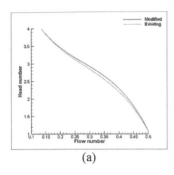
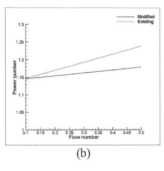
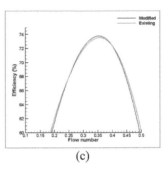

(a)   (b)   (c)

Figure 7.   (a) Flow number vs head number (b) Flow number vs power number (c) Flow number vs efficiency.

Hence, the existing impeller was modified by replacing eight blades with seven blades. Boundary conditions and solution techniques used for the CFD analysis of modified impeller are same as that of the CFD analysis of existing impeller. The comparison of results obtained with modified impeller with existing pump is shown in Figure 7. The modified impeller led to rise in efficiency and head by 0.95% and 2.65% respectively with 3.5% decrease in power consumption.

4   CONCLUSION

In this paper, CFD analysis of three stage mixed flow submersible pump (discharge: 0.03 m³/s, head: 8 m per stage) is carried out with the existing and modified impellers. Among different turbulence models viz. k-ε, SST, BSL, RNG, the results obtained with k-ε model were found most appropriate. The maximum efficiency of pump was found as 57.71% and 73.54% with experimental results provided by manufacturer and CFD respectively. The reason for higher efficiency in CFD could be negligence of mechanical losses, volumetric losses and surface friction along with modelling errors. The predicted efficiency of modified impeller through CFD analysis was 74.24%, which is 0.95% higher than that of existing impeller. The modified impeller led to 2.65% increment in head and 3.5% decrease in power consumption.

REFERENCES

Chaudhari, S.C., Yadav, C.O. & Damor, A.B. 2013. A comparative study of mix flow pump impeller: CFD analysis and experimental data of submersible pump. *International Journal of Research in Engineering & Technology*: 57–64.

Lobanoff, V.S. & Ross, R.R. 1985. Centrifugal pumps design and application, Second ed. Texas: Gulf Publishing Company, Houston.

Manivannan, A. 2010. Computational fluid dynamics analysis of a mixed flow pump impeller. In International Journal of Engineering, Science and Technology: 200–206.

Miner S.M. 2001. 3-D Viscous Flow Analysis of a Mixed Flow Pump Impeller. *International Journal of Rotating Machinery*: 53–63.

Muggli, F.A., Holbein, P. & DuPont, P. 2002. CFD calculation of a mixed flow pump characteristic from shutoff to maximum flow. *Journal of Fluids Engineering*: 798–802.

Patel, N. 2015. Analysis and improvement of mixed flow submersible pump using CFD, M Tech Thesis, Nirma University.

Patel, K. & Ramakrishnan, N. CFD Analysis of Mixed Flow Pump. Jyoti ltd: 1–12.

Si, Q., Yuan, S., Yuan, J., Wang, C. & Lu, W. 2013. Multi-objective optimization of low-specific-speed multistage pumps by using matrix analysis and CFD Method. *Journal of Applied Mathematics.*

Multi-disciplinary Sustainable Engineering: Current and Future Trends – Tekwani, Bhavsar & Modi (Eds)
© 2016 Taylor & Francis Group, London, ISBN 978-1-138-02845-6

Updating bearing coefficients of finite length journal bearings

M. Chouksey
SGSITS, Indore, India

J.K. Dutt & S.V. Modak
IIT Delhi, New Delhi, India

ABSTRACT: The work proposes identifications of eight bearing coefficients of hydrodynamic journal bearings using Inverse Eigen Sensitivity Method (IESM). The closed form expressions for the eight bearing coefficients are only available for short and long journal bearing. Therefore, it is not possible to select eccentricity ratio as the updating parameter for finite length journal bearings. Hence, it becomes important to update the eight bearings coefficients independently for the finite length journal bearings. The work proposes application of Inverse Eigen Sensitivity Method (IESM) to update the bearing coefficients (both direct and cross-coupled stiffness and damping coefficients). The usefulness of the method is also checked in the presence of measurement noise.

Keywords: journal bearing coefficients; model updating; inverse eigensensitivity

1 INTRODUCTION

Finite element model updating techniques are used to adjust selected parameters of finite element models in order to make the models compatible with experimental data. Broadly, the model updating methods are classified into direct and iterative methods. Iterative methods provide flexibility of selecting the updating parameters apart from overcoming the shortcomings of the direct methods. Minimization of an objective function forms the basis in the iterative methods. The iterative methods using eigenvalue and eigenvector are termed as Inverse Eigen Sensitivity Methods (IESM).

Rotor systems are non-self-adjoint systems due to presence of spin speed dependent forces like gyroscopic and circulatory forces. The eigensolution and their derivatives for the equation of motion of Non-Self-Adjoint Systems consist of complex numbers. To obtain the derivatives of eigenvalues and eigenvectors in state space, the work of Plaut and Huseyin (1972; Adhikari and Friswell (2001) may be referred to. This work is based on IESM and uses the sensitivities of eigenvalues in writing the sensitivity matrix. The expressions for sensitivities of eigenvalues are obtained after following the work of Plaut and Huseyin (1972).

Journal bearings are characterized by eight bearing coefficients per bearing to represent stiffness and damping terms (Rao (1996)). Closed form expressions for the bearing coefficients are only available for specific cases based on short and long bearing assumptions. Results of modal analysis can be used to update bearing coefficients of rotor systems (Chouksey et al. (2013), Chouksey et al. (2015), Chouksey et al. (2013)). The work by Chouksey et al. (2012), Chouksey et al. (2012), Chouksey et al. (2012) can be referred to for modal analysis of rotor systems. Recently, Chouksey et al. (2014) proposed an IESM based approach to update the eight bearing coefficients of short journal bearings by using a single parameter i.e. eccentricity ratio. Bearing coefficients of finite length journal bearing (not falling in the category of short or long journal bearings) are generally not known and need to be identified. Identification of these bearing coefficients has been attempted in this work through the process of finite element model updating.

2 THE IESM FOR ROTOR-SHAFT-BEARING SYSTEMS

The equations of motion for a rotor-shaft-bearing system may be written in the state space form as:

$$A(p)\dot{w}(t) = B(p)w(t) + f(t) \tag{1}$$

where, each element of $A(p)$ and $B(p)$ is a function of physical parameters represented by vector 'p', and will further be written as A and B respectively. The work as in Chouksey et al. (2014) can be referred for details of the equations of motion and eigenvalue analysis.

The first order derivative of the rth eigenvalue with respect to j^{th} updating parameter is written below after following the work as in Plaut and Huseyin (1972),

$$\lambda_{r,j} = \varphi_r^T (B_{,j} - \lambda_r A_{,j}) \psi_r \tag{2}$$

The matrices $A_{,j}$ and $B_{,j}$ are obtained by differentiating elements of A and B matrix respectively with respect to the jth updating parameter as:

$$A_{,j} = \begin{bmatrix} 0 & M_{,j} \\ M_{,j} & C_{,j} \end{bmatrix}_{2N \times 2N}, \quad B_{,j} = \begin{bmatrix} M_{,j} & 0 \\ 0 & -K_{,j} \end{bmatrix}_{2N \times 2N} \tag{3}$$

Eigenvalues are generally nonlinear functions of design/updating parameters. Changed eigenvalues as a result of changes in the updating parameters may be expressed as a linear function of changes in the updating parameters by linearizing the expression of the eigenvalues using the Taylor series expansion as given in equation (4)

$$\lambda_x^r \approx \lambda_a^r + \sum_{j=1}^{nu} \left(\lambda_{r,j} \Delta p_j \right) \tag{4}$$

where λ_x^r and λ_a^r represents the experimentally identified (In this work, using the RM) and finite element model eigenvalues respectively corresponding to the rth mode, whereas the term Δp_j represents the change in j^{th} updating parameter. The eigenvalues sensitivity ($\lambda_{r,j}$) may be obtained using equation (2).

Using equation (2), equation (4) can be written for 'm' modes and 'nu' updating parameters. Using vector-matrix notation, equation (4) may be expressed for 'm' eigenvalues after multiplying and dividing by p_j and writing b_j in place of $\Delta p_j / p_j$ as:

$$\Delta q \approx S \times b \tag{5}$$

where Δq is the vector formed by difference of experimentally identified and corresponding finite element model eigenvalues, S is the sensitivity matrix of all the m eigenvalues considered with respect to nu updating parameters the size of which is $m \times nu$ and b is the vector of fractional correction factors for nu updating parameters. For further details about solution of equation (7), the work by Chouksey et al. (2014) can be referred to.

3 JOURNAL BEARING MODEL

Bearings are generally modelled using linear load-deflection relationships to simplify the dynamic analysis. A bearing characteristic is generally expressed with the help of four stiffness and four damping coefficients. The relationship among radial bearing force $\{f_{by} \quad f_{bz}\}^T$, radial displacement $\{y \quad z\}^T$ and radial velocity $\{\dot{y} \quad \dot{z}\}^T$ of the bearing elements is given by:

354

$$\begin{bmatrix} k_{yy} & k_{yz} \\ k_{zy} & k_{zz} \end{bmatrix} \begin{Bmatrix} y \\ z \end{Bmatrix} + \begin{bmatrix} c_{yy} & c_{yz} \\ c_{zy} & c_{zz} \end{bmatrix} \begin{Bmatrix} \dot{y} \\ \dot{z} \end{Bmatrix} = - \begin{Bmatrix} f_{by} \\ f_{bz} \end{Bmatrix} \tag{6}$$

The bearing stiffness and damping coefficients for journal bearing are derived by using the Reynolds equation. The Reynolds equation is a non-homogeneous partial differential equation of two variables (pressure and thickness of oil film) for which the general closed form analytical solution is not available (Srikanth et al. (2009)). However, special cases based on short bearing (L/D << 1, L and D being length and diameter of the bearing respectively) or long bearing assumptions (L/D >> 1) enable to give closed form solutions for the bearing coefficients.

4 NUMERICAL EXAMPLE

The rotor-shaft system (as shown in Fig. 1), which is a slightly modified version of the one given as in Friswell et al. (2010), is considered as an example for the purpose of simulation, as well as illustration. The two-disc steel-rotor-shaft, where the density of the shaft and discs is 7810 kg/m³, is supported on plain journal bearings at its ends.

4.1 Initial Model (IM) and Reference Model (RM)

Two FE models 1) Initial Model or the model to be updated (called hereafter as "IM") and, 2) a Reference Model (called hereafter as "RM") are first created with the purpose that the Initial Model (IM) can be updated by adjusting its system matrices to obtain an Updated Model (UM) such that the dynamic behaviour predicted by Updated Model (UM) matches as closely as possible, the dynamic behaviour predicted by RM. Since the Reference Model (RM) is used here to simulate the experimental dynamic behaviour of the rotor-shaft system, it is built with a much finer mesh. The eigenvalues extracted from the RM are treated as experimental eigenvalues in the process of updating. Random noise is also added to the eigenvalues obtained from the RM to simulate measurement noise present in experimentally identified eigenvalues.

In view of this, the IM is built with 6 two noded beam finite elements of length 0.25 m each whereas, the Reference Model (RM) is constructed with 24 two noded beam finite elements of 0.0625 m each. Four degrees of freedom have been considered at any node in each model. Secondly, to simulate the parametric variations, the eight bearing coefficients of the two journal bearings have been augmented by arbitrary variations about the values used for IM to build the RM. Measurement noise is always unwanted, however unavoidable for any updating process, as this impairs the accuracy of updating. Hence the influence of noise has

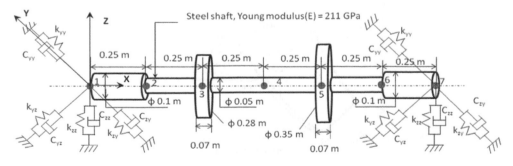

Figure 1. Rotor-bearing system supported on journal bearing at the ends.

also been investigated at last in order to get an idea about the accuracy of updating. This also throws some light on the whole process of modal testing of rotor-shaft system.

4.2   Updating of coefficients of finite journal bearings

This section demonstrates the updating of eight bearing coefficients of journal bearing, simulating the case of finite length journal bearings. The rotor system is considered to be supported on similar journal bearings, so that the bearing coefficients for both of the bearings are assumed to be the same. Table 1 shows the bearing coefficients for the IM along with the percentage deviations to build the RM and the corresponding RM parameters.

The rotor-shaft is assumed to be spinning at 1500 rpm. Based on the difference of real and imaginary parts of initial eight eigenvalues of IM and RM, the objective functions is made, and then updating process is carried out as per equation (10). Fig. 2 shows, the variation of cumulative correction factors after each iteration. The figure shows that the percentage corrections factors converge to finite values as per the deviations introduced in the initial model. Fig. 3(a) and 3(b) depict the percentage errors in the natural frequencies and modal damping factors for the initial eight modes, which shows that errors in both the natural frequencies and modal damping factors almost vanish after about 40 iterations. This shows that the updating procedure is capable of updating the eight coefficients (both direct and cross coupled stiffness and damping coefficients) of the bearing, simulating the case of finite length journal bearings.

The updating of eight coefficients of journal bearing is also checked in the presence of measurement noise. Fig. 4(a) and 4(b) show that even in the presence of measurement noise, the method updated the coefficients in such a way that it reduces the percentage errors in the natural frequencies and modal damping factors considerably.

Hence, the technique of finite element model updating is extended to update coefficients of finite length journal bearings. Recently, Sfyris and Chasalevris (2012) represented coefficients of finite journal bearings in close form in terms of eccentricity ratio. Therefore, the technique of updating eccentricity ratio can also be applied to update the coefficients of finite length journal bearings.

Table 1.   Bearing coefficients for IM and RM.

	K_{yy}	K_{yz}	K_{zy}	K_{zz}	C_{yy}	C_{yz}	C_{zy}	C_{zz}
IMParameters (N/m)	3×10^7	8×10^6	-8×10^6	3×10^7	5000	5000	5000	5000
% deviations	90	70	60	80	50	30	20	40
RMParameters (N/m)	5.7×10^7	13.6×10^6	-2.8×10^6	5.4×10^7	7500	6500	6000	7000

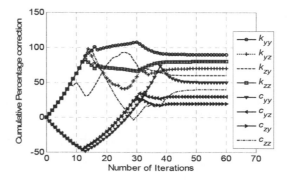

Figure 2.   Cumulative percentage correction for updating parameters.

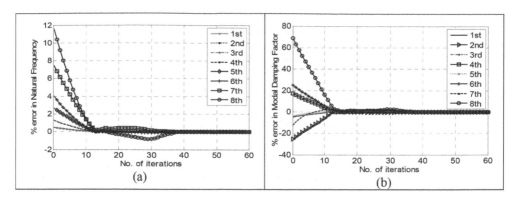

Figure 3. Variation in percentage error for (a) Natural frequencies (b) Modal damping factors.

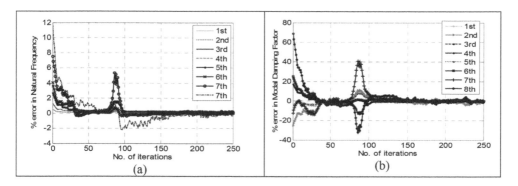

Figure 4. Variation in percentage error (±1% noise) for (a) Natural frequencies, (b) Modal damping factors.

5 CONCLUSIONS

The method of finite element model updating has been applied to update the eight bearing coefficients; both direct and cross coupled stiffness and damping coefficients as in the case of finite length journal bearings. It is shown that the model updating method, based on inverse eigen-sensitivity updates the eight bearing coefficients quite closely (both direct and cross-coupled bearing stiffness and damping coefficients). The effectiveness of the model updating technique to update the eight bearing coefficients has also been shown under the presence of measurement noise. It is found, that the method updates the eight bearing coefficients closely even under the presence of measurement noise.

REFERENCES

Adhikari, S. & Friswell, M.I. 2001. Eigenderivative analysis of asymmetric non-conservative systems. *International Journal for Numerical Methods in Engineering*, 51: 709–733.
Chouksey, M., Dutt, J.K. & Modak, S.V. 2012. *Experimental modal analysis studies for spinning rotor system*. *VETOMAC-VIII*, Gdansk, Poland.
Chouksey, M., Dutt, J.K. & Modak, S.V. 2012. Modal analysis of a flexible internally damped rotorshaft system with bearing anisotropy. *Advances in Vibration Engineering*.
Chouksey, M., Dutt, J.K. & Modak, S.V. 2012. Modal analysis of rotor-shaft system under the influence of rotor-shaft material damping and fluid film forces. *Mechanism and Machine Theory*, 48: 81–93.

Chouksey, M., Dutt, J.K. & Modak, S.V. 2013. Model updating of rotors supported on ball bearings and its application in response prediction and balancing. *Measurement,* 46(10): 4261–4273.

Chouksey, M., Dutt, J.K. & Modak, S.V. 2013. Updating Bearing Stiffness and Damping Coefficients of a Rotor System. *Journal of the Institution of Engineers (India)—Series C.*

Chouksey, M., Dutt, J.K. & Modak, S.V. 2014. Model updating of rotors supported on journal bearings. *Mechanism and Machine Theory,* 71(0): 52–63.

Chouksey, M., Dutt, J.K. & Modak, S.V. 2015. Multi Speed Model Updating of Rotor Systems. *Vibration Engineering and Technology of Machinery*, Springer: 709–719.

Friswell, M.I., Penny, J.E.T. & Garvey, S.D. 2010. *Dynamics of Rotating Machines*, Cambridge University Press.

Plaut, R.H. & Huseyin, K. 1972. Derivatives of eigenvalues and eigenvectors in non-self-adjoint systems. *American Institute of Aeronautics and Astronautics Journal,* 11(2): 250–251.

Rao, J.S., 1996. *Rotor Dynamics*, New age international (P) Ltd.

Sfyris, D., & Chasalevris, A. 2012. An exact analytical solution of the Reynolds equation for the finite journal bearing lubrication. *Tribology International,* 55(0): 46–58.

Srikanth, D.V., Chaturvedi, D.K. & Reddy, A.C.K. 2009. Modelling of large tilting pad thrust bearing stiffness and damping coefficients. *Tribology in industry,* 31(3 & 4): 23–28.

Multi-disciplinary Sustainable Engineering: Current and Future Trends – Tekwani, Bhavsar & Modi (Eds)
© 2016 Taylor & Francis Group, London, ISBN 978-1-138-02845-6

Thermal analysis and designing of spacecraft payloads and control system

Bhargav Desai & V.J. Lakhera
*Department of Mechanical Engineering, Institute of Technology, Nirma University,
Ahmedabad, Gujarat, India*

Prasanta Das
Space Application Centre(SAC), ISRO, Ahmedabad, Gujarat, India

ABSTRACT: Thermo vacuum testing of satellite payloads are done to study the behavior of payloads under space condition. The aim of the present study is to carry out thermal analysis of the panels and design the thermal control system, and to compare it with the data obtained through thermo vacuum test and further optimize the thermal control system design to maintain the temperature gradient of around 20° over the panels to avoid thermal distortion. The thermal simulation of South and North Panel of typical satellite was done using simulation software and the temperature distribution over the panels and the subsystems was obtained. Thereafter, thermal control system design was carried out for providing artificial heating to the panels and it was implemented in the thermo vacuum testing of the panels and the desired gradient of around 20°C to be maintained over the panel was achieved. The temperature results of test and simulation were compared and the difference obtained was ~2°C. In order to reduce the temperature gradient further, new control system design was checked through simulation, and finally two different designs for North and South Panel were finalized, which had around 3–4°C lesser gradient than the older design which can be implemented in further test to be conducted.

1 INTRODUCTION

Thermo vacuum testing of space craft payloads is to be carried out for its acceptance before launch. So, before carrying out thermo vacuum test, thermal analysis of the panels is done to study the temperature distribution throughout the panels.

The aim is to avoid temperature gradient more than 20°C over the panels. Hence, to achieve that, external heating is provided with the help of **IR** lamps and controlling of heat supplied is done with the help of **PID** controllers and **RTD** sensors. The reason behind maintaining uniform temperature throughout the panel was to avoid thermal distortion.

So the objective was to carry out thermal analysis of spacecraft panels and compare its result with that obtained through thermo vacuum test and also improving the thermal designing of controlling system.

2 METHODOLOGY

2.1 *Thermal analysis of north and south panel of a typical satellite*

Thermal modeling and analysis of panels are done to study the temperature distribution over the panels which will be helpful in providing artificial heating to the panels during thermo vacuum test in order to avoid temperature gradient of more than 20°C.

First of all, thermal modeling was done for each of the subsystem mounted on the panels and the panel itself along with heat pipes embedded inside the panels. After that, various boundary conditions were assigned to the panels such as, heat load was given to each of the subsystem, which dissipate heat; radiation to environment was given to outer portion of the panel of −100°C and internal radiation was assigned to the inner portion of the panel and the subsystems. Also, coupling of 1000 W/m² K was given between the subsystems and the panel, 4000 W/m² K for the heat pipes and element source and conductivity of 3.5 W/mK between the two shells of the panel and it was solved. The temperature results are as shown in Figure 1 and Figure 2, in which it is seen that the portion of the panel below the packages is having higher temperature compared to the vacant portion of the panel.

2.2 *Designing of thermal control system used in thermo vacuum test*

After the temperature results were known over both the panels, designing of the control system was carried out by grouping the IR (Infrared) lamps in different zones and providing artificial heating accordingly, as shown in Figure 3.

The analysis was done at −5°C and 55°C because these are the two extreme temperature limits which should not be crossed, as beyond these temperature limits the subsystem functioning may

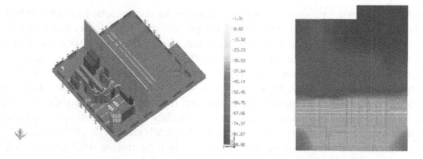

Figure 1.   Thermal modeling and analysis results of south panel.

Figure 2.   Thermal modeling and analysis results of north panel.

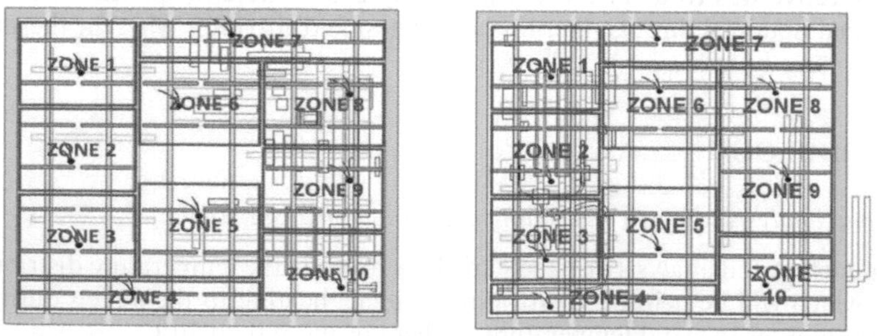

(a)	(b)

Figure 3.   Grouping of IR (Infrared) lamps for north (a) and south (b) panel.

start deteriorating. After applying various heat powers to different groups, the optimum solution obtained was that on North Panel the minimum gradient achieved at −5°C and 55°C panel temperature was 14.42°C and 10.75°C as shown in Figure 4. Similarly, on the South Panel minimum gradient obtained was 20.76°C and 17.87°C as shown in Figure 5. [Research INC].

2.3 Thermo vacuum testing and validation of simulation results

Thermo vacuum testing is done to check the performance of panels in space environment. So, as the results obtained through control system design had temperature gradient less than 20°C, this design was implemented in thermo vacuum test, which is done at −100°C test chamber temperature and vacuum of around 10^(−6) milibar as shown in Figure 6. [ISRO].

As seen in Figure 6, the test cycle is to be followed between −5°C and 55°C, and the condition hot soak and cold soak is the one in which the payloads (subsystems) are in working condition and during the rest of the cycle the payloads are in off condition. After the vacuum test, the test results and simulation results were compared and the output obtained was as shown in Table 1. [STG Department, SAC, ISRO].

It can be seen from the above table that the difference between the test results and simulation results is ~2°C. Hence, the simulation results are accurate and can be validated.

(-5°C)   (55°C)

Figure 4. Temperature results of control system design at −5°C and 55°C of north panel.

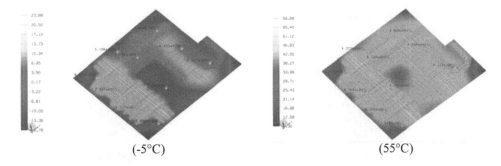

(-5°C)   (55°C)

Figure 5. Temperature results of control system design at −5°C and 55°C of south panel.

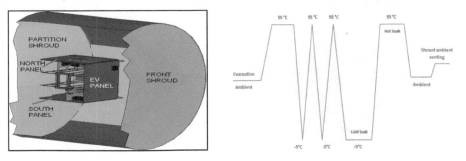

Figure 6. Thermo vacuum test chamber and test cycle.

2.4 *Optimization of control system design*

The control system design was needed to be optimized in order to reduce the temperature gradient over the panels further and hence, different designing of IR lamps were simulated further along with reduced number of IR lamps from forty to thirty six, and also changing the alignment of the lamps in order to provide sufficient heating even at the corners of the panel which was not possible in pervious designs. So in order to reduce the gradient further different design were simulated and the best grouping obtained was as shown in Figure 7.

The above optimized design was obtained after trying one after the other designs by rectifying the drawbacks faced in the previous designs.

By applying various heating power conditions to the optimized groups, the minimum gradient obtained for the North Panel was 12.41°C and 7.93°C for the panel temperature of −5°C and 55°C, as shown in Figure 8. Similarly, for the South Panel, the minimum gradient obtained was of 16.82°C and 12.27°C for the panel temperature of −5°C and 55°C as shown in Figure 9.

Table 1. Validation of simulation results with that of test results.

Panel	Test Results		Simulation Results	
	Cold Soak	Hot Soak	Cold Soak	Hot Soak
North	~13°C	~8°C	14.42°C	10.75°C
South	~19°C	~5°C	20.76°C	17.87°C

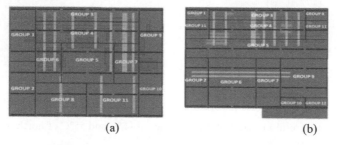

(a) (b)

Figure 7. Optimized grouping of IR lamps for north (a) and south (b) panel.

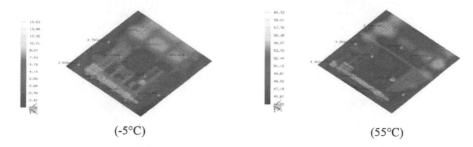

(-5°C) (55°C)

Figure 8. Optimum solutions for north panel at −5°C and 55°C.

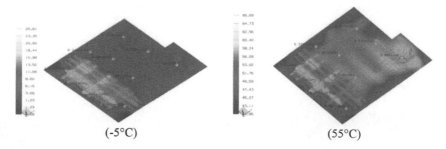

(-5°C) (55°C)

Figure 9. Optimum solutions for south panel at −5°C and 55°C.

Table 2. Comparison of old and new design of control system for north panel.

Design Type	Panel Temperature	Temperature Gradient
Old	−5°C	14.42
New	−5°C	12.41
Old	55°C	10.75
New	55°C	7.93

Table 3. Comparison of old and new design of control system for south panel.

Design Type	Panel Temperature	Temperature Gradient
Old	−5°C	20.76
New	−5°C	16.82
Old	55°C	17.87
New	55°C	12.27

3 RESULTS AND DISCUSSION

The Old and New design that is, the design, used in thermo vacuum test and the one which is the optimum design were compared and the results are as shown in Tables 2 and 3.

From the data mentioned in the table, it is observed that on both the North and the South panel the gradient has reduced ~3–4°C in cold soak and hot soak conditions. Also, the temperature distribution all over the panel is also uniform by making use of lesser number of IR lamps.

4 CONCLUSION AND FUTURE SCOPE OF WORK

- Conclusion:
  1. The thermal analysis and designing of control system of South and North panel was done with the help of simulation software and the results were compared with those of test and the difference in temperature gradient was ~2°C, thereby validating the simulation results.
  2. The design of control system was modified after the thermo vacuum test and the temperature gradient was reduced further by 3–4°C in comparison to the old design. Hence, the new design is optimized and can be implemented in further tests to be carried out.

- Future work:
  1. The testing of South and North panel of a typical satellite is proposed in the thermo vacuum chamber according to the modified design of the control system, in order to maintain the temperature gradient over the panel as per requirement.

REFERENCES

Bhargav Desai, M. Tech. (Thermal Engineering) Thesis, Thermal Analysis and Designing of Spacecraft Payloads and Control System.
Research INC. Providing Control Engineering, Simulate Infrared Heaters Model 5236.
STAG Department, SAC, ISRO, Testing and Evaluation requirement for test instrumentations for IRNSS Thermo vacuum Thermal management.
STG Department, SAC, ISRO, Thermo Vacuum Testing Report of IRNSS 1-D Satellite.

Multi-disciplinary Sustainable Engineering: Current and Future Trends – Tekwani, Bhavsar & Modi (Eds)
© 2016 Taylor & Francis Group, London, ISBN 978-1-138-02845-6

Design improvement of base frame of poucher machine by optimization

Vrajesh T. Makwana & Kaushik M. Patel
Institute of Technology, Nirma University, Ahmedabad, Gujarat, India

Falgun R. Desai
KHS Machinery Pvt Ltd., Ahmedabad, Gujarat, India

ABSTRACT: Poucher machines have a wide scope in packaging products like chocolate, pickle, grains to liquid detergents. From forming of pouch by using the roll of web, filling, and sealing everything is carried out automatically.

In existing design the base frame of the machine assembly is fabricated by welding solid members of rectangular cross-section, which is an over-designed structure. With the help of an optimization base frame the weight material cost and fabrication cost can be reduced drastically which can be helpful in transportation. Hence, in this paper an attempt has been made to optimize the base frame. RHS (Rectangular Hollow Section) of dimensions nearer to existing cross-section has been selected and analyzed, and it is observed that the weight of the base frame has been reduced to 30% without any compromise to the strength of the base frame.

Keywords: frame optimization; FEA; value addition

1 INTRODUCTION

Poucher machine is an exclusive horizontal machine for form, fill, and seal of various sizes of pouch. It produces stand-up re-closeable pouches. Flat roll-stock film is formed into either fin or gusset pouches, which are heat sealed by the individual, temperature controlled seal bars. The dual servo driven feed-roll system, in conjunction with a photo registration sensor, produces pouches of required width. The pouches are then cut with a cutoff knife. Registered pouches are cut and transferred into the filling and closing turret. The individual pouches are opened and filled with product then conveyed to the top seal area. The filled pouches are, then, deflated, closed and top-sealed.

Abbreviation notation	
FEA	Finite Element Analysis
RHS	Rectangular Hollow Section
Pro/E	Pro Engineer
MS	Microsoft

Base frame is the structure of the machine on which various sub-assemblies are mounted. This frame is fabricated by welding rectangular solid members of mild steel material, which is over-designed, as there is no any heavy mounting sub-assembly which gets mounted over the base frame. The base frame has some geometrical constraints that restrict the topology optimization which are of the length of the machine, number of sub assembly, and limitation in use of section other than rectangular hollow section.

2 LITERATURE REVIEW

Farkas & Jármai (Jármai 1999) suggested using cheaper welding technology like SAW instead of SMAW or GMAW, if it is possible. Farkas & Jármai (Farkas 2014) summarized in their article that, it has been shown in previous studies that in the case of square symmetry, the torsional stiffness of cellular plates equals to their bending stiffness. Baotong et al. (Baotong 2014) concluded in their work that together with inspiration from nature and engineering knowledge about machine tool structures a new and practical topology optimization technique can been developed. Heidarpour et al. (Heiderpour 2013) analyzed the data available for the behavior of stub columns at ambient as well as, at the elevated temperatures. And generated one innovative stub column and analyzed it using numerical modelling and concluded that innovative stub column has more buckling strength.

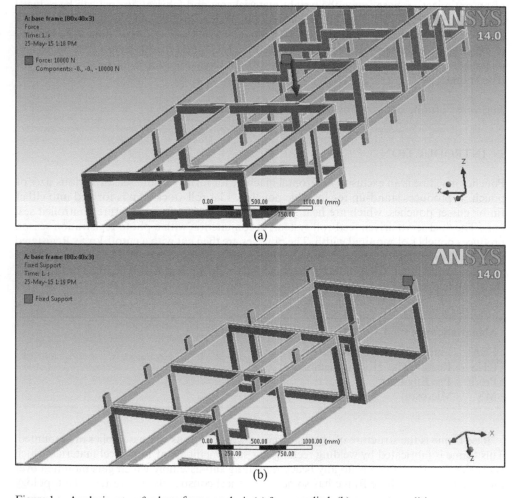

Figure 1.   Analysis setup for base frame analysis (a) force applied, (b) support condition.

3 METHODOLOGY

Pro/E comes with a provision of integration with MS Excel. By using this facility the number of dimensions of CAD model can be modified simultaneously. Thus, a primitive CAD model of base frame has been generated with cross-section dimension similar to the existing one. Then the base frame is analyzed for static structural analysis for given sub-assembly weight. The results are shown in the figures below.

Figure 1 shows how the forces are applied. Here overall weight of the sub-assembly is considered as UDL acting on the top face of the frame structure.

From the results of analysis shown in Figures 2 (a) and (b) it can be concluded that the frame should be further optimized to reduce cost of the base frame assembly.

Now, with help of Pro/E MS excel integration the CAD model cross-section dimensions are modified to the one which are readily available in the market having similar material configuration. The analysis results of two such frames are shown in the Figures 3 and 4.

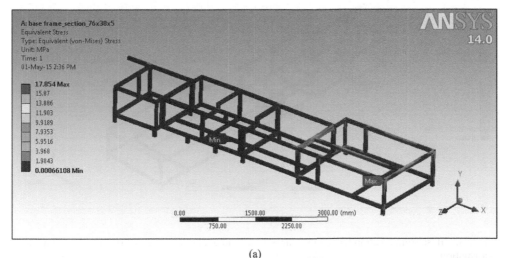

(a)

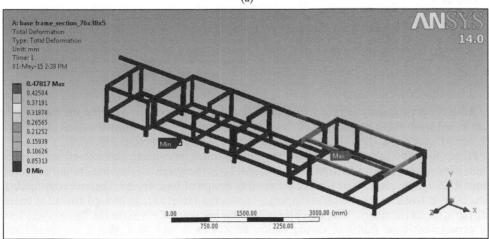

(b)

Figure 2. Base frame section ($76 \times 38 \times 5$ mm) static load analysis results (a) stresses developed (b) deformation.

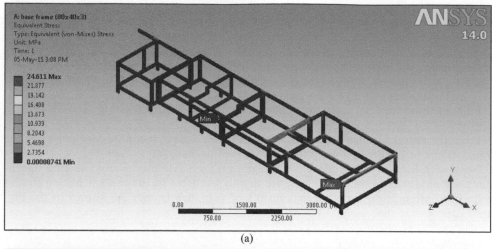

(a)

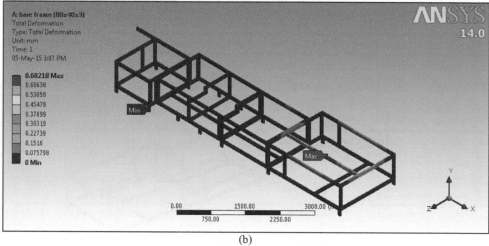

(b)

Figure 3. Base frame section ($80 \times 40 \times 3$ mm) static load analysis results of (a) stresses developed, (b) deformation.

After observing analysis results of Figures 3 and 4, it can be narrated that the base-frame fabricated from RHS is sturdy enough for the application of Pouching machine. The comparative factor of safety for matching cross-section is given in table as following.

Hence, form the Table 1 it is clear that the optimization of base-frame is effective from material utilization point of view and from weight criteria. Here the weight of the base-frame fabricated from RHS weighs almost one third of the existing design of base-frame. Therefore definitely it is a cost effective solution. As far as tapping are concerned rivet nut can be used instead of creating tap in frame members. Also, which is not possible in case of RHS as the thickness is very less. In experiment base frame of configuration $80 \times 40 \times 3$ and $70 \times 30 \times 3$ are considered for optimization because weight of base frame was minimum without compensating the strength.

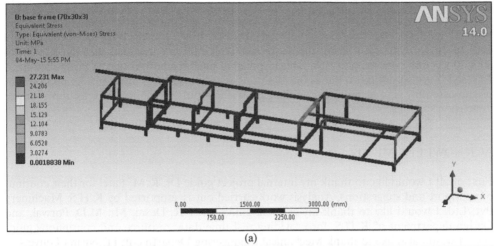

(a)

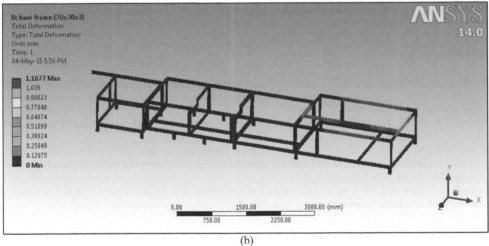

(b)

Figure 4.   Base frame section (70 × 30 × 3 mm) static load analysis results of (a) stresses developed, (b) deformation.

Table 1.   Suitable sections for base frame members and its behavior upon the given load case.

Sr. No.	Section Dimension (B × W × T mm)	Maximum Deformation (mm)	Maximum Stress (MPa)	Factor of Safety
1	78 × 38 × 5	0.47817	17.854	20.5
2	80 × 40 × 3	0.68179	24.611	14.8
3	70 × 30 × 3	1.1677	27.231	13.7

Table 2.   Properties of base frame materials used to carry out analysis.

Young Modulus (N. m²)	2.10e + 11
Poisson Ratio	0.28
Density (Kg. m⁻³)	7860
Thermal Expansion (K)	1.17e − 5
Yield Strength (N. m²)	3.7e + 8

4 CONCLUSION

A CAD model of the base-frame with members of frame having RHS has been created with Pro/E software tool. From the above results, it can be concluded that the RHS with $80 \times 40 \times 3$ mm cross-section is suitable as poucher machine base-frame. A cross-section of $70 \times 30 \times 3$ is also sufficient, but as the mounting point dimensions and considerable amount of sub-assembly is to be mounted on the base-frame it puts restriction on selection of lower dimension sections.

ACKNOWLEDGMENT

First of all I would like to thank my internal project guide Dr. K.M. Patel for their continuous support and suggestions. Analysis work carried out is supported by K.H.S. Machinery Pvt. Ltd. I would like to thank my external guide Mr. F.R. Desai, Mr. M.D. Porwal, and Design department of K.H.S. for providing real time data, resource and continuous guidance. I would also like to thank Mechanical Engineering Department, IT, Nirma University for providing high performance computing resource required for analysis work.

REFERENCES

Baotong, L., Jun, H. & Zhifeng, L. 2014. "Stiffness design of machine tool structures by a biologically inspired topology optimization method", *International Journal of Machine Tools & Manufacture* (84), pp. 33–34.

Farkas, J. & Jármai, K. "Optimization of welded square cellular plates with two different kinds of stiffeners", 2014. *Journal of Constructional Steel Research* (101), pp. 61–65.

Heidarpour, A., Cevro, S., Song, Q. & Zhao, X. 2013. "Behaviour of innovative stub columns utilising mild-steel plates and stainless steel tubes at ambient and elevated temperatures", *Engineering Structures 57 (57)*, pp. 416–427.

Jármai, K. & Farkas, J. 1999, "Cost calculation and optimization of welded steel structures", *Journal of Constructional Steel Research* (50), pp. 115–135.

Multi-disciplinary Sustainable Engineering: Current and Future Trends – Tekwani, Bhavsar & Modi (Eds)
© *2016 Taylor & Francis Group, London, ISBN 978-1-138-02845-6*

Design improvement of radial flow submersible pump using CFD

A.S. Jain
Department of Mechanical Engineering, Vadodara Institute of Engineering, Gujarat, India

R.N. Patel & S.V. Jain
Department of Mechanical Engineering, Institute of Technology, Nirma University, Ahmedabad, Gujarat, India

V.A. Gundale
Vira Pumps, Kolhapur, Maharashtra, India

ABSTRACT: The demand of highly efficient and energy saver submersible pumps force manufacturers to improve the current design and manufacturing processes. In the present study, CFD analysis of 8-stage radial flow submersible pump (discharge: 7 lps and head: 18 m per stage) is carried out using ANSYS CFX software. Steady state simulations were performed in a wide range of discharge varying from 40–180% of rated discharge. As boundary conditions, static pressure is defined at impeller inlet and mass flow rate at impeller outlet. To consider turbulence effects four different turbulence models are used viz. k-ε, SST k-ω, BSL, and SSG. For discretization of momentum and turbulence equations second order upwind scheme was used. SIMPLE scheme was applied for the pressure velocity coupling. To improve the efficiency of pump, the impeller blades are redesigned. The efficiency of modified pump is found as 67.85%, which is around 6.8 % higher than the existing pump.

1 INTRODUCTION

India is an agricultural country. Currently the rain condition has become uncertain due to which farmers have to rely on ground water. The submersible pumps have been proved to be a very good equipment to extract ground water. Average annual electricity consumption by agricultural sector in Gujarat on pumps was 119492 GWh in 2009–10 (Teri, 2012). Overall efficiency of submersible pumps available is about 40% due to which their electricity consumption is more. Indian farmers get electricity at cheaper rates hence, their concern regarding efficiency and more electricity consumption is less. Gujarat Energy Development Agency (GEDA) has done an economic based calculation which compares available pump and efficient pump. It was concluded that if efficient pumps were used in agriculture 34,141 GWh energy could be saved per year (TERI, 2012).

Mehta and Patel (2013) carried out CFD analysis of mixed flow pump impeller by considering backward curved, radial, and forward curved blades. As flow rate increased, pressure decreased in backward curved blades, pressure remained constant in radial blades and pressure increased in forward curved blades. Gundale and Patil (2012) improved the design of radial type vertical submersible pump impeller using CFD. A remarkable increase in the discharge was observed in newly designed impeller. Gundale and Joshi (2013) presented impeller's vane design procedure for radial flow submersible pump. A new vane profile was developed by varying the impeller vane angle such as that of, all other basic dimensions remain unaltered. Muggli et al. (2002) analyzed the behavior of the flow in a vertical semi-axial mixed flow pump by CFD approach and the results were validated with the test data.

In the present study, CFD analysis of eight stage radial flow submersible pump (discharge: 7 lps and head: 18 m per stage) is carried out. As all the stages are similar, analysis of only first stage is carried out. To improve the efficiency new design of impeller was done and analyzed with CFD as a numerical simulation tool.

2 ANALYSIS OF EXISITNG IMPELLER

2.1 Geometry creation and grid generation

The two-dimensional (2D) drawing of impeller is shown in Fig. 1(a). From the 2D drawing, 3D fluid model of was prepared. The 3D fluid model was meshed using meshing module of ANSYS. The impeller was meshed into 322905 tetrahedral elements and 62797 nodes. The fluid model and its meshing are shown in Fig. 1(b & c).

2.2 Boundary conditions and solution techniques

As boundary conditions, static (atmospheric) pressure is defined at inlet of impeller and mass flow rate at outlet of impeller. Steady state mass and momentum conservation equations are solved using Raynold's Averaged Navier-Stokes (RANS) equations in ANSYS CFX software. Impeller was considered as rotating frame and to provide the rotation effect frozen rotor model was used. To consider turbulence effects in fluid, four different turbulence models were used viz. k-ε, Shear Stress Transport (SST) k-ω, Baseline (BSL), Reynolds stress model, and Speziale, Sarkar, and Gatski Reynolds stress model (SSG). Water at standard temperature and pressure was specified as fluid in impeller. Convergence criteria were given as 0.001 for both momentum and continuity equations. SIMPLE scheme was used for the pressure velocity coupling. For momentum and turbulence, second order upwind scheme was used. Reference pressure was kept as 101325 Pa.

2.3 Results and discussions

2.3.1 Grid independency test and selection of turbulence model

Grid independency test was performed by fixing the boundary conditions and varying the grid size to judge the independency of the result with grid size. After around 3 lacs elements the results did not change much with refining grid and deviation in efficiency was less than 1%. Hence, further analysis was done with 3,22,905 grid elements. Among different turbulence models, results obtained with SST k-ω model were found to be closest to the model testing results. SST k-ω model is a combination of original Wilcox k-ω model for near wall region and k-ω model for away from wall region, the combination is made using blending function. Also SST k-ω model is recommended for the flow undergoing separation as it provides higher accuracy in prediction of flow separation and boundary layer simulation (Menter, 1993). Hence, further simulations were performed with SST k-ω model.

2.3.2 Pressure contour and velocity vectors

CFD analysis of impeller was carried out in the wide range of discharge varying from 40–180% of best efficiency point discharge i.e. Q_{bep}. The pressure contours and velocity vectors for different discharge are shown in Figs. 2 and 3 respectively (Jain, 2015). It is observed that at rated and over rated discharge, static pressure variation is uniform but at part load it is non-uniform. The pressure at blade tip is increased which may lead to lower efficiency at part load conditions.

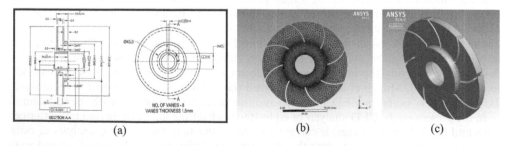

| (a) | (b) | (c) |

Figure 1. (a) 2D drawing (b) 3D fluid model (c) meshing of impeller.

At rated and part load conditions, recirculation is observed near the blade passages which may lead to higher losses and decrease in efficiency. It may be noted that recirculation observed is based on the qualitative assessment of the velocity vectors shown in Fig. 3.

2.3.3 *Validation of CFD results*
The CFD results are compared with the model test results (Fig. 4) in terms of non-dimensional parameters e.g. head number (ψ), discharge number (ϕ), power number (π) defined below.

$$\psi = \frac{g.H}{n^2.D^2} \tag{1}$$

$$\phi = \frac{Q}{n.D^3} \tag{2}$$

$$\pi = \frac{P}{\rho.n^3.D^5} \tag{3}$$

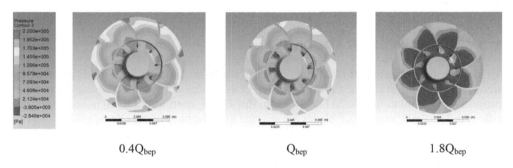

| 0.4Q$_{bep}$ | Q$_{bep}$ | 1.8Q$_{bep}$ |

Figure 2.   Pressure contours in impeller at different discharge.

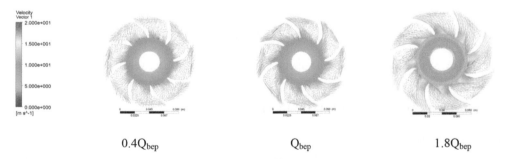

| 0.4Q$_{bep}$ | Q$_{bep}$ | 1.8Q$_{bep}$ |

Figure 3.   Velocity vectors in impeller at different discharge.

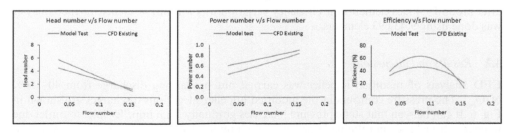

Figure 4.   Validation of CFD results with model test results.

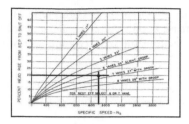

Figure 5.   Chart for selection of number of blades.

where, g is acceleration due to gravity (m/s²), H is head (m), n is rotational speed (rps), D is impeller diameter (m), Q is discharge (m³/s), P is input power (W).

It can be seen that head, power and efficiency predicted by CFD analysis are lower than the model test results provided by the manufacturer. However, the trends of these curves are similar. The average deviation in CFD results is found around 10–12%. This may be due to modeling errors, *as many simplifications were done in the model due to complexity of geometry*, as well as meshing errors. At BEP, the head and efficiency are found as 14.54 m and 46.25% at discharge of 8.79 m³/s.

3   IMPROVEMENT IN BLADE DESIGN

To improve the performance of existing impeller, the basic dimensions of modified impeller were kept the same as that of existing impeller. Only blades were modified using the method proposed by Gundale and Joshi (2013). From this Fig. 5, correspond to 20 m head and specific speed of 1800; number of blades are selected as 6 (Lobanofff and Ross, 1985).

To modify the blade profile "Tangent Circular Arc" method was used, in this method the impeller is divided into concentric rings which may not be equally spaced between inner and outer radius of impeller R_1 and R_2 respectively. The value the of R between any two consecutive rings can be calculated using equation mentioned below and the blade shape formed is actually an arc tangent to both rings.

$$R = \frac{R_2^2 - R_1^2}{2(R_2 \cos\beta_2 - R_1 \cos\beta_1)} \tag{4}$$

where, $\beta_1 = 18°$ and $\beta_2 = 25°$ are blade inlet and outlet angles respectively.

4   ANALYSIS OF MODIFIED IMPELLER

4.1   *CFD methodology*

The boundary conditions and solution technique used were similar to that used for the analysis of existing impeller. To study the effects of grid size variation on results, grid independency test is performed by varying the grid elements between 5 lacs to 28 lacs. It is found that, after around 9.4 lacs elements the efficiency variation was within 1%. Hence, further analysis was done using 9,43,693 elements.

4.2   *Results and discussions*

CFD analysis of modified impeller was carried out by varying discharge from 40–180% of Q_{bep}. The pressure contours in the modified impeller at various discharges are shown in Fig. 6. It can be seen that static pressure increases gradually from impeller inlet to outlet due to energy transfer from rotating impeller. The minimum value of static pressure is observed at leading edge of blades on suction side due to flow acceleration.

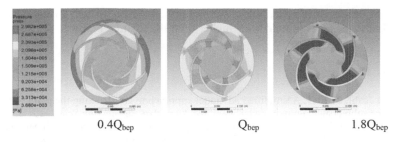

0.4Q_{bep}   Q_{bep}   1.8Q_{bep}

Figure 6.   Pressure contours in modified impeller at different discharge.

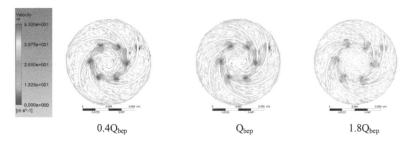

0.4Q_{bep} Q_{bep} 1.8Q_{bep}

Figure 7.   Velocity vectors in modified impeller at different discharge.

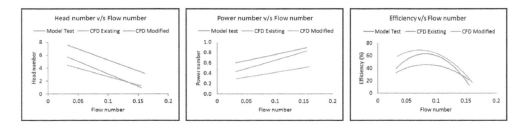

Figure 8.   Comparison of performance curves.

The velocity vectors in modified impeller are shown in Fig. 7 at different discharge. In this case, the recirculation of fluid in impeller blade passages is absent which was there in existing impeller. This may lead to decrease in hydraulic losses and rise in pump efficiency.

The comparison of performance curves of modified impeller with existing impeller and model test results are shown in Fig. 8 (Jain, 2015). It can be seen that, the modified impeller lead to increase in head and efficiency and decrease in power input. At BEP, the head and efficiency are found as 23.51 m and 67.85% at discharge of 8.79 m³/s. The modified impeller led to 5.53 m and 6.8% rise in head and efficiency respectively at best efficiency point compared to existing design.

5 CONCLUSION

In the present study, CFD analysis of radial flow submersible pump (discharge: 7 lps and head: 18 m per stage) was done using ANSYS CFX software. Among different turbulence models, SST k-ω model is found most appropriate for the analysis. The maximum efficiency of existing pump was found as 46.25% and 61% with CFD and model testing respectively. The higher deviations in the results are may be due to modeling and meshing errors. For improving the efficiency impeller blades were modified. The efficiency of modified pump is found as 67.85%, which is around 6.8% higher than the existing pump.

REFERENCES

Gundale, V. & Joshi, G. 2013. A simplified 3d model approach in constructing the plain vane profile of a radial type submersible pump impeller. *Research Journal of Engineering Sciences* 2(7): 33–37.
Gundale, V. & Patil, S. 2012. Improvement in the design of a radial type vertical submersible open well pump impeller using CFD. *International Journal of Engineering Research & Industrial Application* 5(2): 99–108.
Jain, A. 2015. Design improvement in radial flow submersible pump using CFD, M. Tech Thesis, Nirma University.
Lobanoff, V.S. & Ross, R.R. (2), 1985. Centrifugal pumps design and application. Texas: Gulf Publishing Company, Houston.
Mehta, M.P. & Patel, P.M. 2013. CFD analysis of mixed flow pump impeller. *International Journal of Advanced Engineering Research and Studies* 2(2): 15–19.
Menter, F.R. 1993. Zonal Two Equation k-w Turbulence Models for Aerodynamic Flows, *24th Fluid Dynamics Conference: AIAA Paper #93–906.*
Muggli, F.A., Holbein, P. & DuPont, P. 2002. CFD calculation of a mixed flow pump characteristic from shutoff to maximum flow. *Journal of Fluids Engineering*: 798–802.
Teri. 2012. "Promoting energy efficient pumps in industry in India: Addressing the energy and climate change problem", New Delhi: The Energy and Resources Institute, 25. [Project report number 2011IE12].

Multi-disciplinary Sustainable Engineering: Current and Future Trends – Tekwani, Bhavsar & Modi (Eds)
© 2016 Taylor & Francis Group, London, ISBN 978-1-138-02845-6

Chronological design and optimization of modular assembly tool for fabrication of a large size pressure vessel

Umesh Tilwani, Dhaval B. Shah & Shashikant J. Joshi
Department of Mechanical Engineering, Institute of Technology, Nirma University, Ahmedabad, Gujarat, India

ABSTRACT: The storage tanks or pressure vessels are manufactured in a wide range of sizes and shapes depending on the loading and design constraints. The most common shape for the pressure vessel is the cylindrical shape with various types of dished ends. These cylindrical pressure vessels are limited to size in diameter due to restrictions on transportation volume and cost implications for handling. The present paper is about confronting this challenge of fabrication of large diameter and large vertical pressure vessels. The pressure vessel is divided into segments based on the transportation constraints for partial fabrication at the factory and remaining on site. This site fabrication emerges as a requirement of a fixture or assembly tool. The conceptual design of assembly tool for the site fabrication of the complete pressure vessel is proposed. Moreover, during the site fabrication of this large size pressure vessel, the assembly tool/fixture shall be subjected to various types of loadings. The various loading conditions are determined based on the activities to be performed for the site fabrication and the generalized design chronology of the assembly tools is defined. In designing, one of the most important activities to be considered is the optimization for cost saving purpose, with respect to this the topography and topological optimization techniques for the assembly tool are suggested.

Keywords: assembly tool; design chronology; optimization

1 INTRODUCTION

The pressure vessels are extensively used in the oil and gas sector and their size and volume are restricted by certain limitations. One of them is the transportation volume which varied from place to place. Hence, the sizes of vessels sometimes results in them being too long in height, which brings across a requirement of a very meticulous handling of the vessel on site for the installation as shown in the Figure 1. Any type of error may lead to a catastrophic accident ending into a huge lose, as the material used for the fabrication, cost in transportation, also other several resources cost. Moreover, the catastrophe will also have a large impact on the schedule of the project at both ends purchaser and seller. To avoid such consequences a large size pressure vessel can be fabricated partially in the factory and remaining on site fabrication using assembly tools/fixtures which hold the job properly while fabrication on site up till final installation to its location. The segmentation of large pressure vessels shown in Figure 6, describes the scope of the site fabrication and factory manufacturing. While performing site activities for various fixture requirements which help in lifting of radial sectors, cylindrical segments, fixture for fabricating segments of pressure vessels, assembly tools for shifting of the large cylindrical segments to the location of installation.

K. Subramanian and M. Velayutham studied the main factors influencing the seismic load and dynamic analysis results for various structural systems with various zone factors that are compared in various standards (IS 1893, EN 1998-1, UBC 1997, NZS 1170.5). The author concludes that even though the various codes differ in detail, they have some essential common features like seismic risk, spectral content, importance of building, structural behavior,

Figure 1.   Handling of pressure vessel at site.

and soil and foundation type and comparable. Mike Zuckerbot performed the stresses on the lugs and its weld by manual calculations against the force acting on it using the standards. The lifting lug geometry is checked for the tensile stresses, shear stresses due to the oblique load acting due to the lifting plan using slings. Also, the stresses at the welds of the lugs are checked due to the vertical load and in-plane load. Kent L. Richards described the lug analysis in the book of Machine Design. The chapter comprises of the failure analysis of the lugs for tensile strength, shear strength, bearing strength, and fatigue against axial loading, transverse loading, and oblique loads. Zheng Jianxiang et al. introduced a new method for lifting of the steel cylinder of the containment. J. Olearczyk et al. focused on the improvement in traditional lifting design for heavy industrial modules. The FEA was used to analyze the load redistribution due to lifting of three different types of modules under extreme conditions and load redistributions are compared with the results of the newly designed lift frame for a variety of modules and loads. Pecker Alian (2013) presented the French regulations for the seismic protection of the critical industrial facilities.

Present paper focuses on the modular type assembly tool/fixture design for various large size vessels to serve the dual purpose, i.e., fabrication of pressure vessel segments (shells 0–360 and dished ends 0–360) and shifting of segments. Various design constraints are identified and explained in detail for the application and purpose with respect to the fixture and job. The design chronology has been described for the fixture design by analysis method and set in an order which takes care of the optimization part too with minimum number of iterations. The topological and topographical techniques for the optimization have also been suggested for fixture.

2   DESIGN CONSTRAINTS

The fixture will be used for the assembly of the radial sectors of the shell and dished ends as well as for transporting the sub-assemblies, i.e., 0°–360° dished ends and shells to final installation location. The outer dimensions of the fixture will be too large due to the large diametrical dimensions. The fixture needs to be of a modular type in order to create an ease in transportation at site for the fabrication. Also, the fixture can be disassembled and reused for the fabrication of a variety of large diameter vessels.

2.1   SPMT vehicle specification

For transfer of the fabricated dished ends and shells rested on the fixture, typical SPMT vehicles shown in Figure 2 can be used. Stools on which the fixture and the job are resting need to

Figure 2. SPMT vehicle with shell (Courtesy: Sarens).

be placed in the proper way while fabrication is going on. It should be ensured that the SPMT and stools locations do not coincide with each other. The Size of SPMT is referred from the catalogue of their respective companies.

The radial sectors of shells and dished ends are rested on the frame for fabrication on site, and then there shall be few in-situ machines to be placed on the frame. Maximum weights of 30 MT are to be placed on the frame for in-situ machines, welding machines, welding tracks, etc. Height of support stools below frame shall be sufficient to provide free entry of SPMT (Self Propelled Modular Transporter) vehicle in the collapsed condition under the frame/fixture. The frame/fixture height shall be sufficient to withstand manu-facturing and transportation load of the dished ends and shells. Hence, the support stools on which the fixture/frame will be rested while fabrication should have sufficient height to maintain the clearance between the fixture and SPMT, while SPMT is being located at its desired location.

2.2   Seismic loadings

As such there are no codes and standards which define the seismic loading consideration while the pressure vessel is under fabrication stage, but still the fixture is to be designed subjected to the seismic loads. The reason for designing of the fixture against seismic loads is that the fixture is made to be of modular type so that it can be reused for any variety of dimensional pressure vessels also designed against the seismic, and ensures the safety of the capital invested for the frame/fixture as well as for the job being fabricated on it. Other than this, designing against seismic also avoids any delay (Schedule impact) on the project, which may occur if seismic is not taken into consideration.

The seismic loads to be considered are based on the seismic zones of the country. The seismic zone with seismic load is to be considered for that nation and applicable codes & standards. Detailed guidelines are provided by codes and standards describing various meth-odologies of seismic analysis, like time history method, response spectrum method, simpli-fied dynamic analysis etc.

Of the above, the most suitable method for the seismic validation is a simplified dynamic analysis, wherein a static coefficient is determined which is to be applied for the vertical & horizontal acceleration determined for the seismic load. Other than this, the seismic combi-nations are determined with the help of the new marks' rule, which takes into consideration the acceleration. The determined various seismic accelerations are then to be combined with new marks rule with different sign conventions and the coefficients are considered in ASME.

3 DESIGN CHRONOLOGY

The Transportation frame which is one of the required assembly tools which is to be designed for different loading conditions are listed below:

- Design loads to be determined from the reference to the applicable codes and standards.
- Seismic loads determined by the standard of the particular location of the site.
- Transport loads while transferring the dished ends and shells from the temporary workshop at the site to the final location of the installation.

In the subsequent sections the values of the design loads have been assumed as 1.5 g vertical and 0.15 g in-plane loading. Whereas, for the seismic loading the value assumed is 2.16 g vertical and 0.2 g in-plane. Also, the combination of the seismic loads has been considered as below:

i. Seismic combination 1: + 100% vertical + gravity + 40% horizontal
ii. Seismic combination 2: − 100% vertical + gravity + 40% horizontal
iii. Seismic combination 3: + 50% vertical + gravity +108% horizontal

Based on the above loading conditions a sequence of activities has been listed below for designing the transport frame which can be used for fabrication and transportation of both.

I. Concept Design of Modular Frame/Fixture for the assembly of dished ends and shells with location of support stools on which the fixture is rested, also location of SPMT on which the dished ends and the shells will be transferred to the actual location of the installation using the same fixture. The location of stools shall be with respect to SPMT Vehicle movement under the fixture. The frame shall be made of modules (maximum size as feasible), and the connection between modules shall be of bolted type.

II. Validation of frame on support stools with support stool locations including support structures on the frame and complete Base Section rested on it.

- Design load (Vertical) = 1.5 g
- In Plane Load (Horizontal) = 0.15 g

III. Seismic Validation of frame with complete Base Section and In-situ Machines

- Seismic Load Combination 3 Total Load = 2.08 g (Vertical) + 0.233 g (Horizontal)

IV. Optimization of the Transport Frame Conceptual Design for assembly of Base Section with Support Stools and Transportation from site workshop to installation place on SPMT vehicles.

V. Validation of the frame including support structures and complete Base Section rested on it for movement on SPMT vehicles confirming payload capacity of the available SPMT vehicle.

- Design load (Vertical) = 1.5 g
- In Plane Load (Horizontal) = 0.028 g

According to the Report EQ001 – Equipment Description—KAMAG 2400 ST
Speed of the SPMT is 1 km/h Speed = 1000/3600 = 0.2778 m/sec
Maximum acceleration = 0.2778/1 = 0.2778 m/sec2 (for achieving 1 Km/h velocity in 1 sec)
Maximum Load applicable = Maximum acceleration/Gravity = 0.2778/9.81 g = 0.028 g
Total Load = 1.5 g (Vertical) + 0.028 g (Horizontal)

a. Validation of SPMT with respect to pay load capacity

- Design load (vertical) = 1.0 g
- In Plane Load (Horizontal) = 0.028 g

Total Load = 1.0 g (Vertical) + 0.028 g (Horizontal)

VI. Validation of the frame on support stools with support stool locations including support structures on the frame and complete Base Section rested on it.

- Design load (Vertical) = 1.5 g
- In Plane Load (Horizontal) = 0.15 g

Total Load = 1.5 g (Vertical) + 0.15 g (Horizontal)
a. Validation of all Support Stools with respect to the above analysis.

- Vertical Load = Maximum reaction forces on Stool
- In-Plane Load = Maximum reaction forces on Stool

VII. Seismic Validation of the frame with complete base section and in-situ machines

- Seismic Load Combination 1 Total Load = 3.16 g (Vertical) + 0.0864 g (Horizontal)
- Seismic Load Combination 2 Total Load = 1.16 g (Vertical) + 0.0864 g (Horizontal)
- Seismic Load Combination 3 Total Load = 2.08 g (Vertical) + 0.233 g (Horizontal)

a. Validation of Stools with respect to the above seismic analysis

- Vertical Load = Maximum reaction forces on Stool
- In-Plane Load = Maximum reaction forces on Stool

VIII. Validation of frame and support structure with base section for buckling on support stools

- Design load (vertical) = 1.5 g
- In Plane Load (Horizontal) = 0.15 g

Total Load = 1.5 g (Vertical) + 0.15 g (Horizontal)

IX. Transport frame conceptual design for assembly of top lid with support stools and transportation from site workshop to installation place on SPMT vehicles.

X. Validation of frame including support structures and complete top lid rested on it for movement on SPMT vehicles confirming payload capacity of available SPMT vehicle.

- Design load (Vertical) = 1.5 g
- In plane load (Horizontal) = 0.028 g

Total Load = 1.5 g (Vertical) + 0.028 g (Horizontal)
a. Validation of SPMT with respect to pay load capacity

- Design load (vertical) = 1.0 g
- In plane load (horizontal) = 0.028 g

Total Load = 1.0 g (vertical) + 0.028 g (horizontal)

XI. Validation of frame on support stools with support stool locations including support structures on the frame and complete top lid rested on it.

- Design load (vertical) = 1.5 g
- In plane load (horizontal) = 0.05 g

Total Load = 1.5 g (Vertical) + 0.05 g (Horizontal)
Validation of all the support stools with respect to the above analysis.

- Vertical Load = Maximum reaction forces on Stool
- In-plane Load = Maximum reaction forces on Stool

XII. Seismic Validation of frame with complete top lid and in-situ machines

- Seismic load combination 1 total load = 3.16 g (vertical) + 0.0864 g (horizontal)
- Seismic load combination 2 total load = 1.16 g (vertical) + 0.0864 g (horizontal)

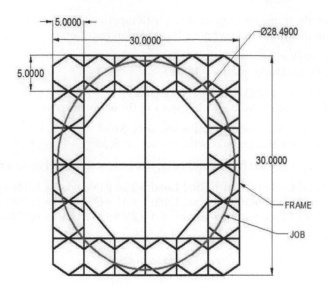

Figure 3.   Conceptual design of frame/assembly tool.

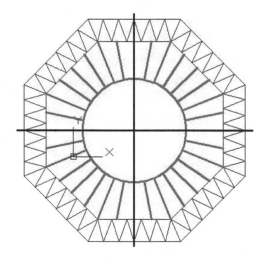

Figure 4.   Optimized frame for topological surface.

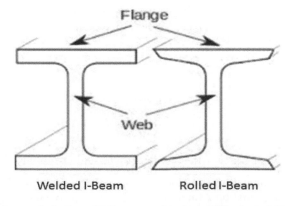

Figure 5.   Cross section optimization of I-Beams used in frame (Topological Phenomenon).

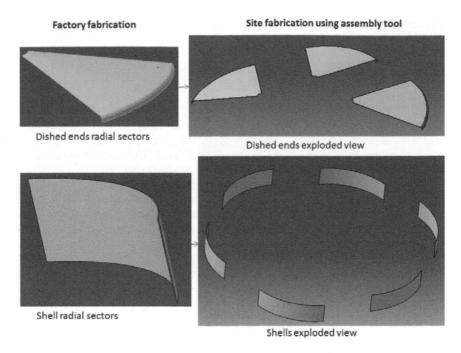

Factory fabrication

Dished ends radial sectors

Site fabrication using assembly tool

Dished ends exploded view

Shell radial sectors

Shells exploded view

Figure 6.   Segmentation of shells and dished ends of large dia pressure vessel.

- Seismic load combination 3 total load = 2.084 g (vertical) + 0.233 g (horizontal)

Validation of the stools with respect to the above seismic analysis

- Vertical Load = Maximum reaction forces on the stool
- In-Plane Load = Maximum reaction forces on the stool

XIII. Detailed analysis and validation of bolted joints used for the assembly of the frame structure

- Vertical load = Maximum reaction force at particular location for design load case
- In-Plane load = Maximum reaction force at particular location for design load case

4   OPTIMIZATION OF THE BASE SECTION TRANSPORT FRAME

There are basically two types of techniques topography and topological optimization. In the present work, for the design of transport frame, both types of optimization have been applied.

Topographical optimization of frame: observing the conceptual design of the transport frame in the Figure 3 the scope of optimization on the surface based are determined. The four corners of the frame can be chopped at 45° forming an octagonal frame; sample is shown in the Figure 4.

Topological optimization of frame: assuming the frame is built from the I-sections. I-sections are manufactured by two types:

1. Rolled I-sections
2. Welded I-sections: – Plate and Girders

Refer to Figure 5 for the difference between rolled I-sections and welded I-sections.

The topological optimization has been performed by considering the welded I-sections (Plate & girders). This gives a wide range of I-sections for selection, since the flanges and web of the I-sections are to be made from the plates available from the rolled plates. The topological optimization gives the liberty to select a beam with a large number of cross sections with various dimensions.

5 CONCLUSION

A new concept using appropriate fixture /assembly tool for fabrication of the pressure vessels with large sizes is proposed. The proposed concept suggested partial fabrication at factory and remaining fabrication on site using the modular fixture. The same modular fixture can be reused for various dimensions of the large pressure vessel also being of modular type it can be transported to the site required, which again leads to cost saving.

The design chronology has been developed for the development and design of the fixture/assembly tools which are to be used for the manufacturing and fabrication of the large size pressure vessels on site. The design chronology takes in to consideration the conceptual design and the preliminary validation of the concept to find the scope of optimization if any. These steps defines a systematic approach to the fixture design and also takes care that the fixture doesn't get over designed to avoid extra costing for manufacturing of the fixture as well as reduces the mass of the fixture providing sufficient strength. The same assembly tool can be conceptualized so that the fabrication of the dished end and the shell both are taken care of using a single fixture.

REFERENCES

ASME Section II part D.
ASME Section III Division N Dynamic analysis procedure for Seismic EN 1998 (Euro code 8).
ASME Section VIII Div. 2.
Alain, P. 2013.Overview of Seismic Regulations for French Industrial Facilities, International Conference on Seismic Design of Industrial Facilities, RWTH Aachen University.
IS 800.
Richards, Kent L. Machine Design handbook Chapter 12 Lug analysis.
Subramanian, Dr. K. & Velayutham, M. Influence of seismic zone factor and the international codal provisions for various lateral loads resisting systems in multi storey buildings, Indian Society of Earthquake Technology Golden Jubilee symposium, Department of Earthquake Engineering Building IIT Roorkee.
Westover, L., Olearczyk, J., Hermann, U., Adeeb, S. & Mohamed, Y. Analysis of rigging assembly for lifting heavy industrial modules.
Zheng Jianxiang, Qiu Guozhi, Lu Xiaojun, Wu Shaopeng and Gong Jinghai, Study on lifting method for steel cylinder of the containment of AP1000 Nuclear Plant.
Zuckerbrot, Mike. Stress analysis of lifting lugs, Fermilab ES & H manual.

Multi-disciplinary Sustainable Engineering: Current and Future Trends – Tekwani, Bhavsar & Modi (Eds)
© 2016 Taylor & Francis Group, London, ISBN 978-1-138-02845-6

A comparison of recently developed meta-heuristic optimization methods for improving ride comfort of a bio-mechanical quarter car model

B. Gadhvi, V. Savsani, F. Doshi, R. Doshi, G. Patel & M. Chauhan
Pandit Deendayal Petroleum University, Gandhinagar, Gujarat, India

ABSTRACT: The present study comprises Human-Vehicle-Road (HVR) model, consisting of a biomechanical representation of the driver which is used for analysis of vehicle suspension to improve ride comfort. The Mass-Spring-Damper (MSD) model having four degrees of freedom (dof) is devised using SIMULINK with the help of Newton's laws of motion. The suspension acceleration function from this model is used as an objective function for optimization algorithms. The optimized values of spring stiffness (k_s) and damping co-efficient (c_s) are obtained by the iterative process of the employed algorithms. For a comprehensive result, 2 recently developed optimization techniques namely Grey Wolf Optimizer (GWO), Firefly Algorithm (FFA) are used along with Genetic Algorithm (GA)—a standard well developed optimization technique. The thesis provides a two-fold conclusion-first for the best values of spring stiffness and damping co-efficient to maximize ride comfort and second for comparison of the 3 optimization methods for the given problem.

1   INTRODUCTION

With the onset of the 21st century, came the times of massive industrialization and rapid urbanization. Technology and automation were at people's disposal to increase their productivity and improve their living standards. People in an effort to reach out to larger masses were required to travel long distances. As a result, highways were laid down but not without its inherent road surface irregularities. Passenger cars became the need of the day and people were seen spending long and tiresome hours travelling in them.

Engineers and designers in an effort to reduce vehicle vibrations over human body have come to a conclusion that very little can be done to improve the road surface roughness. Hence large amount of research has been directed in designing a good suspension system that would produce minimum vibration levels over varying road profiles.

Biomechanical model consists of integrating human body properties along with the sprung and un-sprung mass-spring-damper system model of the vehicle. According to this approach, the human body is represented as various parts (lower body, upper body and head) with characteristic values of mass, spring stiffness and damping coefficient values. The integration helps in representing it as a single four degrees of freedom biomechanical quarter car model. The biomechanical model gives us values of variables like acceleration at each and every stage of the body (head, pelvis, lower body) and helps determine how much exposure is considered risky.

The current research paper employs the meta-heuristics optimization approach. It is randomized search technique which uses large set of feasible solutions to find good solutions with less computational effort than conventional algorithms (Huang & Arora 1979, Paeng & Arora 1989, Tamboli & Joshi 1999). Meta-heuristics does not guarantee an exact global optimum solution but provides more convincing and better results for unknown problems like the one employed in the present study (Blum & Roli 2003).

2 BIO-MECHANICAL MODEL OF QUARTER CAR

The quarter car model in the current study comprises four lumped masses namely sprung mass (m_s), un-sprung mass (m_u), mass of the lower body (m_{st}) and mass of the head (m_h). Also the suspension system of the vehicle in this case is passive system with spring stiffness (k_s) and damping coefficient (c_s). Moreover the tire, the seat and the thorax & pelvis of the human body are also assumed to have stiffness and damping properties (Qassem 1996, Wan & Schimmels 2003). The present study limits the selection criteria by employing constants from an already published study (Mitra and Benerjee, 2013) for ease of comparison and verification. The vertical displacement of the sprung mass is denoted by 'z_s' and the road profile is denoted by 'z_r'.

Using Newton's laws of motion the equations of motion of the model are obtained which are stated as follows:

$$m_h \ddot{z}_h + k_h(z_h - z_b) + c_h(\dot{z}_h - \dot{z}_b) = 0 \tag{1}$$

$$m_b \ddot{z}_b + k_h(z_b - z_h) + c_h(\dot{z}_b - \dot{z}_h) + k_b(z_b - z_s) + c_b(\dot{z}_b - \dot{z}_s) = 0 \tag{2}$$

$$m_s \ddot{z}_s + k_b(z_s - z_b) + c_b(\dot{z}_s - \dot{z}_b) + k_s(z_s - z_u) + c_s(\dot{z}_s - \dot{z}_u) = 0 \tag{3}$$

$$m_u \ddot{z}_u + k_s(z_u - z_s) + c_s(\dot{z}_u - \dot{z}_s) + k_t(z_u - z_r) + c_t(\dot{z}_u - \dot{z}_r) = 0 \tag{4}$$

The equation of motion obtained here is a set of four linear, second order, ordinary differential equation. These equations are modeled in SIMULINK. For obtaining the solution of the equations SIMULINK uses the 'ode45' function of MATLAB which is based on Runge-Kutta fourth order method.

The sinusoidal road excitation (z_r) has been modeled as follows (Mitra & Benerjee 2013).

$$z_r(x) \begin{cases} = 0 & when\ t < \dfrac{d}{v} \\[2mm] = h\sin\left[\dfrac{\pi v}{L}\left(t - \dfrac{d}{v}\right)\right] & for\ \dfrac{d}{v} \le \left(t < \dfrac{d}{v}\right) \le \dfrac{d+L}{v} \\[2mm] = 0 & when\ t < \dfrac{d+L}{v} \end{cases} \tag{5}$$

where t = time.

Table 1. Value of constants. (Mitra and Benerjee, 2013).

Parameter	Symbol	Corresponding value
Sprung mass	m_s	240 kg
Unsprung mass	m_u	36 kg
Mass of upper body and head	m_h	20 kg
Mass of lower body and seat	m_b	45 kg
Spring stiffness of thorax and pelvis	k_p	45005.3 N / m
Damping co-efficient of thorax and pelvis	c_p	1360 N·s / m
Spring stiffness of seat cushion	k_c	20000 N / m
Damping co-efficient of seat cushion	c_c	1649.3 N·s / m
Spring stiffness of tyre	k_t	160000 N / m
Damping co-efficient of tyre	c_t	800 N·s / m

Table 2. Road profile parameters.

Parameter	Symbol	Corresponding value
Height of the bump	h	0.3 m
Distance of approach	d	10 m
Length of bump	L	1 m
Velocity of vehicle	v	10 m / s

Table 3. Optimization parameters.

GWO		FFA		GA	
Parameter	Value	Parameter	Value	Parameter	Value
No. of search agents	60	Population size	60	Population size	120
No. of iterations	100	No. of iterations	100	Crossover fraction	0.80
		Value of alpha	0.2	Generation	100
		Value of beta	0.5	Selection function	Tournament
		Value of gamma	1	Crossover function	Heuristic
				Mutation function	Adapt feasible

3 OPTIMIZATION

Ride comfort can be measured by various approaches like subjective testing based on SAE norms (Strandemar, K. 2005), minimization of sprung mass acceleration (Gadhvi & Savsani 2014), vibration analysis (ISO 2631) and many more. The present study uses vertical acceleration of sprung mass as an indicator of ride comfort. Table 3 shows the optimization parameters employed to obtain the optimized values of spring stiffness (k_s) and damping coefficient (c_s). In the present study recently developed optimization techniques namely Grey Wolf Optimization (Mirjalili et al. 2014) and Firefly Algorithm (Yang 2010, Pal et al. 2012) are employed along with a standard well developed technique—Genetic Algorithm (Man 1996, Gundogdu 2007).

4 RESULTS AND DISCUSSION

In order to find the best possible spring stiffness coefficient and spring damping values to isolate the passenger from road undulations and for improved ride comfort, we have performed 10 optimization runs for each of the three algorithms (each run included 100 generations). The lower and upper bounds for the suspension stiffness and damping coefficient chosen are:

2500 N/m $< k_s < 20,000$ N/m and 500 N·s/m $< c_s < 3000$ N·s/m.

The effectiveness of all three algorithms is gauged by comparing the best, worst, mean and standard deviation of optimized values obtained in each run is done for GWO, FFA and GA. Table 4 shows the most optimized value of all 10 runs for each algorithm. Figure 2 shows the convergence of all three methods: GA, FFA and GWO. The graph indicates how fast the objective function value converges to the most optimized value. It should be noted that the objective function values are averaged at each generation for all 10 runs for all three optimization methods. Figure 3 shows the graph of sprung mass acceleration i.e. the objective function value, as the quarter car passes over the described road profile.

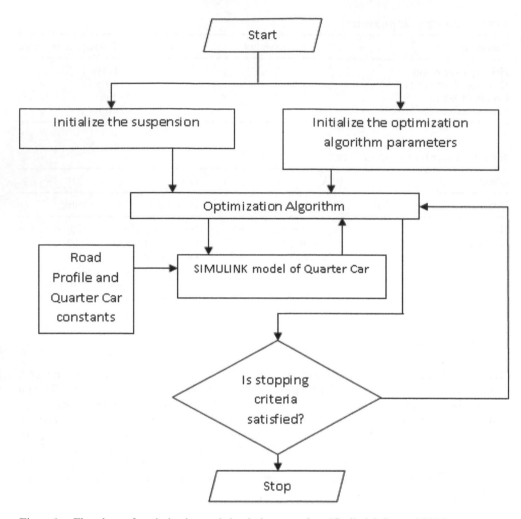

Figure 1. Flowchart of optimization and simulation procedure (Gadhvi & Savsani 2014).

Table 4. Objective function (sprung mass acceleration) values of GWO, FFA and GA.

Run	GWO	FFA	GA
1	6.6375	8.024362	6.624
2	6.6203	7.166225	6.683
3	6.6368	7.806469	6.62
4	6.6235	7.748824	6.619
5	6.6326	8.350107	6.625
6	6.6251	7.65516	6.628
7	6.6288	7.742206	6.683
8	6.6245	8.386131	6.651
9	6.6338	6.710201	6.636
10	6.6233	8.037534	6.63

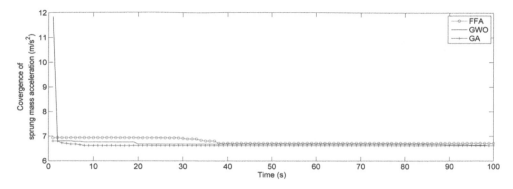

Figure 2. Convergence of optimization algorithms for given problem.

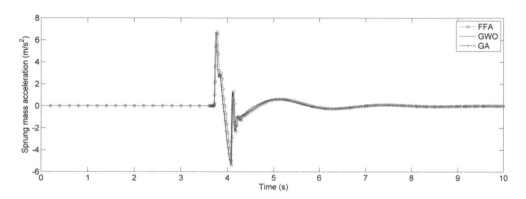

Figure 3. Sprung mass acceleration vs time.

5 CONCLUSION

Thus by employing a material which provides values of spring stiffness and damping coefficients closest to 2503.32 N/m and 500.0995 N·s/m respectively, ride comfort can be maximized. Moreover for the given problem three optimization techniques reveal following trends:

Order of convergence (fastest to slowest): GA > GWO > FFA
Minimum objective function result (highest to lowest): FFA > GWO > GA
Standard deviation of iterations (largest to smallest): FFA > GWO > GA

Hence it can be seen that GA has provided best results for the given problem. However results of GWO and FFA are very close to the most optimized result and this shows that all 3 techniques are equally capable for optimization of similar multi variable problems. Moreover other criteria like time required in obtaining the most optimized result can also be used as a basis for comparison for the optimization methods employed.

REFERENCES

Blum, C. & Roli, A. 2003. Metaheuristics in Combinatorial Optimization: Overview and Conceptual Comparison. *ACM Computing Surveys (CSUR)* 35(3): 268–308.
Gadhvi, B. & Savsani, V. 2014. Passive Suspension Optimization Using Teaching Learning Based Optimization and Genetic Algorithm Considering Variable Speed Over a Bump. In *ASME (American Society of Mechanical Engineers) 2014 International Mechanical Engineering Congress and Exposition* (pp. V04AT04A018-V04AT04A018).

Gundogdu, O. 2007. Optimal seat and suspension design for quarter car with driver model using genetic algorithms. *International Journal of Industrial Ergonomics* 37(4): 327–332.

Huang, M.W. & Arora, J.S. 1979. Applied optimal design: mechanical and structural systems.

Man, K.F., Tang, S.K. & Kwong, S. 1996. Genetic Algorithms: Concepts and Applications. *IEEE transactions on industrial electronics* 43(5): 519–534.

Mirjalili, S., Mirjalili, S.M. & Lewis, A. 2014. Grey Wolf Optimizer. *Advances in Engineering Software* 69: 46–61.

Mitra, A.C. & Benerjee, N. 2013. Ride comfort and Vehicle handling of Quarter Car Model Using SIMULINK and Bond Graph. *Proceedings of the 1st International and 16th National Conference on Machines and Mechanisms (iNaCoMM2013)*, IIT Roorkee, India.

Paeng, J.K. & Arora, J.S. 1989. Dynamic response optimization of mechanical systems with multiplier methods. *Journal of Mechanical Design* 111(1): 73–80.

Pal, S.K., Sing, A.P. & Rai, C.S. 2012. Comparative Study of Firefly Algorithm and Particle Swarm Optimization for Noisy Non-Linear Optimization Problems. *International Journal of Intelligent Systems and Applications (IJISA)* 4(10): 50.

Qassem, W. 1996. Model prediction of vibration effects on human subject seated on various cushions. *Medical engineering & physics* 18(5): 350–358.

Strandemar, K. 2005. On objective measures for ride comfort evaluation. *Master's thesis.*

Tamboli, J.A. & Joshi, S.G. 1999. Optimum design of a passive suspension system of a vehicle subjected to actual random road excitations. *Journal of sound and vibration* 219(2): 193–205.

Wan, Y. & Schimmels, J.M. 2003. Improved vibration isolating seat suspension designs based on position-dependent nonlinear stiffness and damping characteristics. *Journal of dynamic systems, measurement, and control* 125(3): 330–338.

Yang, X.S. 2010. *Nature-inspired metaheuristic algorithms.* Luniver press.

Multi-disciplinary Sustainable Engineering: Current and Future Trends – Tekwani, Bhavsar & Modi (Eds)
© 2016 Taylor & Francis Group, London, ISBN 978-1-138-02845-6

The optimization of process parameters of grinding machine in a bearing manufacturing plant

T.A. Bhatt
B H Gardi College of Engineering and Technology, Rajkot, Gujarat, India

A.M. Gohil
Institute of Technology, Nirma University, Ahmedabad, Gujarat, India

S.D. Dhadve
SKF Technologies (I) Pvt. Ltd., Ahmedabad, Gujarat, India

ABSTRACT: This paper represents the use of design of experiments method to optimize the process parameters of grinding machine in a bearing manufacturing industry. Grinding is a very important cutting process that achieves the required dimensions within close tolerance values. Design of experiments is carried out on an internal grinding machine to optimize the output of outer ring of cylindrical roller bearing. Experiments are carried out using fractional factorial method to reduce the number of runs because of the machine availability constraints. The objective of this study is to; reduce the rejection rate of outer ring in ovality and roughness and at the same time reducing the cycle time. Minitab 16 software is used for statistical analysis. Result is analyzed by Analysis Of Variance (ANOVA) technique to achieve optimum ovality and surface roughness. A mathematical model is also developed using regression analysis to get a relationship between responses and predictors. Optimum set of process parameter is selected from model and suggested to the company to get better quality and reduced cycle time.

1 INTRODUCTION

Higher level of rejection rate during manufacturing is not acceptable in any company. Higher scrap level costs the company and it leads to loss of material, time, and efforts added to convert the raw material into final product. Bearing manufacturing company in India is facing a problem of higher scrap due to non-conformance of the product with the quality requirements. There is a need to find the process parameters which optimizes the performance of the machine and leads to lower rejection rate.

Design of Experiments (DOE) is a technique to find out the individual or interactive effects of many influencing factors that could affect the output. Prior to a few years it was accepted that quality control alone offers sufficient confidence to meet customer fulfillment. In today's scenario, it is perceived that examinations utilizing engineering outline procedures or statistically planned tests (DOE) give significantly more stupendous open doors to quality change and expanded profit (Kivak 2014).

DOE is a way of conducting experiments with the following characteristics:

- Structured procedure.
- Simultaneous study of many experimental factors.
- Fewer experimental runs: smallest number of experiments to get the largest amount of information.
- The approach of the experimentation is well considered.
- Universally applicable for any kind of operation or process (grinding, honing, heat treatment etc.)

2 OBJECTIVE OF THIS WORK

The main objective of this paper is to optimize the process parameters in internal grinding machine and to reduce rejection rate of bearing in roundness and roughness. Cylindrical Roller Bearing (CRB) line is manufacturing new type of outer ring. In one of the batch production of that ring, 25 rings got rejected from a total of 300 rings due to not conforming to the quality requirements. This is an intolerable rejection rate. It shows not only the waste of raw material but also a waste of all the additional value given to the raw material by different processes at various stages. Also it is a cost to the company. So there is a need to identify process parameters which gives optimum setting to achieve the quality established by the company.

So the objectives of this work is to study the effect of different machining parameters on work piece material and implement an enhancement procedure that could give ideal machining parameters to minimize machining time, reduce and/or eliminate ovality problem and to attain required surface finish. The methodology adopted for the same is shown in Fig. 1.

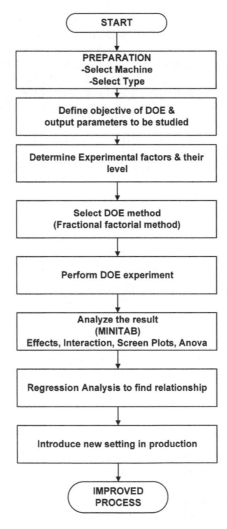

Figure 1. DOE methodology.

3 FRACTIONAL FACTORIAL METHOD

In a full factorial design all combinations of the experimental factors (2 levels each) are used. The main advantage of full factorials is that all main effects and interactions can be analyzed independently from each other. Number of experiment in full factorial design is 2 k. Where, k is the number of factors. The disadvantage of a full factorial design is that higher numbers of runs are required when numbers of factors are large (Tsao 2013).

As in a bearing manufacturing plant there is a continuous production going on to meet the customer demands, company cannot afford to do the experiments on full factorial design. To reduce number of experiments two level fractional factorial method is used. In Fractional factorial method number of experiment is 2(k-1). There are five process parameters affecting the grinding process. So the total number of experiment needs to performed is 16 (Loukas 1998).

3.1 Control parameters and range selection

There are many factors which affect the quality of grinding process is shown in Fig. 2. To select the parameters and its levels for experimentations, several exploratory experiments were conducted to determine important control factors. Out of several available controllable input parameters on the grinding process, some of main parameters are selected with maximum feasible range.

Feed rate is taken as one of the process parameter and is radial cutting depth per unit revolution that cutting tool advances into work piece. Fine feed rate and rough feed rate are two different rate of grinding to achieve smooth and coarse surface finish respectively. Dressing of the grinding wheel is another process parameter. Two levels chosen are intermediate re-sharpening (two time re-sharpening of grinding wheel) and interval re-sharpening (one time re-sharpening of grinding wheel).

3.2 Input parameter (Predictor)

The process variables selected for the experimentation are listed in Table 1.

3.3 Output parameter (Response)

Roundness and roughness are selected as output parameters of the experiments which is important from the bearing life requirement.

Figure 2. Control parameter affecting grinding process.

Table 1. Process parameter.

Symbols	Process variable	Unit	Upper level (+)	Lower level (−)
N_w	Work head RPM	RPM	250	100
N_s	Spindle RPM	RPM	6000	3000
V_1	Fine feed rate	μ/sec	2	1
V_2	Rough feed rate	μ/sec	8	3
D	Dressing interval	sec	Intermediate	Interval

Table 2. Fractional factorial factor observation table.

Work Head Speed (N_w)	Grinding wheel Speed (N_s)	Rough Feed Rate (V_1)	Fine Feed Rate (V_2)	Dressing Interval	Roughness (R_a) (In μm)	Roundness (Ovality) (In μm)	Cycle Time (CT) (In sec)
100	3000	3	1	Intermediate	0.38	7	259
250	3000	3	1	Interval	0.62	6	241
100	6000	3	1	Interval	0.52	6	241
250	6000	3	1	Intermediate	0.34	5	259
100	3000	8	1	Interval	0.54	7	200
250	3000	8	1	Intermediate	0.46	6	218
100	6000	8	1	Intermediate	0.38	8	218
250	6000	8	1	Interval	0.59	4	200
100	3000	3	2	Interval	0.62	5	217
250	3000	3	2	Intermediate	0.52	6	235
100	6000	3	2	Intermediate	0.43	7	235
250	6000	3	2	Interval	0.64	4	217
100	3000	8	2	Intermediate	0.46	9	194
250	3000	8	2	Interval	0.72	5	176
100	6000	8	2	Interval	0.64	10	176
250	6000	8	2	Intermediate	0.47	4	194

3.3.1 *Roundness (Ovality)*

Roundness means how closely the shape of bearing is nearer to circle. Ovality means the value of out of roundness. Ovality is measured manually by the universal gauge apparatus. To get the ovality, ring is rotated about its axis and by measuring the difference between the readings, required value is obtained. Maximum tolerance value for this outer ring is 8 micron.

3.3.2 *Roughness*

It shows the texture of the bearing track surface or irregularity of surface. It is measured by Mitutoyo MWA machine. It plays an very important role in bearing because roller is in contact with this surface. Maximum tolerance limit for that is 0.7 micron.

3.3.3 *Cycle time*

Cycle time is the total time from the beginning to the end the process. It includes the loading time, machining time, and unloading time.

4 DESIGN OF EXPERIMENTS

The experiments are performed on internal grinding machine for outer ring. Sixteen experiments are carried out. Table 2 gives the summary of input and output parameters of the experiments. To perform experiments in systematic manner, Minitab16 software is used for

DOE and ANOVA analysis. Minitab16 software helps to identify significant parameters. Various features of the software not only help in experiments but also enhance the analysis capabilities by statistical approach. Minitab calculates response tables and generates main effects and interaction plots (Kanlayasiri 2007).

4.1 Effect of process parameter on ovality

In order to see the effect of process parameters on the ovality, experiments are conducted using fractional factorial method. The experimental data is given in Table 2. Fig. 3 and Fig. 4 shows the significant factors affecting ovality and effect of each parameter on the ovality. Residual plots are valuable for evaluating information for the issues like non normality, non-irregular variety, non-steady fluctuation, higher-order relationship, and outliers. It could be seen in Fig. 5 that the residuals follow straight line in the normal probability plot. Approximate symmetric nature of histogram shows that the residuals are regularly dispersed. Residuals have steady change as they are scattered haphazardly around zero in residuals versus the fitted qualities. Since residuals display no agreeable example, there is no mistake because of time or information accumulation request (Davidson 2008). Surface plot for the ovality verses fine feed rate and work head speed is shown in Fig. 6.

4.2 Analysis of result

From the Fig. 3 it is seen that work head speed is the most significant parameter affecting ovality. Other parameters like grinding wheel speed and fine feed rate is also a signifi-

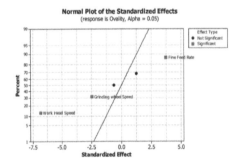

Figure 3. Effect of process parameter on ovality.

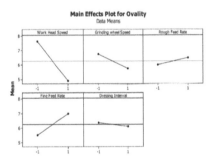

Figure 4. Main effect plot for ovality.

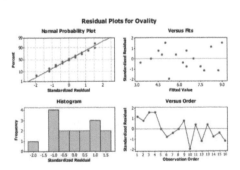

Figure 5. Residual plot for ovality.

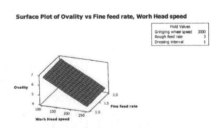

Figure 6. Counter plot for ovality.

Table 3. Estimated effects and coefficients for ovality.

Term	Effect	Coef	SE Coef	T	P
Constant		6.25	0.202	31.01	0
Work head speed	−3	−1.375	0.202	−6.82	0
Grinding wheel speed	−1	−0.5	0.202	−2.48	0.033
Rough feed rate	1	0.25	0.202	1.24	0.243
Fine feed rate	2	0.75	0.202	3.72	0.004
Dressing interval	−0	−0.125	0.202	−0.62	0.549

S = 0.806226   PRESS = 16.64
R-Sq = 87.25%   R-Sq(pred) = 67.37%   R-Sq(adj) = 80.88%

Table 4. Analysis of variance for ovality.

Source	DF	Seq SS	Adj SS	Adj MS	F	P
Main effects	5	44.5	44.5	8.9	13.69	0
Work head speed	1	30.25	30.25	30.25	46.54	0
Grinding wheel speed	1	4	4	4	6.15	0.033
Rough feed rate	1	1	1	1	1.54	0.243
Fine feed rate	1	9	9	9	13.85	0.004
Dressing interval	1	0.25	0.25	0.25	0.38	0.549
Residual error	10	6.5	6.5	0.65		
Total	15	51				

DF-Degree of freedom SS-Sum of square.
MS-Mean square variance.
F-ratio of variance of a source to variance error.
P<0.05, shows significance of a factor at 95% confidence level.

cant parameter. Surface plot for two main significant factors is shown in Fig. 6. From that optimum value for each parameter can be found to achieve low ovality. Estimated Effects and Coefficients for ovality are shown in Table 3. Analysis of variance for ovality is shown in Table 4.

4.3 *Effect of process parameter on roughness*

Figs. 7 and 8 show the significant factors affecting roughness and effect of each parameter on the roughness. Residual plot is shown in Fig. 9. Since residuals exhibit no clear pattern, there is no error due to time or data collection order. Surface plot for the roughness, fine feed rate, and work head speed are shown in Fig. 10.

4.4 *Analysis of result*

From Fig. 7 it is seen that dressing interval is the most significant parameter affecting the roughness. Other parameters like grinding wheel speed, work head speed, and fine feed rate is also an significant parameter. Surface plot for two main significant factors is shown in Fig. 10. From that we can find out the optimum value for each parameter to achieve low roughness. Estimated effects and coefficients for roughness are shown in Table 5. Analysis of variance for roughness is shown in Table 6.

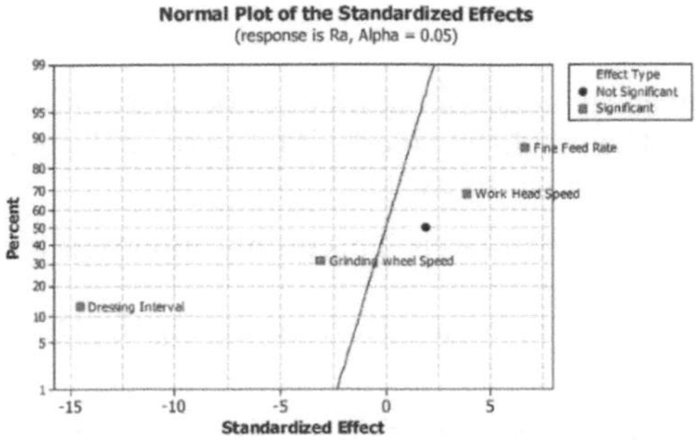

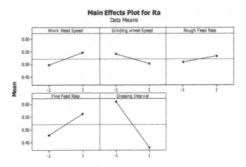

Figure 7.   Effect of process parameter on roughness.

Figure 8.   Main effect plot for roughness.

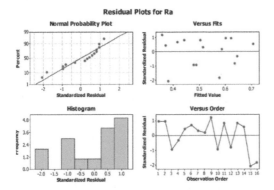

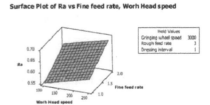

Figure 9.   Residual plot for roughness.

Figure 10.   Counter plot for roughness.

Table 5.   Estimated effects and coefficients for roughness.

Term	Effect	Coef	SE coef	T	P
Constant	0.04875	0,52063	0.006256	83.22	0
Work head speed	−0.0388	0.02437	0.006256	3.9	0.003
Grinding wheel speed	0.02375	−0.01937	0.006256	−3.1	0.011
Rough feed rate	0.08375	0.01187	0.006256	1.9	0.087
Fine feed rate	−0.1813	0.04187	0.006256	6.69	0
Dressing interval	0.04875	−0.09063	0.006256	−14.49	0
S = 0.0250250	PRESS = 0.016032				
R-Sq = 96.59%	R-Sq(pred) = 91.26%		R-Sq(adj) = 94.88%		

Table 6.   Analysis of variance for roughness.

Source	DF	Seq SS	Adj SS	Adj MS	F	P
Main effects	5	0.177231	0.177231	0.035446	56.6	0
Work head speed	1	0.009506	0.009506	0.009506	15.18	0.003
Grinding wheel speed	1	0.006006	0.006006	0.006006	9.59	0.011
Rough feed rate	1	0.002256	0.002256	0.002256	3.6	0.087
Fine feed rate	1	0.028056	0.028056	0.028056	44.8	0
Dressing interval	1	0.131406	0.131406	0.131406	209.83	0
Residual error	10	0.006262	0.006262	0.000626		
Total	15	0.183494	0			

5   REGRESSION ANALYSIS

Regression analysis is used to build up a mathematical model to examine the behavior of the experimental factors within the operating region. The best fitting model is taken to be the one closest to the data in the least squares sense. This may then be used to:

1. Interpolate a response by using a setting within the DOE-levels. This is only applicable for quantitative experimental factors.
2. Interpolate a response by using settings which have not been performed during the DOE experiment in a fractional factorial design.

Due to the fact that only two level or linear-effects-designs are used in this document, the regression analysis must assume a first order mathematical model. The estimated values from such an analysis give an indication of the direction for improvement (Janardhan 2012). Minitab 16 is used for regression analysis, from that following mathematical model for ovality, roughness, cycle time is obtained.

$$C \qquad (1)$$

$$Roughness = 0.521 + 0.0244\,(Nw) - 0.0194\,(Ns) + 0.0119\,(V1) + 0.0419\,(V2) - 0.0906\,(D) \qquad (2)$$

$$Cycle\ time = 218 - 20.5\,(V1) - 12.0\,(V2) + 9.00\,(D) \qquad (3)$$

6   OPTIMUM SET SELECTION

Optimum value for each process variable can be found out from Figs. 4 and 8. So it concluded that, ovality is minimum for the higher level of work head speed, higher level for grinding wheel speed, lower level for rough feed rate, lower level for fine feed rate, and intermediate dressing stroke (Jirapattarasilp 2012).

Roughness is minimum for the lower level of work head speed, higher level for grinding wheel speed, lower level for rough feed rate, lower level for fine feed rate, and intermediate dressing stroke.

Individual optimal set for ovality and roughness is shown in Table 7.

In both case cycle time is 259 sec. So it requires the optimum set to get low ovality and low roughness value with lower cycle time. From the regression analysis equation the optimum set obtained is as shown in Table 8.

6.1   *Analysis of optimized parameter*

Comparison of existing and optimized parameters for cycle time result is shown in Table 9. From that it seen that, total production time for batch of 250 rings requires 795 min with existing parameters and 833 min in new optimized process parameter. So it is 38 min less than time required in new process parameters.

But in existing process parameters 15 rings are scrapped due to lower quality in ovality and roughness out of 250. So the effective time for the one cycle with existing parameters is 203 sec, which is higher than cycle time in optimized process parameters. Also as shown in Table 8, value for the ovality and roughness is less than existing parameters. So chances of rejection of rings are very less.

- Rough feed rate has minimum effect on ovality and surface roughness.
- Intermediate dressing has significant effect on surface roughness but it also increases cycle time around 20 seconds per cycle.
- For better surface finish work head speed should be high.

Table 7. Optimized parameter for ovality and roughness.

Optimal set for	Controlled factor	Optimal set	Value
Ovality	Work head speed	250	4.35 (tol 8)
	Grinding wheel speed	6000	
	Rough feed rate	3	
	Fine feed rate	1	
	Dressing interval	2	
Roughness	Work head speed	100	0.3324 (tol 0.7)
	Grinding wheel speed	6000	
	Rough feed rate	3	
	Fine feed rate	1	
	Dressing interval	2	

Table 8. New and old process parameter.

Existing	Work head speed	150 RPM	
Parameter	Grinding wheel speed	5000 RPM	
	Rough feed rate	6 μ/sec	Ovality = 6.4 μ
	Fine feed rate	2 μ/sec	Roughness = 0.65 μ
	Dressing cycle	Interval	CT = 191 sec
New optimized	Work Head speed	250 RPM	
Parameter	Grinding wheel speed	6000 RPM	
	Rough feed rate	8 μ/sec	Ovality = 4.7 μ
	Fine feed rate	1 μ/sec	Roughness = 0.57 μ
	Dressing cycle	Interval	CT = 200 sec

Table 9. Optimized process parameter.

Total no of rings	250 sec
No of scrap	15
Total cycle time as per old parameter	250 × 191 = 47750 sec = 795 min
Total cycle time as per optimized parameter	250 × 200 = 50000 sec = 833 min
Total increase In cycle time by new parameter	833–795 = 38 min
Effective time for one cycle in old parameter	47750/235 = 203.2 sec
Cycle time in new optimized parameter	200 sec

- From the ANOVA analysis dominant factors affecting the surface roughness and ovality are identified. From the regression analysis relation between the input and out parameter calculated. Optimum parameter with less cycle time is obtained.
- Fractional factorial is very effective optimization technique for optimum results with very less number of experiments.

REFERENCES

Chang S.H., Teng T.T. & Ismail N. 2011, *Screening of factors influencing Cu(II) extraction by soybean oil-based organic solvents using fractional factorial design*, Journal of Environmental Management, 92: 2580–2585.

Davidson M.J., Balasubramanian K. & Tagore G.R.N. 2008, *Surface roughness prediction of flow-formed AA6061 alloy by design of experiments*, Journal of materials processing technology, 202:41–46.

Janardhan M. & Krishna G. 2012, *Multi-objective optimization of cutting parameters for surface roughness and metal removal rate in surface grinding using response surface methodology*, International Journal of Advances in Engineering & Technology, 3: 270–283.

Jirapattarasilp K. & Kuptanawin C. 2012, Effect of Turning Parameters on Roundness and Hardness of Stainless Steel: SUS 303, AASRI Procedia, 3:160–165.

Kanlayasiri K. & Boonmung S. 2007, *Effects of wire-EDM machining variables on surface roughness of newly developed DC 53 die steel: Design of experiments and regression model*, Journal of Materials Processing Technology, 192–193: 459–464.

Kivak T. 2014, *Optimization of surface roughness and flank wear using the taguchi method in milling of Hadfield steel with PVD and CVD coated inserts*, Measurement 50:19–28.

Loukas Y.L. 1998, *A computer-based expert system designs and analyzes a $2^{(k-p)}$ fractional factorial design for the formulation optimization of novel multicomponent lipsomes*, Journal of Pharmaceutical and Biomedical Analysis, 17:133–140.

Suard S, Hostikka S. & Baccou J. 2013, *Sensitivity analysis of fire models using a fractional factorial design*, Fire safety journal, 62:115–124.

Tsao H.S.J. & Patel M.H. 2013, *An intuitive design pattern for sequentially estimating parameters of a 2^k factorial experiment with active confounding avoidance and least treatment combinations*, Computers & Industrial Engineering, 66:601–613.

Multi-disciplinary Sustainable Engineering: Current and Future Trends – Tekwani, Bhavsar & Modi (Eds)
© 2016 Taylor & Francis Group, London, ISBN 978-1-138-02845-6

Development of a swinging bucket mechanism for a shot blasting machine

M.R. Malli
A D Patel Institute of Technology, New V.V. Nagar, Gujarat, India

D.A. Jani
Department of Mechanical Engineering, ADIT, New V.V. Nagar, Gujarat, India

V. Rajpura
VP R&D Department, Indabrator Division, Nesco Ltd., Anand, Gujarat, India

ABSTRACT: Shot blasting is a widely used process for cleaning components after casting and forming processes. During the shot blasting process, it is essential to expose all the surfaces of the components to shots to ensure desired cleaning. The most conventional way to achieve is by use of a mill-construction arrangement of the shot blasting chamber. In the present work, an alternative mechanism with "Swing Bucket" has been proposed to replace the mill construction mechanism. The proposed mechanism is expected to overcome disadvantages of mill-construction like frequent failures of flights precluding the operation and wearing of moving components in extremely abrasive environment resulting in higher maintenance cost and loss of production. In addition, the proposed mechanism offers its benefits without compromising functional requirements shot blasting process. Design of bucket was based on targeted pay load, volume, and operation parameters with an intention to keep changes in the existing machine to minimum. After the bucket was designed, a geometrical model was prepared in Creo 1.0® and same was analysed for stresses in ANSYS Workbench 14.0®. An analysis was also performed in LS-Dyna® to check toppling of various shapes.

Keywords: shot blasting machine; swinging bucket mechanism

1   INTRODUCTION

One of the important stages of cast products manufacturing is finishing to remove residues and obtain required surface condition. Among surface treatment methods, mechanical methods seem prevalent, particularly the abrasive (or shot blasting) methods. The shot blasting is a cold surface treatment that involves projecting beads on the workpiece. Their key advantages include low energy demands, high quality of treated surfaces, good potentials for process automation, and the use of shot blasting units made of more durable materials, work safety and environment-friendly features. In shot blasting processes, a stream of abrasive medium with the required kinetic energy is generated and propelled onto the surface to be treated *Wronaa et al. (2012)*. It is possible with the shot blasting operation to obtain excellent cleaning and surface preparation for subsequent finishing operations *(http://www.wirelab.com)*.

In the recent years, the researchers have focused on various aspects of shot blasting process and machine construction. This includes customization of blasting chamber to control the coverage area and impact density *Badreddine et al. (2014)*. Such a design of chamber has been reportedly providing feedback on the shot dynamics. Also, mechanical and electronic systems have also been proposed to measure velocities of shots in the chamber which was otherwise difficult *Hribernik et al. (2003)*.

The proposed system uses a small microphone encapsulated in a metal housing, placed under a metal cap that acts as a membrane. It is used for the detection of shot impacts. The vibration of membrane resulting from the impact of particles is transmitted to the microphone and induces electrical signals *Hribernik et al. (2006)*. The velocity measurement mechanism was further optimized with the use of small microphone encapsulated in a metal housing and placed under the metal cap acting as a membrane.

A micro-shot blasting mechanism has also been developed and reported to be used for precision components like cutting tools. Tests prior to and following the blasting process were conducted to ascertain any improvements resulting from the process *Kennedy et al. (2005)*.

Verities of shot blasting machines are available depending upon specific applications, and geometry of components. This includes belt type machine (used for small size component), hanger type machine (medium size components suspends on hanger while operation is performed), and rotating table type machine (parts like heavy gear or pulley is bolted to a rotating table during the process).

Despite continuous technology addition to the field, certain aspects remained untouched. One of the major drawbacks of existing mechanism i.e. mill construction that falls in belt type category is a frequent failure of flights due to trapping of components and blasts in the space between consecutive flights. This not only causes frequent breakdown and increase in maintenance cost but also reduces productivity. Many times this also requires replacement of complete belt if the damage is sufficiently high.

The present work is based on the development of an alternative mechanism that can replace the mill—construction mechanism. Though the work is reported in context to a specific machine, authors expect the mechanism to be generic and can be extended to other machines with similar construction.

The existing machine as shown in Figure 1 works with mill construction arrangement. Figure 1 depicts working principle of mill-construction where, flight bars are fitted on the belt with fasteners. The spacing between flights ensures rolling action and tumbling of components exposing its various surfaces to shots. The mechanism has many drawbacks as noted during the user survey and personal communications by the first author. Major drawbacks include, trapping of thin sections between consecutive flights, wearing of the moving components like chain and chain-link due to operating in the extremely abrasive environment sometimes may be severe enough to preclude further operation. Other issues include large maintenance period causing loss of production and higher frequency of such incidences, because of which not only the production suffers but also the life of components is compromised. The statistical data from the survey of existing shot blasting machine users also suggested that the average working life of a mechanism is three months and a failure is experienced after that. Each such failure causes substantial loss of production as it requires numerous time replace the damaged parts. Hence, there was an acute need for new a system or mechanism that not only fulfills functional requirements of the existing mechanism but also has to overcome drawbacks discussed.

It was also intended to reduce the number of parts to reduce the cost and improve system reliability. The major challenge was to design a system that can still be used by making minimum changes in the existing machine. Hence, subsystems should also be designed in such a way that the overall structure of the existing machine can be used with minimum modifications

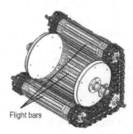

Figure 1. Milling construction *(http://www.sfecindia.net)*.

2 METHODOLOGY

The idea of the new system evolved from the fact that, when a component is kept in a swinging enclosure, it tends to roll or slide inside it. If small obstacles are offered to components in such an enclosure, components topple. This hypothesis encouraged the development of the new mechanism. Moreover, as the new design has to have fewer moving parts as compared to existing mechanism like mill-construction, it is expected to have higher reliability.

The conceptual shape of such bucket is shown in Figure 2a. The actual mechanism was developed to satisfy requirements prescribed in section 2.1 as desired by the NESCO Ltd. (Indbrator division) from here onwards referred as the company. In this mechanism, a bucket is constructed by bending a sheet into a polygon. The bucket is free to swing with desired included angle. Using the concept, it was decided to develop a bucket of an open polygonal shape. The polygonal shape would help components to topple. The opening at one of the faces would facilitate loading—unloading and servicing.

2.1 *Design requirements*

The system proposed in the present work was developed to address some requirements as Pay load mass 1500 kg, Permissible over load is 50% of pay load, Max Filled volume capacity 0.5 m³, Per single work piece 80 kg, Number of blast wheel 2, angle of flow of blast wheel 49⁰, height of blast wheel from loaded material is 0.8 m, Number of oscillation per minute 20 and factor of safety is desired greater than or equal to 2 as desired by company.

The mechanism development was completed in stages. First, bucket shape and size were determined to satisfy fill volume requirement. That was followed by design analysis of the check accomplishment of functional and structural requirements.

2.2 *Fill volume*

Fill volume is the volume of the bucket to be occupied by components during the process. From the data given in the section 2.1, it was concluded to satisfy following requirements for an acceptable initial design: (1) fill volume of at least 0.5 m³ (2) distance between blast wheel outlet, and upper faces of components should be around 0.8 m and (3) distance between end plates of bucket should not exceed 2 m to ensure complete coverage by shots and (4) bucket should be able to operate within a space of 2.5 m × 3 m × 3 m so as to use it with the structure of existing machine.

As the section of the bucket is to remain constant along its length, the expected area to be filled was estimated about 0.25 m². Initially, it was thought to have a bucket with flat faces, but soon it was realized that such a construction may not topple components with blunt or curved faces. Hence, the bucket with corrugation at the bottom, as shown in Figure 2b, was developed. The corrugation also resulted in increased strength of the bucket as discussed later.

Figure 2b is the magnified view of blacken section shown in Figure 2a. As shown with the blacken section, the bucket cross-section can be discretized using trapeziums and triangles (As numbered 1, 2, 3, 4, 5, 6 in Figure 2b). The total area accounted was approximately 0.3 m², as desired. The height of the bucket was decided on from the requirement to ensure

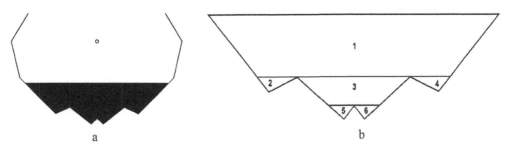

Figure 2a.   Bucket shape and fill volume.   Figure 2b.   Filling area of bucket.

the distance between blast-wheels and height of the filled materials. Thus, the final dimension of bucket are: width 1825 mm, height 1236 mm, opening face of the bucket is 1452 mm, the height of blast wheel from loaded material 800 mm.

Using the constraints discussed, a bucket was developed, and a 3D model of the same was prepared. To ensure the complete coverage of components by shots, length of the bucket was restricted to 2.0 m. This facilitates, maximum permissible material filling volume of 0.6 m³ which more than the desired value of 0.5 m³.

2.3 Oscillatory movement of bucket

In Figure 3, 1-2-3-4-1 indicates one oscillation of the bucket. The new bucket has to be compatible with the existing set-up in which angle of flow blast wheels is 49°. Hence, if bucket oscillates with an amplitude of more than 50° during a cycle, shots will fall outside the bucket and operation will suffer. Thus, during one cycle the bucket will cover 200°. The exposure time of the components to shots can be controlled by number of cycles per minute and run time. It was decided to design system of 20 cycles per minute. However, this can later be adjusted as per the requirements enforced by the application.

For producing the oscillatory motion, a drive mechanism is required. The required 20 oscillations per minute will need a drive mechanism to provide 12 rpm ([20 × 200/360] = 4000/360 = 12 RPM).

2.4 Bucket-material

Due to the good wear resistance and availability, a ferrous alloy with Manganese was the chosen for the bucket material. The material selected was IS-276 Grade I with Manganese (11–14%), Carbon (1.05–1.35%) and yield strength of 300 MPa. As desired, considering factor of safety as 2, permissible stress was reduced to 150 MPa *(Indian Standard ICS 77.140.180, Aug-2000)*

2.5 Design analysis

As per design parameter, payload of the bucket should be 1500 kg, bucket mass was estimated to be 1015 kg, mass of accumulated shots was assumed to be 600 kg, 500 kg mass was considered for other mounting devices and 50% (750 kg) overload allowance was considered. Hence, the total dead load was computed to be 4365 kg (43 kN).

2.5.1 Centrifugal force consideration
Centrifugal force is likely to induce due to the oscillatory motion of the bucket. Hence, it was calculated and added to dead load for all loading conditions. For a loaded bucket, the total mass is equivalent to 4365 kg. Centre of the mass is approximately located at radius 0.7 m from the axis of rotation. As the bucket is expected to rotate at 12 RPM, the centrifugal load is equivalent to 4845 N. Now this load is added to the dead load of 42820 N, resulting in a total load of 48 kN.

2.5.2 Load calculation as per static condition of bucket
For the analysis, the bottom section of the bucket was treated as a beam and checked for three loading conditions viz. Uniformly Distributed Load UDL, Linearly Varying Load LVL_1, and LVL_2. Loads estimated above were used for the analysis.

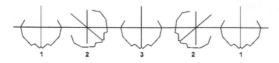

Figure 3. Bucket oscillation.

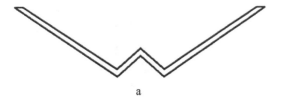
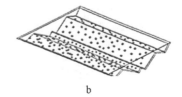

a

b

Figure 4a.   Bucket bottom (W section).   Figure 4b.   Bucket bottom with through holes.

- UDL – Resembles the uniform distribution of material along the length.
- LVL$_1$ – Depicts the situation of material loading, when majority load is on one side and reduces towards another side.
- LVL$_2$ – Majority of the material filled is concentrated at centre and reduces towards ends.

Stresses were evaluated at three locations for each case: Longitudinal ends of the bottom plate (A and B) and Centre of the plate (C).

Consider the bottom section of the bucket as shown in Figure 4a. The analysis was initially performed considering 12 mm thickness. However, as 12 mm thick section was too conservative, the final section had only 6 mm thickness. Thus, section modulus of 6 mm thick W section was estimated to be $Z = 1.079 \times 10^6$ mm^3.

Thus, the highest bending stress for 6 mm thick section was estimated to be 15 N/mm^2 among all three loading conditions.

2.5.3   *Perforation of the bottom portion*

For the targeted machine, shots are expected to impact at an approximate rate of 140 kg/min. Each cleaning cycle has lasts for 4 to 6 min. If the removal of shots is not arranged, they may all be accumulated at the bottom, increasing the load on the bucket. Hence, it was decided to provide slots or holes at the bottom of the bucket. The arrangement should facilitate the exit of the used shots. This helps to avoid increase of load during the operation. Shots accumulated this way may also be reused.

The option of providing slots was not considered as it might have offered trapping of components with thin sections.

As shown in Figure 4b, a number of small (20 mm diameter) holes provided for releasing the shots during the process.

3   FEA OF THE BUCKET

FEA analysis of bucket was done using ANSYS 14.0® Workbench. For the analysis, the bucket was subjected to a load of 48 kN. The load was distributed on the inner faces of 'W' section of the bottom portion. The two holes provided on end-plates (for mounting of the bucket on the shaft) were fixed. Hex dominant mesh was used. FE analyses were done for the corrugated as well as for the bucket with the non-corrugated bottom. Figure 5a shows that maximum equivalent stress value is 55.63 MPa which is very less than the permissible stress value 150 Mpa as a factor of safety is desired 2. Figure 5b shows that the maximum deformation at the bottom portion of the bucket was 0.67 mm which can be negligible for the large component.

4   TOPPLING ACTION

Though it is not possible to include all types of geometries to confirm the toppling action, the analysis was carried out with few shapes to check if components topple during the swing of the bucket. The analysis was done using LS-Dyna® software, for components with circular, rectangular and triangular cross-sections. During the analysis of components of triangular

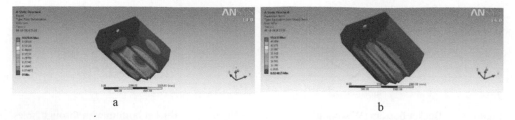

| a | b |
| Figure 5a.   Stresses in the bucket. | Figure 5b.   Deformation of the bucket. |

and rectangular cross-sections, aspect ratios of 1, 20 and 50 were considered. Each component was simulated in the bucket for various orientations and at different locations for one cycle. For all the cases, toppling was confirmed by tracing the path of a node on it. Hence, the shape of the bucket was confirmed to serve the purpose.

5   CONCLUSION

In this work, it has been proposed to replace the existing mill construction in the shot blasting chamber by a swinging bucket mechanism. The purpose of the swinging bucket mechanism is to topple the components during the shot blasting operations so that all the surfaces of the component get exposed to the shots. Corrugated bottom of the bucket helps in toppling components with various shapes, round edges and large aspect ratios. To allow used shots to fall out of the bucket, the perforated pattern is provided at the bottom. Such removal of shots helps to avoid cumulative increase of load during the process. The proposed bucket design was analysed with certain approximations and was also analysed using standard FEA solver. To perform software based analysis, the Static Structural module of ANSYS Workbench 14.0 was used. The induced von Mises stress for the worst loading case (analytically identified as LVL_2 case) in the bucket was noted to be 55.63 MPa. The stress was well below the permissible limit of 150 MPa for the bucket material.

Drive mechanism for the bucket to swing and supporting structure modifications are being accomplished with the view to make minimum changes in the other segments of the machine. This permits the use of the majority of the structure and associated components for the new mechanism also. Though the simulations were done to ensure the toppling of the components inside the bucket, experimental validations are yet to be done for the same.

REFERENCES

Badreddine, J., Remy, S., Micoulaut, M., Rouhaud, E., Desfontaine, V., Renaud, P., 2014, CAD based model of ultrasonic shot peening for complex industrial parts. *Advances in Engineering Software* 76: 31–42.
General Introduction to Shot Blasting, http://www.wirelab.com/Cym%20Shot%20Blasting%20Info.pdf.
Hribernik, A., Bombek, G., Markoc, I., 2003, Velocity measurements in a shotblasting machine. *Flow Measurement and Instrumentation* 14: 225–231. http://www.sfecindia.net.
Hribernik, A., Bombek, G., 2006, Improved method for shot particle velocity measurement within a shotblasting chamber. *Flow Measurement and Instrumentation* 17: 99–10.
Indian Standard Austenite-Manganese Steel Casting Specification (fifth revision) ICS 77.140.180, Aug–2000.
Kennedy, D.M., Vahey, J., Hanney, D., 2005, Micro shot blasting of machine tools for improving surface finish and reducing cutting forces in manufacturing. *Materials and Design* 26: 203–208.
Malli, M., 2014, Development of swinging bucket mechanism in blasting chamber of shot blasting machine. M.E., *Thesis, A.D. Patel Institute of Technology (GTU).*
Wrona, R., Zyzak, P., Ziólkowski, E., Brzezinski, M., 2012, Methodology of Testing Shot Blasting Machines in Industrial Conditions. *Archives of Foundry Engineering 12: 97–104.*

Multi-disciplinary Sustainable Engineering: Current and Future Trends – Tekwani, Bhavsar & Modi (Eds)
© 2016 Taylor & Francis Group, London, ISBN 978-1-138-02845-6

Converting the chaotic motion of the double pendulum into linear motion by changing its masses

Pratik N. Prajapati, Jay K. Airao & Himanshu K. Yadav
L.D.R.P. Institute of Technology and Research, Gandhinagar, Gujarat, India

ABSTRACT: The present paper deals with the linear and nonlinear dynamic analysis of a double pendulum. This paper mainly deals with converting the non-linear motion of the pendulums, which have the same length, into linear motion by changing the values of the mass of the pendulums. The paper contains an experimental analysis, along with a simulation by Pro-E (Creo) software. A three-step procedure has been carried out for dynamic analysis. In the first step, the equation of motion has been developed by the Lagrangian method and the authors have derived new relations between the masses of the pendulums in order to convert the non-linear motion of the pendulum into a linear motion in terms of the initial displacement. In the second step, an analysis has been carried out by different analysis tools, like time displacement plot and phase portrait. The authors have found different nonlinear behaviors like periodic, period 2T, quasiperiodic, multi-periodic, and chaotic. Finally, in the third step, the results have been verified experimentally.

1   INTRODUCTION

A double pendulum consist of two pendulums having point masses. The variables that are present in a double pendulum are the two lengths of the rigid (massless) rods l_1 and l_2, the two masses m_1 and m_2, and the two angles between the vertical massless rods θ_1 and θ_2, as shown in Fig. 1 (a and b). The double pendulum is a complex system. The dynamics of a majority of multi-link robotics manipulators are simply the dynamics of a large number of coupled pendula. In case of EOT crane, for certain types of payloads and riggings, the system can behave like a double pendulum. M. Z. Rafat et al. have investigated a variation of the simple double pendulum, in which the two point masses are replaced by square plates and they show how the behavior varies from regular motion at low energies, to chaos at intermediate energies and back to regular motion at high energies. Carl W. Akerlof has analyzed the double pendulum with distributed masses, while considering all the mountings and bearings used for a double pendulum system, and he compared the actual dynamical nonlinear behavior of a real physical double pendulum with a mathematical simulation. The practical occurrences of the nonlinear phenomena, such as periodic, sub-harmonic, quasi-periodic and chaotic, are explained in the books of different authors (A.H. Nayfeh and B. Balachandran, F.C. Moon, H.S. Steven, J.M.T. Thompson and H.B. Stewart). The authors want to give more importance to all these books because a different tool to identify the nonlinear behavior of the system has been explained very nicely with plots of experiments. M.L. Kansara et al. have numerically analyzed the nonlinear behavior of 2 DOF spring-mass systems by developing a MATLAB program.

So, based on literature survey the authors have decided to derive the relative equations between masses to convert the chaotic motion of the system into a nearly linear motion.

2 PROBLEM FORMULATION

A simple double pendulum system is shown in Fig. 1. It comprises a mass m_1 suspended by rod of length l_1 from a fixed pivot and mass m_2 suspended by a rod of length l_2 from m_1. Take θ_1 and θ_2 as the angle made by mass m_1 and m_2 with respect to the vertical line respectively. The distance of masses m_1 and m_2 from the vertical line at moment is x_1 and x_2 respectively. $\dot{\theta_1}$ and $\dot{\theta_2}$, $\ddot{\theta_1}$ and $\ddot{\theta_2}$ are the tangential velocity and the acceleration of mass m_1 and m_2 respectively.

2.1 *Equation of motion*

The approaches of using the Euler-Lagrange method have been used by H.K. Yadav et al.
Using the Euler-Lagrange equation:

$$\frac{d}{dt}\frac{\partial(L)}{\partial \dot{x}} - \frac{\partial(L)}{\partial x_i} + \frac{\partial(L)}{\partial x_i} = 0 \tag{1}$$

Total potential energy is:

$$T = \frac{1}{2}(m_1 + m_2)l_1^2\,\dot{\theta_1^2} + m_2 l_1 l_2 \dot{\theta_1}\dot{\theta_2} + \frac{1}{2}m_2 l_1^2\dot{\theta_2^2} \tag{2}$$

Total kinetic energy is:

$$U = \frac{1}{2}(m_1 + m_2)gl_1\theta_1^2 + \frac{1}{2}m_2 gl_2\theta_2^2 \tag{3}$$

Now, the Lagrangian, $L = T - U$ gives the two lagrangian equations and by applying the small angle approximation, the two equations of motion are:

$$\ddot{x_2} = g\frac{x_1}{l_2} - g\frac{x_2}{l_2} \tag{4}$$

$$\ddot{x_1} = x_2\frac{m_2 g}{m_1 l_2} - x_1\left(\frac{m_2 g}{m_1 l_2} + \frac{g}{l_1} + \frac{m_2 g}{m_1 l_1}\right) \tag{5}$$

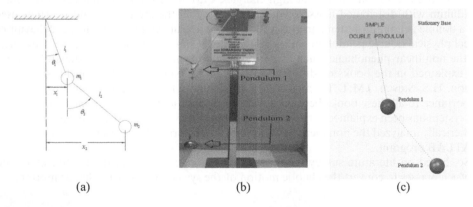

(a) (b) (c)

Figure 1.   (a) Simple double pendulum line diagram (b) Experiment demonstration (c) Preo-E (Creo) geometrical model.

2.2 Solution of the equation of motion

For the principle mode of vibration, let the general solution of eq. (4) and eq. (5) be a simple harmonic motion, and A and B are the amplitudes of pendulums 1 and 2 respectively. From the book "Mechanical vibration" by S.S. Rao, the solution is given below,

$$\omega_{n1}^2 = \left(\frac{(m_1 + m_2) + \sqrt{m_1 m_2 + m_2^2}}{m_1 l} \right) g \tag{6}$$

$$\omega_{n2}^2 = \left(\frac{(m_1 + m_2) - \sqrt{m_1 m_2 + m_2^2}}{m_1 l} \right) g \tag{7}$$

2.3 Derivation for finding the relation between value of the mass for linear motion

$$m_2 = m_1 \frac{A^2}{B(B - 2A)} \tag{8}$$

where, A and B are the initial maximum displacement of masses m_1 and m_2 respectively.

3 RESULTS AND DISCUSSION

In the present paper, the authors have considered two cases. In both cases, the lengths of both pendulums are the same and equal to 0.2 m. By simulation of the present system through Pro-E (Creo) software, the time displacement plots and phase portrait plots are generated. The authors have analyzed the system for two different modes. In mode 1, both pendulums are initially displaced in the same direction and in mode 2 in opposite direction. The two cases are summarized in the following Table 1.

If both masses are chosen randomly, then the system behaves nonlinearly. The authors have analyzed two cases. In case 1, the authors have taken the value of mass m_1 is 1 kg and mass m_2 is 1.5 kg randomly. For case 1, the system behaves nonlinear. Now in case 2, in order to convert the system into a linear one, the value of mass m_2 is calculated from eq. (8) and it is equal to 0.249 kg for the value of mass m_1 is 1 kg. Similarly, the value of mass m_1 can be calculated by eq. (8) to get the linear system for a particular value of mass m_2.

Table 1. Parameters of double pendulum.

	l_1 (m)	l_2 (m)	θ_1 (degree)	θ_2 (degree)	A (m)	B (m)
Mode 1	0.2	0.2	20	50	0.0684040	0.221613
Mode 2	0.2	0.2	20	50	0.0684040	−0.0848049

Table 2. Description of cases.

Length of pendulum (m)	Case-1 Masses for nonlinear analysis (kg)		Case-2 Masses for linear analysis (kg)	
	Mass 1	Mass 2	Mass 1	Mass 2
l1 = l2 0.2	1	1.5	1	0.249

From the time displacement graph of mass m_1 (Fig. 2(a)), it can be observed that a number of cycles have different amplitudes. So, it can be inferred that mass m_1 has a chaotic motion. The same can be confirmed with the help of phase plot (Fig. 2(c)), which has many crossing loops. While from Fig. 2(b & d), it can be analyzed that mass m_2 behaves like a nonlinear period 1T motion. Now, in Fig. 2(e), one can observe mainly two peaks in the case of mass m_1 of mode 2. Similarly, in Fig. 2(g) mainly two dark loops are observed, which indicate that the mass m_1 has the period 2T motion. While from Fig. 2(f & h), it can be observed that mass m_2 behaves chaotic, because in Fig. 2(f) there are many peaks and in Fig. 2(h) there are many crossings and overlapped loops.

3.1 Case 1

3.1.1 Mode 1

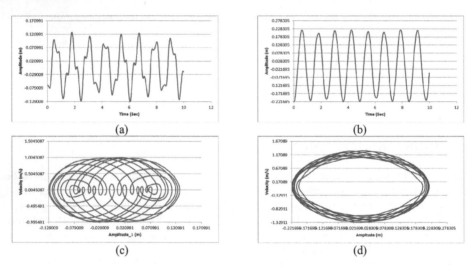

(a) (b) (c) (d)

3.1.2 Mode 2

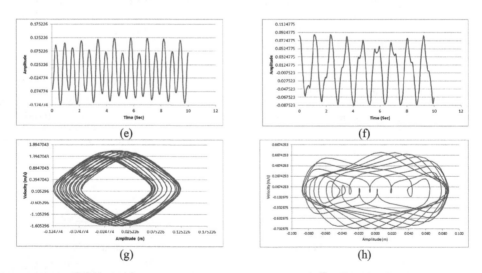

(e) (f) (g) (h)

Figure 2. Time displacement plot for mode 1 (a) for mass m_1 (b) for mass m_2, mode 2 (e) for mass m_1 (f) for mass m_2 and Phase portrait plot for mode 1 (c) for mass m_1 (d) for mass m_2, mode 2 (g) for mass m_1 (h) for mass m_2.

Now, from the time displacement graph of mass m_1 (Fig. 3(a)), it can be concluded that motion of mass m_1 is of the linear type, because there are not many peaks that can be observed. The same can be confirmed with the help of phase plot (Fig. 3(c)), which has only one loop. While from Fig. 3(b & d), it can be analyzed that mass m_2 behaves like a linear motion. Similarly, it can be observed for mode 2 that mass m_1 and mass m_2 behave linearly. But, due to limited space, the authors could not display the plots in the paper.

Similarly, all the results are observed practically by a setup of the double pendulum (Fig. 1(b)).

3.2 Case 2

3.2.1 Mode 1

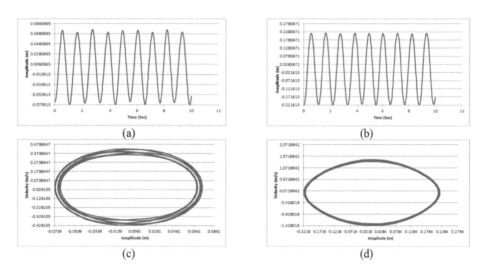

(a) (b)

(c) (d)

Figure 3. Time displacement plot for mode 1(a) for mass m_1 (b) for mass m_2, Phase portrait plot for mode 1(c) for mass m_1 (d) for mass m_2.

From Fig. 2(c & d) mass m_1 and m_2 have multi periodic and period 2T behavior respectively. From Fig. 2(g & h) mass m_1 and m_2 have 2T periodic and multi-periodic behavior respectively. From Fig. 3(c & d) mass m_1 and m_2 have both having linear behavior.

4 CONCLUSION

Based on above analysis, some conclusions can be drawn:

1. For case 1, the system behaves like a nonlinear system.
2. If both masses are chosen randomly, then the system behaves nonlinearly. In case 1, the system behaves like period 1T, period 2T and chaotic. From the newly derived eq. (8) the system can be converted into the linear system. Also, this behavior is verified practically.

REFERENCES

Akerlof Carl W., 2012 September 26, "*The Chaotic Motion of a Double Pendulum*".
Kansara M.L., Gohil J.A., Yadav H.K., 2014, "*Nonlinear dynamics analysis of self-excited vibration in case of two degree of freedom system*", Proceedings of international conference on theoretical, applied, computational and experimental mechanics, IIT Kharagpur, pp. 530.

Moon F.C., 1987, *"Chaotic Vibrations"*, John Wiley & Sons, New Jersey.

Nayfeh A., Balachandran B., *"Applied Nonlinear Dynamics Analytical, Computational, and Experimental Methods"*, 1995, Wiley, New York.

Rao. S.S. 2010, *"Mechanical Vibration"*, New Delhi, Pearson publication.

Rafat M.Z., Wheatland M.S., and Bedding T.R., 2006, *"Dynamics of a double pendulum with distributed mass"*, School of Physics, University of Sydney, NSW, Atraia.

Strogatz S.H., 1994, *"Nonlinear Dynamics and Chaos"*, New York, perseus books.

Thompson J.M., Stewart H.B, 2002, *"Nonlinear Dynamics and Chaos"*, UK, John Wiley & Sons.

Yadav H.K., Upadhyay S.H. and Harsha S.P., 2013, *"Study of effect of unbalanced forces for high speed rotor"*, Procedia Engineering, Elsevier. 64, pp.593–602.

Multi-disciplinary Sustainable Engineering: Current and Future Trends – Tekwani, Bhavsar & Modi (Eds)
© 2016 Taylor & Francis Group, London, ISBN 978-1-138-02845-6

Modal analysis of geared rotor system using finite element method

Ankur Saxena
Department of Mechanical Engineering, Indian Institute of Technology, Indore, India

Manoj Chouksey
Department of Mechanical Engineering, SGSITS, Indore, India

Anand Parey
Department of Mechanical Engineering, Indian Institute of Technology, Indore, India

ABSTRACT: This paper presents a comprehensive study of modal analysis for a geared rotor system supported on rolling element bearings. The effect of coupling due to the contact of pinion and gear has been found out on the modal properties. The finite element technique is used to model the geared rotor system. The effect of translatory inertia, rotary inertia, gyroscopic moment, shear deformation etc. have been considered in modelling the rotor-shaft continuum, whereas the gear and pinion are modelled using inertia element. The analysis includes both transverse as well as torsional degrees of freedom of the gear rotor shaft. Detailed modal analysis has been carried out to investigate the influence of coupling due to mesh stiffness on natural frequencies and mode shapes.

1   INTRODUCTION

Modal analysis is used to have an idea about the modal characteristics of system, viz. primarily modal natural frequencies, damping factors and mode shapes (Ewins 1999). However, modal analysis of rotor system differs from modal analysis of structures due to additional forces induced by rotor spin like gyroscopic effect, asymmetric nature of bearing as well as tangential forces (Friswell 2010). Many studies about modal analysis of rotor systems have been reported in literature, e.g. to see the effect of internal damping, gyroscopic effect, and rotor faults (Lee and Jei 1988; Chouksey et al., 2012). Many researchers built the Finite Element (FE) model of rotor system where discs are considered as point element and represented by its inertial properties. Lee and Jei (1988) investigated continuous rotor bearing system by considering the effects of rotary inertia and gyroscopic moment to see the effects of asymmetry of the rotor system on the modal parameters. Chouksey et al. (2010) carried out modal analysis of a rotor shaft system to see the effect of rotor-shaft material damping on the modal and frequency response characteristics. Further, Chouksey et al. (2012) investigated the influences of internal rotor material damping along with fluid film forces generated in journal bearings on modal behavior of the rotor shaft system. Such studies include modal analysis of a single rotor system supported on bearings. However, modal analysis of geared rotor system calls for combined analysis of the driving and driven rotors as both are directly coupled.

Kahraman et al. (1992) developed a FE model of a geared rotor system with flexible bearings considering transverse and torsional degrees of freedom. The authors also studied the effect of bearing compliances on system dynamics using the FE model. Choy et al. (1991) analyzed a multi stage gear transmission system by considering it as an enclosed structure and reported the results of transient and steady state vibrations due to torque variations, speed changes, rotor imbalances and gear box support motion. Lim and Singh (1991) developed a dynamic model of geared rotor system by using lumped parameter and dynamic finite element techniques and used it to predict the vibration transmissibility through bearings and mounts, casing vibration motion, and dynamic response of the internal rotating system. However, they did not take into account the gyroscopic effect. Pedersen et al. (2010) applied a theory of modal analysis to study spur gear pair considering time varying gear mesh stiffness. They concluded that vibration related to higher frequency mode is more sensitive to the time variant mesh stiffness.

It is very important to find out the effect of coupling of the driving and driven gear on the modal characteristics of the rotor system. However, it is found that there are very few studies about modal analysis of geared rotor systems. In view of this, present work attempts modal analysis of geared rotor system to investigate the effect of the coupling due to mesh stiffness on the modal behavior of the rotor system.

2   FINITE ELEMENT MODEL

Finite element method has been used in this work for the modelling and analysis of the geared rotor system. The effect of translatory inertia, rotary inertia, gyroscopic moment, shear deformation etc. have been considered in modelling the rotor-shaft continuum, whereas the gear and pinion are modelled using inertia element. The analysis includes both transverse as well as torsional degree of freedom of the geared rotor shaft.

Firstly, the driving and driven rotor systems are discretized and assembled to get the mass and stiffness matrix. This is followed by the assembly of the system matrices for the driving and driven gear by including the coupling matrix. The coupling in actual is between the transverse displacement and the angular displacement of the gear and pinion. In this work, the co-ordinate axis are considered in such a way that one of the principal axis (Y axis) is taken along the pressure line. This facilitates consideration of the forces only along the Y direction, as there will not be any force component due to mesh stiffness along the Z direction. The coupling matrix in this work has been considered after following Kahraman et al. (1992), and is given by:

$$[K_m] = \begin{bmatrix} k_m & k_m r_p & -k_m & -k_m r_g \\ k_m r_p & k_m r_p^2 & -k_m r_p & -k_m r_p r_g \\ -k_m & -k_m r_p & k_m & k_m r_g \\ -k_m r_g & -k_m r_p r_g & k_m r_g & k_m r_g^2 \end{bmatrix} \qquad (1)$$

where, $[K_m]$ = mesh stiffness matrix, k_m = mesh stiffness coefficient, r_p, r_g = base circle radii of pinion and gear, respectively. The corresponding displacement vector for the above matrix is

$$\{q\} = [y_p \ \theta_p \ y_g \ \theta_g]^T \qquad (2)$$

where, y_p, y_g = coordinates in the direction of pressure line of the pinion and gear respectively, θ_p, θ_g = angular displacement of the pinion and gear respectively.

Adding the mesh stiffness matrix given in equation (1) to the stiffness matrix of the uncoupled geared rotor system yields the total stiffness matrix of the coupled geared rotor system. After assembling the governing equations for all the elements by incorporating the boundary conditions and including the coupling effect of spur gear pair, the equations of motion of the geared rotor system are obtained as follows:

$$M\ddot{q}(t) + Kq(t) = f(t) \qquad (3)$$

where, $M = M_{trs} + M_{rot}$; $K = [K_{brg} + K_{sh} + K_m]$

In the above 'M' is mass matrix and includes mass components due to translatory (M_{trs}) and rotary inertia (M_{rot}). The symbol 'K' is the overall stiffness matrix and is the summation of stiffness matrix due to bearing flexibility (K_{brg}), shaft bending (K_{sh}) and gear coupling matrix (K_m) respectively. The work by Chouksey et al. (2012) can be referred to for details of modal analysis.

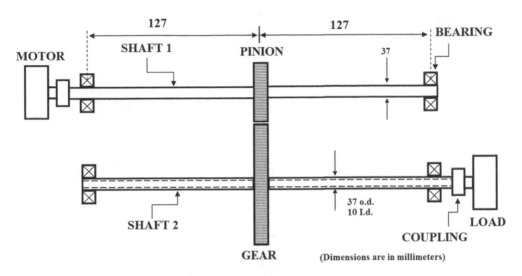

Figure 1.   A schematic diagram of geared rotor system.

Table 1.   Parameters of the geared rotor system.

Parameter		Value
Moment of inertia	Driven gear, I_g	0.0018 kg · m²
	Driving gear, I_p	0.0018 kg · m²
Mass	Driven gear, m_g	1.84 kg
	Driving gear, m_p	1.84 kg
Length of shaft	Driven gear, L_g	0.254 m
	Driving gear, L_p	0.254 m
Diameter of shaft	Driven gear, d_g	0.037 m
	Driving gear, d_p	0.037 m
Base circle radius	Driven gear, r_g	0.0445 m
	Driving gear, r_p	0.0445 m
Mesh stiffness coefficient, k_m		1.0×10^8 N/m
Bearing stiffness coefficient, k_{xx}		1.0×10^9 N/m

3   ILLUSTRATIVE EXAMPLE

The geared rotor-shaft system as shown in Fig. 1 has been considered for the purpose of sim-ulation as well as illustration after following Kahraman et al. (1992). The geared rotor system consists of two rotor shafts connected by gear and pinion. The driving shaft is connected through motor and the driven shaft is connected to load. Each rotor-shaft is supported by a pair of roller bearings. The parameters of the system are mentioned in Table 1. Both the gears are mounted at the center of shaft of lengths L_p, L_g respectively.

4   SIMULATED RESULTS

In this section, the simulated results of the modal analysis of the geared rotor system has been presented. The work investigates the influence of mesh stiffness due to contact of driving and driven gears on the natural frequencies and mode shapes.

Initially, the results of modal analysis are given for the uncoupled rotor system, i.e. by considering zero value of mesh stiffness. The results are given for a spin speed of 1000 rpm. Table 2 shows the results of natural frequencies and mode shapes of the uncoupled geared

Table 2. Natural frequencies and mode shapes of uncoupled geared rotor system.

Mode No.	Natural frequency (Hz)	Driving Rotor	Driven Rotor
1	662.8 (B)		
2	663.3 (F)		
3	669 (B)		
4	669.5 (F)		

Table 3. Natural frequencies and mode shapes of coupled geared rotor system.

Mode No.	Natural frequency (Hz)	Driving Rotor	Driven Rotor
1	499.5 (B)		
2	663.2 (F)		
3	665.9 (B)		
4	669.5 (F)		

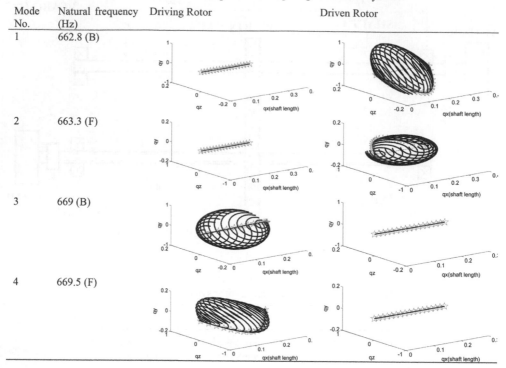

rotor system. The first two natural frequencies are obtained as 662.8 Hz. and 663.3 Hz. which are natural frequencies of the driven rotor. It may be noted that both the driving and driven rotors are fully symmetric and are supported on isotropic bearings. However, the initial two natural frequencies are not equal, and this is due to the gyroscopic effect. The analysis of the mode shapes corresponding to the two natural frequencies shows that the first mode is backward mode, whereas the second mode is forward mode. It may be seen that the mode orbit at each node is drawn open and the starting point of which is marked with a star. This facilitates the identification of forward and backward modes. It may also be seen that there is no motion in the driving shaft in these modes, which is attributed to the uncoupling of the two rotors. Similarly, the 3rd and 4th natural frequencies are obtained as 669 Hz. (B) and 669.5 Hz. (F), and are the natural frequencies of the driving rotor. The corresponding mode shapes are also drawn.

Further, the effect of the coupling due to mesh stiffness has been found out. Table 3 shows the results of natural frequencies and mode shape of the coupled geared rotor system by considering mesh stiffness (k_m = 1e8 N/m). It is seen that the coupling due to mesh stiffness significantly altered the first natural frequency, which in this case comes out to be 499.5 Hz. It has been observed that due to coupling the mode shapes of driving and driven rotors also gets coupled. This has been depicted in the mode shapes of the coupled geared rotor systems as given in Table 3.

5 CONCLUSIONS

The work studied the influence of coupling due to mesh stiffness of the gear and pinion contact on the natural frequencies and mode shapes of the geared rotor system. The results of modal analysis of the uncoupled and coupled geared rotor systems have been compared. The gyroscopic effect has been considered in the model and the forward and backward modes for the geared rotor system are studied. It is concluded that the coupling significantly affects the first natural frequency of the rotor system and in this case it is found that the first natural frequency reduced by 24.6%. Further it is seen that the due to mesh stiffness the mode shapes of the gear and pinion also gets coupled, which is reflected in the plots of mode shapes of the coupled rotor system.

REFERENCES

Chouksey, M., Dutt, J.K., and Modak, S.V., 2010, Influence of rotor-shaft material damping on modal and directional frequency response characteristics. *Proceedings of ISMA-2010*, Katholieke Universiteit, Leuven, Belgium, 1543–1557.
Chouksey, M., Dutt, J.K., and Modak, S.V., 2012, Modal analysis of rotor-shaft system under the influence of rotor-shaft material damping and fluid film forces. *Mechanism and Machine Theory*, vol. 48, pp. 81–93.
Choy, F.K., Tu, Y.K., Savage, M., and Townsend, D.P., 1991, Vibration signature and modal analysis of multi-stage gear transmission. *Journal of the Franklin Institute*, vol. 328(2), pp. 281–298.
Ewins, D.J., 1999, *Modal testing: theory and practice* (Vol. 6). Letchworth: Research studies press.
Friswell, M.I., 2010, *Dynamics of rotating machines*. Cambridge University Press.
Kahraman, A., Ozguven, H.N., Houser, D.R., and Zakrajsek, J.J., 1992,Dynamic analysis of geared rotors by finite elements. *Journal of Mechanical Design*, 114(3), 507–514.
Lee, C.W., and Jei, Y.G., 1988, Modal analysis of continuous rotor-bearing systems. *Journal of Sound and Vibration*", vol. 126(2), pp. 345–361.
Lim, T.C., and Singh, R., 1991, Vibration transmission through rolling element bearings. Part III: Geared rotor system studies. *Journal of sound and vibration*, 151(1), 31–54.
Pedersen, R., Santos, I.F., and Hede, I.A., 2010, Advantages and drawbacks of applying periodic time-variant modal analysis to spur gear dynamics. *Mechanical Systems and Signal Processing*, vol. 24(5), pp. 1495–1508.

Multi-disciplinary Sustainable Engineering: Current and Future Trends – Tekwani, Bhavsar & Modi (Eds)
© 2016 Taylor & Francis Group, London, ISBN 978-1-138-02845-6

Computational investigation of chatter in hard turning

Mayur S. Ghormade & D.H. Pandya
L.D.R.P. Institute of Technology and Research, Gandhinagar, Gujarat, India

ABSTRACT: Chatter is one of the major problems in hard turning. Machine tool chatter should be avoided for a better machine life, improved surface finish and increased tool life. The chatter in hard turning is reduced by introducing the material damper in the cutting tool system. In the present study, an analytical model was simulated for three different shim materials, namely carbide, aluminum, and brass, using the MATLAB program. The damping ratio of cutting tools with three different shims was obtained by using the frequency response function and the half bandwidth method. Stability region for different shim materials from their respective damping ratio values was identified.

1 INTRODUCTION

Chatter is a problem of instability in the hard metal cutting process. The cutting process is accompanied by violent vibrations with loud noise, which results in low surface finish and thus poor quality of machined components. The vibration amplitude has a direct influence on the tool life. Higher vibration amplitudes result in substantially reduced tool life. (Kayhan & Budak, 2009). Self-excited vibration is an undesirable phenomenon in machining. It has adverse effects on surface finish and metal removal rates (Tlusty & Ismail, 2000), which also has reduced tool life (Altintas & Budak, 2000). There are four major types of chatter: regenerative chatter, mode coupling, interrupted cutting, and thermo-mechanical chatter. The regenerative shown in Figure 1 is due to the interaction between the cutting force and the workpiece surface undulations produced by preceding tool passes. The regenerative type is found to have the most adverse effect on the production rate in most machining processes (Nijmeijer et al 2003). Davies and Balachandran (1996) studied chatter caused by interrupted cuts, as shown in Figure 2, which led to impact oscillations, a form of forced machine tool vibration. Kuster and Gygax (1990) presented a new model and illustrated its good accuracy in the estimation of chatter stability when comparing the experimental cutting test

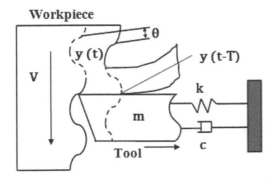

Figure 1. Regenerative chatter.

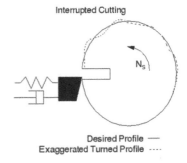

Figure 2. Interrupted cut.

with simulation. Kim et al. (2006) introduced a mechanical damper to reduce tool vibrations during the high-speed milling process. Highly damped tools have a lower tendency to chatter. Thus, Rivin and Kang (1989) substantially increased the damping of a lathe tool by using a sandwich of steel plates and hard rubber viscoelastic material to form a laminated clamping device to hold the tool. Kelson and Hsueh (1996) also achieved an increase in stability by redesigning the tool holder for added stiffness and damping. Aouici et al. (2013) conducted an experimental investigation on the effect of the cutting speed, feed, and depth of cut on hard turning, and also developed a mathematical model to study the effect on surface finish.

The main objective of the present study was to find the stability limit of the cutting tool with different material dampers and three different shim materials during the hard turning (45 HRC) operation.

2 MATHEMATICAL MODELING

Basic and important concepts and equations of the structural dynamics, which are presented in the following sections, are used to develop the MATLAB program, with the discussion on a single degree of freedom system. A viscously damped single degree of freedom system model is shown in Figure 3. We assume that any increment of force (P) due to the regeneration effect occurring in the y-direction continues to act in the β-direction and that is the only force acting on the system.

If the principal mode (x) is inclined to an angle α in the direction of normal (y) and to the generated surface, the motion along the direction of y is related to that along the direction of x by

$$y = x \cos \alpha$$

Thus, the chip thickness variation is given by

$$y = y(t) - y(t - T) \quad y = [x(t)] - x(t - T)] \cos \alpha \tag{1}$$

The equation of motion along the direction of x is given by

$$m\ddot{x} + c\dot{x} + kx = P_x(t) = P(t)\cos(\alpha - \beta) \tag{2}$$

Solving the equation of motion for x(t),

$$x(t) = \frac{P(t)\cos(\alpha - \beta)}{k} \frac{1}{1 - \left(\dfrac{\omega}{\omega_b}\right)^2 + j2\zeta\left(\dfrac{\omega}{\omega_b}\right)} \tag{3}$$

If the coupling coefficient between the force along the direction of y and z is given by r, the above equation can be rewritten as follows:

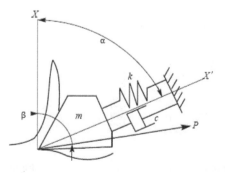

Figure 3. Single degree of freedom cutting system.

$$x(t) \left\{ \frac{K}{\cos\alpha\cos(\alpha-\beta)} \right\} \left[(1-r^2) + j2\xi r \right] = -r[x(t)] - x(t-T)] \tag{4}$$

Let

$$\text{Coupling coefficient } u = \cos\alpha\cos(\alpha-\beta) \tag{5}$$

$$\text{Cross receptance } \phi = \frac{u}{k} \frac{1}{(1-r^2) + j2\xi r} \tag{6}$$

Combining Equation (4) with Equation (6), we obtain

$$\frac{x(t)}{\phi} = -r[x(t)] - x(t-T)]$$

Thus,

$$\frac{|x(t)|}{|x(t-T)|} = \frac{r}{\left(\frac{1}{\phi}\right) + r} = \frac{\phi}{\phi + \left(\frac{1}{r}\right)} = q \tag{7}$$

q< 1, the system is stable, i.e., the amplitude does not build up. However, when q = 1, the system is at the threshold of stability. Here, we assume 'r' as real and solve the stability criterion with only the real part of ϕ.

The governing equations of machine chattering can be derived from the general equation of vibration and the regenerative chatter equations. In orthogonal cutting, the cutting force P(t) is proportional to the cutting area (the product of the chip width or depth of cut a_p and thickness h), which is given by

$$P(t) = k_s a_p h = k_s a_p [x(t-T) - x(t)] \tag{8}$$

where T is the time interval between the previous and current cuts.

Substituting Equation (8) into the general equation of vibration, we obtain

$$m\ddot{x} + c\dot{x} + kx = P(t) \tag{9}$$

The depth of cut is determined from Equation (8):

$$a_p = -\frac{1}{2k_s G} \tag{10}$$

where G is the real part of the frequency response function (FRF).

Solving Equations (8) and (9) together, the depth of cut a_p is dependent on the frequency f of machine vibration or chatter through the frequency ratio r. For each chatter frequency generated on a machining system, there is a corresponding critical chip width (minimum depth of cut) a_p. The cutting process is stable when its depth of cut is less than the critical value, but unstable otherwise.

3 RESULTS AND DISCUSSION

The implementation of MATLAB in chatter suppression considers design parameters such as mass stiffness of the tool, rake angle damping of the tool, and speed. The values of the parameters can be predetermined by the simulation and analysis of the required model.

The above-mentioned parameters are very well expressed in the form of equations given in the above section. In this section, the effect of the mode orientation angle and the cutting speed on the stability of the cutting tool system is discussed.

3.1 *Effect of the mode orientation angle on stability*

From Equation (5), the effect of the mode orientation angle α on cross receptance is simulated using the MATLAB program. The function u (coupling coefficient) for β = 60° is plotted in Figure 4(a). From Figure 4(b), it can be observed that for α = (β/2) and α = [(π/2) + β], the coupling coefficient u becomes zero and an unconditional stability is achieved. This is also clearly shown in Figure 4(c) where a stability plot for different values of α and keeping β invariant shows that the stability limits are affected by the mode orientation angle α. The limit of stability, connoted by the coupling coefficient r*, is minimum for α = (β/2). From Figure 4(b) and 4 (c), it can be indicated that the permissible stability value r* is minimum at α1 = 30°. Figure 4(c) shows the effect of the mode coupling angle α on the stability limit for finding various damping ratio values and keeping β invariant. The higher damping ratio results in a higher stability region. The stability limit is increased when the damping ratio values increase.

3.2 *Effect of the cutting speed on stability for various damping ratio values*

From Equation (5), the effect of the mode orientation angle α on cross receptance is simulated using the MATLAB program. The function u (coupling coefficient) for β = 60° is plotted in Figure 4(a). From Figure 4(b), it can be observed that for α = (β/2) and α = [(π/2) + β], the coupling coefficient u becomes zero and an unconditional stability is achieved. This is also clearly

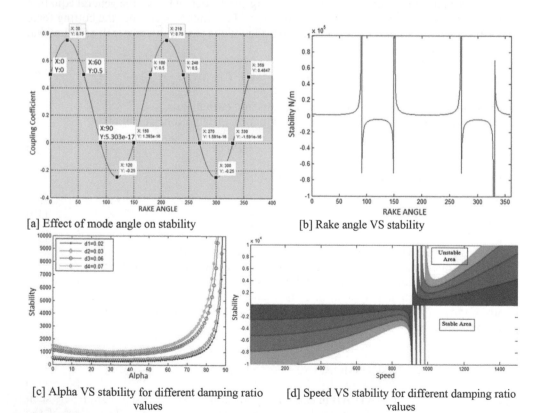

[a] Effect of mode angle on stability

[b] Rake angle VS stability

[c] Alpha VS stability for different damping ratio values

[d] Speed VS stability for different damping ratio values

Figure 4.   Stability analysis.

shown in Figure 4(c) where a stability plot for different values of α and keeping β invariant shows that the stability limits are affected by the mode orientation angle α. The limit of stability, connoted by the coupling coefficient r*, is minimum for α = (β/2). From Figure 4(b) and 4(c), it can be indicated that the permissible stability value r* is minimum at α1 = 30°. Figure 4(c) shows the effect of the mode coupling angle α on the stability limit for finding various damping ratio values and keeping β invariant. The higher damping ratio then leads to a higher stability region. The stability limit is increased when the damping ratio values increase.

3.3   *Prediction of the damping ratio with harmonic analysis of the cutting tool using FEA*

The damping ratio is predicted using FEA for different shim materials, as it is the prerequisite for mathematical modeling. The principle followed in the design phase is to enhance the damping capability of the tool by replacing the conventionally used material interface of carbide shim with brass and aluminum shim interfaces. The evaluation criteria are the dynamic characteristics, frequency, and damping ratio of the machining system. The design of the insert, tool holder, and shim was implemented using the PRO-E software as per the original dimension. The tool holder specification is PCLNR 2020 K12 WIDAX and for tool insert, it is CNMG 120408 MJ. The assembly model of the tool holder is shown in Figure 5(a).

The assembly model of the tool holder is exported to ANSYS to perform the harmonic analysis. The 8-node brick element solid 185 is used. The material property of the holder, insert, and shim is defined using linear isotropic models. After defining the material property, the part of the model is meshed. The holder is coarsely meshed, and the shim and insert are finely meshed, as shown in Figure 5(b). The harmonic analysis module is then used to find the responses of the tool holder when it is subjected to the force or displacement-controlled harmonic excitation.

3.4   *Frequency response function*

A harmonic force is applied to the top surface of the holder and nearby the insert, as shown in Figure 5(b). The frequency response is calculated over the frequency range covering the first mode of the tool. The damping ratio is predicted by using the half-power bandwidth method. The bandwidth is the frequency difference between the upper and lower frequencies for which the power is dropped to half of its maximum value.

The damping ratio is given by

$$\xi = -\frac{f_2 - f_1}{2f_r}$$

The terminology of the half-power bandwidth is explained in Figure 5(c). From Figure 6(a), it can be seen that the amplitude peak is 0.022202 mm and the corresponding frequency is 172 Hz. Using the half bandwidth method, the damping ratio for the carbide shim tool holder

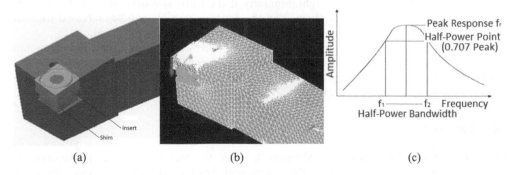

(a)   (b)   (c)

Figure 5.   (a) Tool assembly; (b) Meshed assembly; (c) Half-power bandwidth.

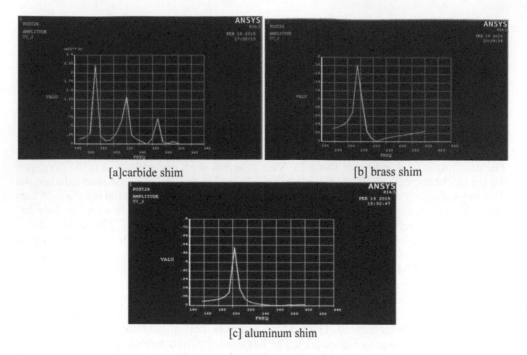

[a] carbide shim   [b] brass shim

[c] aluminum shim

Figure 6.   Frequency response for different shim materials.

can be calculated as 0.01162. Similarly, from Figure 6(b), it can be seen that the amplitude peak of brass shim is 0.181833 mm, the corresponding frequency is 195 Hz and the damping ratio is 0.0205. From Figure 6(c), it can be seen that the amplitude peak of aluminum shim is 0.537885 mm and the corresponding frequency is 202.5 Hz. Using the half bandwidth method, the damping ratio for aluminum shim is found to be 0.025.

4   CONCLUSION

From the above mathematical modeling and simulation, the stability limit of the cutting tool with different dampers can be predicted using the MATLAB program. The permissible stability value r* is minimum at $\alpha = \beta/2 = 30$. Moreover, between $60 < \alpha < 180$, the self-excited vibrations due to mode coupling would not occur. The higher damping ratio lifts the stability lobe upwards and widens the stability region. However, the boundary region moves downwards for the lower damping ratio and the size of the stable region becomes smaller and narrower. By increasing the stiffness of the system, the natural frequency is increased and the boundary curve moves to the right. Similarly, if the stiffness is decreased, then the curve moves to the left. This finding confirms that the dynamic stability can be increased without any modification in the structure and the weight of the existing machine tool.

Aluminum has the highest damping ratio value followed by brass and carbide. The aluminum shim shows the peak amplitude at a much higher frequency when compared with the brass and carbide shims. Therefore, among the three shim materials investigated, a significant improvement in the damping capacity is exhibited by aluminum when compared with carbide shim.

REFERENCES

Aouici1, H. Yallese, M.A. Belbah, A. Mfameur, & Elbah, M., 2013, Experimental Investigation Of Cutting Parameters Influence On Surface Roughness And Cutting Forces In Hard Turning of X38CrMoV5-1 with CBN tool. *Sadhana—Indian Academy of Sciences 38(3):429–445.*

Altintas, Y. & Budak, E., 1995. Analytical Prediction of Stability Lobes in Milling. *Annals of the CIRP 44:357–362.*

Davies, M.A. & Balachandran, B., 1996. Impact Dynamics in the Milling of Thin-Walled Structures. *Journal of Nonlinear Dynamics 22(4):375–392.*

Faassen, R.P.H. van de Wouw, N. Oosterling, J.A.J. and Nijmeijer. H., 2003. Prediction of regenerative chatter by modelling and analysis of high-speed milling. *International Journal of Machine Tools & Manufacture 43:1437–1446.*

Kayhan, M. & Budak, E., 2009. An Experimental Investigation of Chatter Effects on Tool Life. *Proc. IMechE, Part B: Journal of Engineering Manufacture 223 (B11): 1455–1463.*

Kelson Z.Y. & Hsueh W.C., 1996. Suppression of Chatter in Inner—Diameter Cutting. *JSME International Journal of Series C: Dynamic Control Robot, Design and Manufacturing 39(1):25–32.*

Kuster, F. & Gygax, P.E., 1990. Cutting Dynamics and Stability of Boring Bars. *CIRP Annals—Manufacturing Technology 39(1):361–366.*

Rivin, E.I. & Kang H., 1989. Improvement of Machining Conditions for Slender Parts by Tuned Dynamic Stiffness of Tool. *International Journal of Machine Tools Manufacturing 29:361–376.*

Tlusty, J. & Ismail, F., 1982. Basic Nonlinearity in Machining Chatter. *C.I.R.P. Annals. 10:229–304.*

Multi-disciplinary Sustainable Engineering: Current and Future Trends – Tekwani, Bhavsar & Modi (Eds)
© 2016 Taylor & Francis Group, London, ISBN 978-1-138-02845-6

Design and analysis of a 100 T capacity bottom pulley hook block for a double girder EOT crane

Naresh D. Chavda
A D Patel Institute of Technology, New V. V. Nagar, Gujarat, India

Y.D. Patel & D.A. Jani
Department of Mechanical Engineering, ADIT, New V. V. Nagar, Gujarat, India

P. Jha
Department of Design, Anupam Industries Ltd., New V. V. Nagar, Gujarat, India

ABSTRACT: Electric Overhead Traveling (EOT) cranes are widely used in industries as a material handling equipment. The bottom pulley hook block is one of the most critical parts for the effective functioning of the crane. Indian standards (IS 4137 and IS 3177) for cranes provide guidelines for the bottom pulley hook block up to 16 falls using a single axle. The use of a higher number of falls facilitates high-speed operations, reduced end approach length, and clearance. However, an increased number of falls will enforce a longer axle. This problem was resolved using double axles. A detailed design of a bottom pulley hook block for a 100 T capacity for a general-purpose double girder EOT crane is presented in this paper. Two alternatives were used: one with single axle-8 pulleys and another with double axle-8 pulleys. Stress analysis of hook and other critical components was also carried out using ANSYS. Selection of second alternative with double axle 8- pulleys was based on cost criteria as it saves upto 8% as compared to first alternative.

Keywords: EOT crane; hoisting mechanism; bottom pulley hook block

1 INTRODUCTION

Electric Overhead Traveling (EOT) cranes are widely used in industries as a material handling equipment. Crane is equipped with a hoist, wire ropes or chains, and sheaves to lift and transport materials from one place to another in a confined space. The bottom pulley hook block is one of the most critical parts for the effective functioning of the crane. Shufang et al. (2009) presented a parametric design for an overhead traveling crane based on a holographic model. Holography is a technique to measure the deformation and displacement of the 3D object. They improved the design efficiency up to 30%. Huang et al. (2010) studied a rapid variant design technique of the crane based on parameterized templates. It is the technique of adapting existing design specifications to satisfy new design goals and constraints. Combined with the modular technology and the product configuration technology, the concept of parameterized templates is introduced to the variant design of products to solve the problems. This new design method is applied to the design case of gantry cranes so that it improves the variant design efficiency and reduces the design cost. Sharma et al. (2013) worked on the design of the hoisting mechanism of an EOT crane using IS:3177. The dimensions of the main components are determined after the system is designed. The stress and deflection are calculated at critical points using ANSYS and optimized. Load per wire for wire ropes is then found. Dimensions for the pulley, rope drum, power requirement, and ratings for the motor brakes used in the hoist mechanism are also calculated. Chauhan et al. (2012) presented the work on the analysis of stress and strain conditions in the power structure of the overhead crane. They developed a methodology based on computer-aided solving of complex static indeterminate structures. This methodology presents an optimized model of the crane hook determined after the analysis of the whole overhead crane. Torres et al. (2010) described the terms and conditions that are to be considered in the design, manufacture, use, and control of lifting hooks. Parmanik (2010) discussed the history of the crane, various types of the crane, application, and a model design of the various parts of the EOT crane using the IS standard. Ezzell (2008) attempted to increase the hoisting efficiency of a crane and hoisting systems by increasing the number of running wire ropes through encased sheaves attached to the encasement.

In this paper, the design of a 100 t capacity bottom pulley hook block for a double girder EOT crane with 16 falls is presented using two approaches. Moreover, critical components are further analyzed using a high-end computer software.

2 DESIGN CALCULATIONS

Basic inputs as well as the design calculation of the bottom pulley hook block components for the double girder EOT crane are detailed below (Table 1).

2.1 *Design inputs*

The design of the components of the bottom pulley hook block is presented in this section. The design is governed by the following inputs: lifting capacity 100 t, number of falls 16, and RPM of the pulley 10. Two alternative arrangements (a configuration with eight pulleys and single axle arrangement, and a design with eight pulleys and double axle arrangement) for the bottom pulley hook block were considered to satisfy functional requirements (Figures 1 and 2).

It was found that the maximum Bending Moment (BM) was 12,051.58 Nm at point B in the case of the first alternative, as shown in Figure 3, which was higher than the value (10,987.2 Nm) at point B using the second alternative, as shown in Figure 4. Using the maximum BM, it can be stated that the axle diameter required for the first alternative is greater than that for the second alternative.

Table 1. Design calculation of the bottom pulley hook block components.

Description	Dimension value for alternative-I	Dimension value for alternative-II
Wire rope diameter (mm)	28	28
Pulley diameter (mm)	900	900
Pulley bearing size (mm)	130 (6226-z)	110 (NJ222ECJ)
Bearing life	10298.03 rev	1224857 rev
Axle diameter (mm)	130	101(Cal)
Side plate support thickness (mm)	30	30
Hook bed diameter (mm)	428	250
Hook nominal diameter (mm)	113	180
Height of hook head (mm)	60	280
Hook bearing diameter (mm)	190	190
Hook nut (mm)	80	80
Cross head length/height of locknut (mm)	680	300
Cross head shaft diameter (mm)	240	420×25

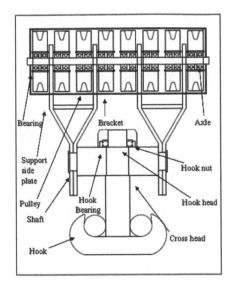

Figure 1. First alternative design arrangement.

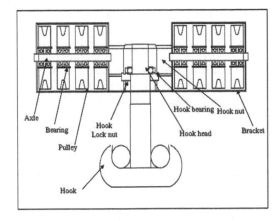

Figure 2. Second alternative design arrangement.

A B C   D E F G H I   J K L
P = 12.051.58 Nm

Figure 3. BM diagram of the first alternative.

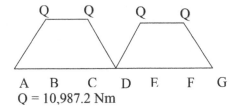

A   B   C   D   E   F   G
Q = 10,987.2 Nm

Figure 4. BM diagram of the second alternative.

Table 2. Cost calculation.

Sr. no.	Components	Material	Weight (kg)	Qty	Total cost (Rs.) of the existing design	Total cost (Rs.) of the alternative design
1	Hook	20C15 IS 4367	1127	1	6,92,000	6,92,000
2	Hook bearing	Steel	10	1	20,000	20,000
3	Pulley	C.S Gr.280–520 N IS:1030	920	8	2,55,200	2,55,200
4	Pulley bearing	Steel	13/7	16	2,32,000	1,36,000
5	Axle	42CrMo4	145/104	1/2	18,125	24,250
6	Cross head/locknut	E250 IS:2026 Gr:BR	940/120	1	50,000	15,000
7	Frame structure	E250 IS:2026 Gr:BR	1700	1	85,000	85,000
Total cost					13,34,325	12,27,450

3 COST CALCULATION

Among the other advantages, the cost is also one of the criteria for the selection of the bottom pulley hook block. Cost calculation carried out on the basis of the interaction with company personnel and comparison between the two alternative designs is presented in Table 2. It was found that the design with two axles had less weight, reduced height of the bottom pulley hook block and rope drum diameter, and also improved lifting height. In addition, it was expected to be more economical.

4 FINITE ELEMENT ANALYSIS OF BOTTOM PULLEY HOOK BLOCK COMPONENTS

The FEA of bottom pulley hook block components such as axle, hook, and pulley was done using ANSYS software. The properties of the components are given in Table 3. The meshing of the axle was carried out using the SOLID 185 element. The axle was loaded at four points of pulley mounting, and the fixed boundary condition was applied at the points of support. In the present case, a load of 1,37,273 N was applied at each point, as shown in Figure 5. The maximum total deformation of an axle was 0.05 mm, as shown in Figure 6. The maximum working stress was 200 MPa, which was less than the allowable stress of 490 MPa. Thus, the proposed design was suitable for a selected material property.

A structural 10-noded solid 187 element was used for the meshing of the hook. The meshed model had 19,844 nodes and 11,291 elements. Figure 7 shows the loading and boundary conditions applied to the present case. The hook shank was constrained in all directions. The inner curvature of the hook was subjected to a load of 10,000 kN. A maximum deformation of 4.55 mm was observed at the bottom of the hook. The deformation was within a limit as per IS 15660. The maximum stress of 270.3 MPa was induced at the shank, as shown in Figure 8. It was less than the allowable stress of 420 MPa.

The meshing of the pulley was carried out using 8-noded brick solid 45 elements. The inner portion of the hub was constrained, and the load was applied on the periphery of the rim as per belt tensions in the tight and slack side. On that basis, cylindrical support was applied at the center surface of the pulley where the axle was fixed. In the present case, the

Table 3. Material properties.

Part	Name	Properties
Axle	Name of the material	42CrMo4 (Steel)
	Mass density	7830 kg/MA^3
	Young's modulus	210 GPa
	Poisson's ratio	0.2880
Hook	Name of the material	20C15 (Low-alloy Steel)
	Mass density	7.85 g/cm^3
	Young's modulus	210 GPa
	Poisson's ratio	0.2880
Pulley	Name of the material	280–520 N (Carbon Steel)
	Mass density	7.85 g/cm^3
	Young's modulus	210 GPa
	Poisson's ratio	0.2880

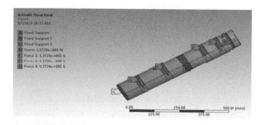

Figure 5. Loading condition for the axle.

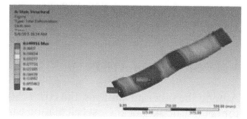

Figure 6. Total deformation of the axle.

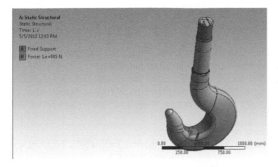

Figure 7.   Loading condition for the hook.

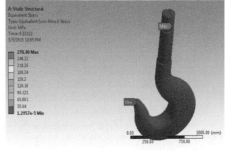

Figure 8.   Equivalent stress on the hook.

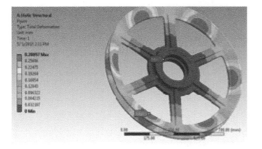

Figure 9.   Total deformation of pulley.

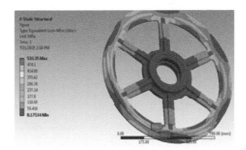

Figure 10.   Equivalent stress on pulley.

total load of 1,25,000 N was applied on the periphery of one pulley rim. The maximum total deformation was 0.2 mm between the two arms of the pulley for the applied rope tension, as shown in Figure 9. The maximum stress was found to be 533.35 MPa at the corner between the arm and rim of the pulley, as shown in Figure 10. The allowable stress for the carbon steel grade 280–520 N was 620 MPa.

5   CONCLUSION

The rapid industrial development in many parts of the world has made the use of overhead cranes a crucial component of every major industry. The design of 100 T capacity bottom pulley hook block for double girder EOT crane was carried out using thumb rules and relevant Indian Standard. The design of two different arrangements (one with a single axle and 8 pulleys, another with a double axle and 8 pulleys) was presented along with their comparison. The selection of the bottom pulley hook block was based on cost as well as feasibility criteria. Furthermore, MATLAB codes were developed for the design of the hook, pulley, and axle that were used for iterative theoretical calculations. The design analysis of the hook, pulley, and axle using ANSYS software confirmed the results within the limit. The results indicated that the proposed design improved the material as well as reduced the cost up to 8% by using the double axle arrangement for the bottom pulley hook block.

REFERENCES

Chauhan, N., Bhatt, P., 2012, Improving the durability of the E.O.T. crane structure by finite element analysis, and optimize the hook material for improving its solidity. International conference on modelling optimization and computing. Procedia Engineering 38: 837–842.

Huang, Y., Cao, Z., Wang, J., Pu, K., Wang, W., 2010, Simulation verification for the Lifting of the Overhead Crane. International Conference on Computer Application and System Modelling (ICCASM)15: 216–219.

IS 3177:1999, Indian Standard code of practice for Electric Overhead Travelling cranes and Gantry cranes other than steel work cranes (second revision). Bureau of Indian Standards.

IS 4137:1985, Indian Standard code of practice for heavy duty Electric Overhead Travelling cranes include special service machine for use in steel work (first revision). Bureau of Indian Standards.

IS 15560:2005, Indian Standard point hook with shank up to 160 tones specification (second revision). Bureau of Indian Standards.

Parmanik, R., 2010, Design of Hoisting Arrangement of E.O.T. Crane.

Phillip S. 2008. Failure mode effect analysis. Crane Safety Associates of America, Inc. at NIS Technology service group.

Sharma, S., Khan, T., Md, P., Kumari, K. ,2013, Computer aided analysis and design of hoisting mechanism of an EOT crane. International journal of mechanical engineering and robotics research 2(3): 89–95.

Torres, Y., Gallardo, J., Domínguez, J., Jiménez, E., 2010, Brittle fracture of a crane hook. Engineering Failure Analysis 17: 38–47.

Wu, S., Wang, Z., Wang, Y., Wang, X., 2009, Research and application of parametric design for Overhead Travelling Crane based on the Holographic Model. Technology and Innovation Conference. 1–5.

432

Multi-disciplinary Sustainable Engineering: Current and Future Trends – Tekwani, Bhavsar & Modi (Eds)
© 2016 Taylor & Francis Group, London, ISBN 978-1-138-02845-6

Analytical and finite element analysis of dovetail joints in aeroengine disc

Ranjan Kumar
RVSCET, Jamshedpur, India

Vinayak Ranjan
ISM, Dhanbad, India

H. Nilesh Gupta
GIT, Gandhinagar, India

ABSTRACT: In the design of aeroengine blade/disc assemblies, the desire for weight reduction must be balanced against the safety and reliability of these assemblies. Typically, a dovetail joint is implemented to attach the blades to compressor disc in aeroengine. The contact region is highly prone to fretting wear attributed to the relative motion of the blade and the disc results in surface damage. In this paper, reliability of the analytical approach has been proven first and, the complexity in developing exact formulation for geometry of disc-blade interface is clearly shown. Analytical methods for computation of stresses is found inadequate for the analysis. Therefore, finite element model is developed for dovetail joint in aeroengine compressor. Attention of the study is devoted to analyse the stress distribution in the interface region of the blade and disc in a dovetail joint.

Keywords: finite element method; aeroengine; contact stress; dovetail joint

1   INTRODUCTION

An aeroengine operates under extremely severe and complex conditions of temperature and stress, where there are numerous different cases of the failure of aero engine components. One of the important factor concerning mechanical integrity of aero engine is the interface region between the blade and the rotor disc. A dovetail joint is adopted in fastening blades to discs in compressor because, they provide adequate areas of contact over which large contact loads can be accommodated. The joint between a blade and the disc represents the most critical load path within the assembly. In the majority of cases, cracks are generally initiated in such regions due to the fretting action at the blade disc interface. The loads associated with these regions are the self-generated centrifugal forces of the disc and associated blades, the bending loads due to the gas pressure, and the thermal loads. Interface conditions (friction, surface roughness, residual stress), and the detailed geometry of the joint, determine the severity of the resulting stress. Through the stress analysis of the turbine rotor, failure location and the maximum stress point can be determined. The stress analysis of the dovetail-rim region of aeroengine discs has received the attention of several investigators. Papanikos and Meguid (Papanikos, P., and Meguid, S. A., 1994) have presented finite element stress analysis of the critical geometrical features and interface conditions of different dovetail configuration with the prediction of the direction of potential fatigue cracks. Chan and Tuba (Chan S., and Tuba, I. S., 1971) analysed the effect of blade-disc clearance and friction using a two-dimensional model. It was shown that changes in the coefficient of friction had little effect on the maximum stress in fillet region and load distribution by each flank. Variation in clearance changes between the blade root and disc, had significant effects on the stress distribution. Papanikos et al. (Papanikos P, Meguid S A, Stjepanovic Z.,1998) developed 2D and 3D FE models to examine the effect of geometric features in the dovetail joint under centrifugal loading. The results reveal that the maximum stress occured just below the contact point for various geometries, and that the stress distribution through the thickness of the 3D model revealed a much higher stress level at the middle of the disc than that of at the disc surfaces. Vale (Vale, T. O, Villar G. d. C., and Menezes, J. C., 2012) also performed a comparable study using a refined FE model for dovetail attachments. It was concluded that minimizing clearance between the teeth of the blade and disc reduced the maximum contact stress. Fretting fatigue analysis for the dovetail joint was also modeled by a number of researchers (Kheto, M. K., Babu, N. C., and Madan, J., 2009, Enright, M. P., Chan, K. S., Moody, J. P., Golden, P. J., Chandra, R., and Pentz, A. C., 2010). Fretting is a surface damage phenomenon, in which the relative cyclic motion between two contacting surfaces accelerates the initiation of cracks. Patrick and Sam (Patrick J. Golden., Sam Naboulsi., 2012) investigated fretting fatigue in a turbine engine fan disc, which is a significant driver of fatigue and failure of discs. Sinclair and Cormier (Sinclair, G. B., and Cormier, N. G., 2002) used FE analysis to identify the contact shears due to friction as a major source of peak tensile hoop stress at the edge of contact during loading. They identified a pinching mechanism that occurs with sufficient friction during unloading thereby explaining the counter-intuitive response of an increase in normal contact stress during unloading. The analysis was carried out under steady state conditions using Ansys software. Of particular interest to this study, is the work of Kenny et al. (B. Kenny, E. Patterson, M. Said, K. Aradhya., 1991) and Nurse and Patterson (A.D. Nurse, E. Patterson, 1993). Their work was concerned with stage two fatigue crack growth paths in fir-tree fixtures. Since no allowances were made for contact elements in their model, approximate contact pressure was assumed at the interface between the blade and the disc. The numerical analysis at dovetail joints was also treated by Boddington et al. (P.H.B. Boddington, K. Chen, C. Ruiz). In their work, a technique was developed to model the relative motion at the interface of the assembly, and thus accounted for contact at the interface. In the present work, comprehensive three-dimension model of realistic disc geometries is developed using modelling software and commercial finite element package ANSYS V14.0 has been used as a tool for analysis. Accurate modelling of the blade No attempt was made for accurate modelling of the blade, except insofar as providing the necessary centrifugal loading at the interface. A dove-tail interface for a sector is considered and contact stress changing with the flank angle and friction coefficient is analysed at compressor blade-disc interface.

2 PROBLEM FORMULATION

2.1 *Rotating disc with central hole*

The mathematical model has been developed assuming thin rotating disc. Problem formulation is based on axisymmetric hypothesis of disc i.e. stress and strain component are independent of θ coordinate. The axial stress is neglected and hence σ_r and σ_θ are the only principal stress [20].

We have stress-strain relation from Hook's law,

$$\varepsilon_r = \frac{1}{E}(\sigma_r - \mu\sigma_\theta) \tag{1}$$

$$\varepsilon_\theta = \frac{1}{E}(\sigma_\theta - \mu\sigma_r) \tag{2}$$

Considering the inertia force due to rotation of disc in radial direction,

$$R_r = \rho\omega^2 r$$

We have equilibrium equation for rotating disc,

$$\frac{\partial \sigma_r}{\partial r} + \frac{1}{r}\frac{\partial \tau_{r\theta}}{\partial \theta} + \frac{1}{r}(\sigma_r - \sigma_\theta) + R_r = 0 \qquad (3)$$

$$\frac{\partial \tau_{r\theta}}{\partial r} + \frac{1}{r}\frac{\partial \sigma_\theta}{\partial \theta} + \frac{2\tau_{r\theta}}{r} = 0 \qquad (4)$$

where, σ_r = radial stress, σ_θ = circumferential stress

Due to symmetry $\tau_{r\theta}$ vanishes and σ_r, σ_θ becomes independent of θ and hence equation (3) reduced to,

$$\frac{\partial \sigma_r}{\partial r} + \frac{1}{r}(\sigma_r - \sigma_\theta) + \rho\omega^2 r = 0 \qquad (5)$$

$$\frac{\partial(r\sigma_r)}{dr} - \sigma_\theta + \rho\omega^2 r^2 = 0 \qquad (6)$$

$$\sigma_\theta = \frac{d}{dr}(r\sigma_r) + \rho\omega^2 r^2 \qquad (7)$$

The strain displacement relation for the axisymmetric problem are given by

$$\varepsilon_r = \frac{du}{dr}$$

$$\varepsilon_\theta = \frac{u}{r} \Rightarrow u = r\varepsilon_\theta \qquad (8)$$

$$\frac{du}{dr} = \varepsilon_r = \frac{d}{dr}(r\varepsilon_\theta)$$

Combining the above relationship with equation (1) and equation (2) Then,

$$\frac{1}{E}(\sigma_r - \mu\sigma_\theta) = \frac{1}{E}\left[\frac{d}{dr}(r\sigma_\theta - \mu r\sigma_r)\right] \qquad (9)$$

Put $F = r\sigma_r$, in equation (7) and substitute in equation (9)

$$r^2\frac{d^2F}{dr^2} + r\frac{dF}{dr} - F + (3+\mu)\rho\omega^2 r^3 = 0 \qquad (10)$$

The solution of above differential equation is

$$F = cr + \frac{C_1}{r} - \left(\frac{3+\mu}{8}\right)\rho\omega^2 r^3 \qquad (11)$$

From $F = r\sigma_r$ and (9) get

$$\sigma_\theta = C + \left(\frac{C_1}{r^2}\right) - \left(\frac{3+\mu}{8}\right)\rho\omega^2 r^2 \qquad (12)$$

$$\sigma_r = C - \left(\frac{C_1}{r^2}\right) - \left(\frac{3+\mu}{8}\right)\rho\omega^2 r^2 \qquad (13)$$

435

The above equations are the general Lame's expressions for tangential and radial stresses for rotating disc. These analytical solutions of stresses in rotating circular discs holds good for a disc, whose thickness is uniform and small in comparison with the radius. In a turbine disc, the solid disc contains a central bore for the shaft, meaning a free boundary exists at both inner and outer radii. The boundary conditions necessary to solve for the constants and can be determined using this criterion. The radial stresses at the free boundaries must equal zero and are given as:

At r = a, $\sigma_r = 0$, and r = b, $\sigma_r = 0$.

where 'a' is inner radius and 'b' is outer radius of rotating disc with central hole. Applying initial boundary condition to equation (12) & equation (13)
We can find C and C_1

$$C = \left(\frac{3+\mu}{8}\right)\rho\omega^2(b^2 + a^2) \tag{14}$$

$$C_1 = -\left(\frac{3+\mu}{8}\right)\rho\omega^2 b^2 \tag{15}$$

Substitute the above constant in equation (12) and equation (13), we get the radial Stress for the hollow disc is

$$\sigma_r = \left(\frac{3+\mu}{8}\right)\rho\omega^2\left(b^2 + a^2 - \left(\frac{a^2b^2}{r^2}\right) - r^2\right) \tag{16}$$

$$\sigma_\theta = \left(\frac{3+\mu}{8}\right)\rho\omega^2\left(b^2 + a^2 + \left(\frac{a^2b^2}{r^2}\right) - \left(\frac{1+3\mu}{3+\mu}\right)r^2\right) \tag{17}$$

The expressions above do not account for the centrifugal effect of blades. For such cases, the boundary conditions must be adjusted accordingly.

2.2 Rotating disc with attached blades

Previous modelling has only considered a free boundary at the outer disc edge. The disc with blades can easily be modelled analytically by imposing a non-zero radial stress at the outer radii. The pull of the blades on the outer surface of the disc imparts this stress which can be averaged over the entire disc outer surface as:

$$\sigma_b = \sigma_{r(r=b)} = \frac{m_b\omega^2 r_m N}{2\pi bt} \tag{18}$$

where 'm_b' the mass of each blade, 'r_m' is the mean radius from the center of the disc to the centroid of the blade, 'N' is the number of blades, 'b' is the outer radius of the disc and 't' is the thickness of the disc. Using the general equations (12) and (13) with the new boundary conditions at the outer radius and $\sigma_r = 0$, at r = a, the new constants C and C_1 can be determined for a purely mechanical loading:

$$C = \frac{b^2\left[\dfrac{\rho(\upsilon+3)\omega^2}{8}(b^2 - a^2) + \sigma_b\right]}{b^2 - a^2} + \frac{\rho a^2\omega^2(\upsilon+3)}{8} \tag{19}$$

$$C_1 = \frac{a^2b^2\left[\dfrac{\rho(\upsilon+3)\omega^2}{8}(b^2 - a^2) + \sigma_b\right]}{b^2 - a^2} \tag{20}$$

3 FINITE ELEMENT MODELLING OF DISC-BLADE ASSEMBLY

Design of the attachments between compressor or turbine blades and their discs is critical in order to transfer the large loads generated due to centrifugal, thermo-mechanical actions and other gas loads between the two components. The centrifugal force that is transferred through the contacts in a blade disc attachment results in high stress region near the contacts. Dove-tail joints are commonly used for the disc blade attachments in compressors for disc blade attachments. The contact region is highly prone to fretting wear attributed to the relative motion of the blade and the disc resulting in surface damage. Analytical solution for such a complex problem is practically impossible. Even though certain techniques are developed to analyze analytically but, it is limited to uniform thickness disc only. Finite Element Method (FEM), which is the numerical technique, is one solution to such a complex problem. FEM is a powerful tool determining the stresses and deformation of complex structures like aero-engine.

3.1 Modelling of dovetail joint

Rotor consists of a number of contact pairs of disc and blade. It would be complex and time consuming to study all the interfaces together. In view of symmetry of geometry and loading, only one sector of disc/blade interface is considered for analysis purpose. No attempt was made to accurately model the blade, except insofar as providing the necessary centrifugal loading at the interface. Figure 1 shows the dimensions disc and blade while, Figure 2 shows the disc blade assembly. A thickness of 10 mm is considered for all the models.

3.2 Material properties for compressor disc/blade

The material properties used for modelling the disc and blade are Nickel based superalloy Inconel 720 and Inconel 718, respectively. The material has high tensile strength, endurance strength, creep strength and rupture strength In this work, the following value for disc material is used Young's modulus = 220 GPa, Poisson's ratio = 0.3, density = 8510 Kg/m³ and for the material of blade Young's modulus = 210 GPa, Poisson's ratio = 0.3, density = 8230 Kg/m³.

All dimensions in mm

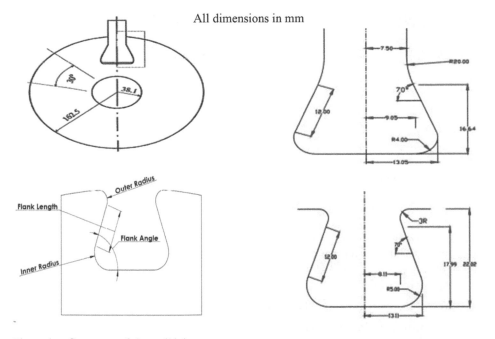

Figure 1.   Geometry of dovetail joint.

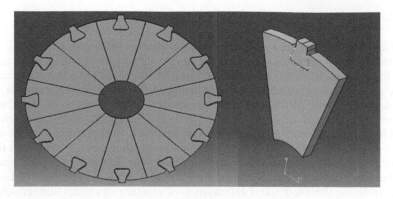

Figure 2.   Full disc assembly and single blade/disc sector of dovetail joint.

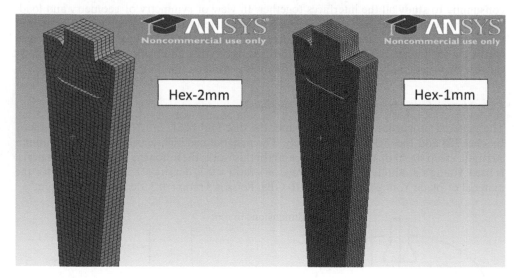

Figure 3.   Types of mesh for dovetail joint.

3.3   *Mesh*

There are several methods of discretisation for finite element mesh in non-linear contact stress analysis. A free meshing routine was used due to necessity to model a complex geometry with large transitions in the stress field. In addition a convergence requirement is satisfied using mesh convergence test. Two types of meshes have been studied. The meshes with Hex 2mm and 1mm of size are used for mesh convergence test as shown in Figure 3. In this paper, for analysis purpose mesh with Hex 2mm is used after convergence requirement to save computational time.

3.4   *Loads and boundary conditions and contact elements*

The disc is constrained in all directions at bore region, while blade is allowed to have a translation radial degree of freedom. The cyclic symmetry tool is effectively used which greatly reduced the computational time, compared to a full disc/blade model that would require a much computational effort. The model is subjected to centrifugal loading with specific constant speed of 1000 rpm. The surface to surface contact is defined between the blade and disc interface. The outer surface of the blade is (along flank length) considered as the contact surface element and the inner surface of disc sector (along flank length) is considered as target element. The contact region between the blade and disc is recognized automatically when geometry is transferred from design modeler to simulation as bonded contact in Ansys. Further frictional conditions between the blade and disc is considered and coefficient of friction is varied from 0.05 to 0.25. Augmented Lagrange multiplier contact formulation method is used to obtain solutions in Ansys. The contact region is considered as a non-linear contact conditions. The contact elements used are CONTA174 and TARGE170. These contact element allow for modelling of gap and friction condition at interface.

4   RESULTS AND DISCUSSION

In the present paper, the effect of coefficient of friction on contact stresses in the interface region of blade & disc is investigated for 70° flank angle, as shown in Figure 4. Whereas, the Figure 5 shows the effect of variation of flank angle on contact stresses for a coefficient of friction value of 0.25. The von-Mises stresses are calculated for comparison & analysis. From Figure 6, it is clear that as friction coefficient increases, the maximum contact stress in interface region decreases. This is because of the disc, which becomes more firm in assembly due to increased friction. It is further observed from Figure 6, that the rate of decrease of contact stress reduces with increase in coefficient of friction above 0.25. It is also evident from Figure 7, that contact stress increases with increase in flank angle of dovetail joint. This is because the blade assembled in thedisc becomes untwisted and runs away easily with increase in flank angle. It is noteworthy from Figure 7, that the rate of increase of contact stress above 60° flank angle almost remains same.

4.1   *Effect of coefficient of friction (μ)*

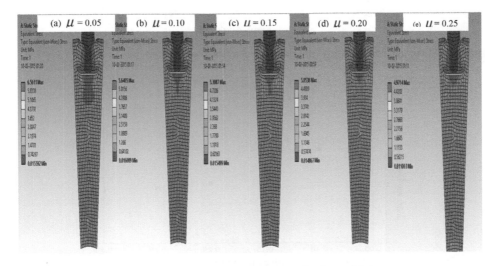

Figure 4.   Effect of coefficient of friction on contact stresses for dovetail joint.

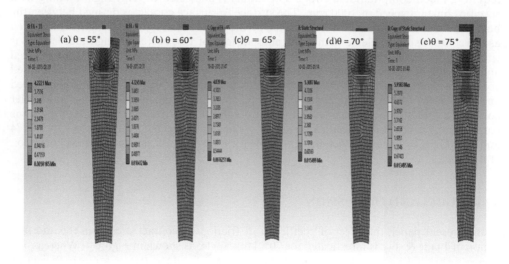

Figure 5.   Effect of flank angle on contact stresses for dovetail joint.

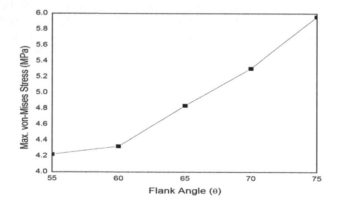

Figure 6.   Maximum contact stress variation with coefficient of friction.

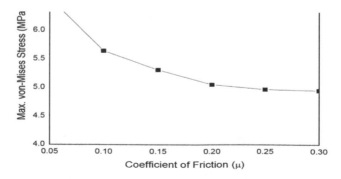

Figure 7.   Maximum contact stress variation with flank angle.

5 CONCLUSION

1. Present investigation gives an insight into the understanding of the influence of contact stress acting on the blade-disc interface, which is very essential in determining the structural integrity of an aeroengine disc.
2. Finite element model has been developed for compressor disc-blade interface region and contact stress value with varying flank angle and coefficient of friction is examined.
3. The maximum contact stress decreases with the increase of friction coefficient and increases with the increase of flank angle.
4. The maximum contact stress appears at the fillet below the flank where the fracture failure can easily emerge.

REFERENCES

Boddington, P.H.B., Chen, K. and Ruiz, C., The numerical analysis of dovetail joints, Comput. Struct. 20 (1985) 731–735.

Chan S. and Tuba, I.S., 1971, A finite element method for contact problems of solid bodies. II. Application to turbine blade fastenings, International Journal of Mechanical Sciences, 13(7), pp. 627–639.

El-Sayed, A.F., 2008, Aircraft propulsion and gas turbine engines, Taylor & Francis.

Enright, M.P., Chan, K.S., Moody, J.P., Golden, P.J., Chandra, R. and Pentz, A.C., 2010, Probabilistic Fretting Fatigue Assessment of Aircraft Engine Disks, Journal of Engineering for Gas Turbines and Power, 132(7), pp. 072502.

Hill, P.G. and Peterson, C.R., 1992, Mechanics and thermodynamics of propulsion, Addison-Wesley, Reading, Mass.

Kenny, B., Patterson, E., Said, M. and Aradhya. K., 1991, Contact stress distributions in a turbine disc dovetail type joint—A comparison of photoelastic and finite element results, Strain Vol. 27, pp. 21–24.

Kheto, M.K., Babu, N.C. and Madan, J., 2009, Fretting fatigue analysis in dovetail joint of compressor through numerical simulation, SAS Technical Journal, 8(1), pp. 71–76.

Kumar Ranjan, Ranjan V., Kumar Subodh and Srivastava J.P., 2014, An investigation of critical stresses on a gas turbine disc rotating at different speeds by analytical and finite element method, Proceeding of International symposium on aspects of mechanical engineering & technology for industry, NERIST, Vol. 1, pp. 202–211.

Mattingly, J.D., 2006, Elements of propulsion gas turbines and rockets, AIAA education series, American Institute of Aeronautics and Astronautics.

Meguid, S.A., 1988, Engineering Fracture Mechanics, Elsevier Applied Science, London, pp. 287–295.

Mustapha, H. and Zelesky, M., 2003, Axial and radial turbines, Concepts NREC, White River Junction.

Nurse, A.D. and Patterson, E. 1993, Experimental determination of stress intensity factors for cracks in turbine discs, Fatigue Fract. Eng. Mater. Struct. Vol. 16, pp. 315–325.

Papanikos, P. and Meguid, S.A., 1994, Theoretical and experimental studies of fretting initiated fatigue failure of aero engine compressor discs, Fatigue and Fracture of Engineering Materials and Structures, 17(5), pp. 539–550.

Papanikos, P., Meguid, S.A. and Stjepanovic, Z., 1998, Three-dimensional nonlinear finite element analysis of dovetail joints in aeroengine discs, Finite Elements in Analysis and Design, Vol. 29: 173–186.

Patrick J. Golden. and Sam Naboulsi., 2012, Hybrid contact stress analysis of a turbine engine blade to disk attachment, International Journal of Fatigue, Vol. 42, pp. 296–303.

Sinclair, G.B. and Cormier, N.G., 2002, Contact Stresses in Dovetail Attachments: Physical Modeling, Journal of Engineering for Gas Turbines and Power, Vol. 124(2), pp. 325–331.

Srinath, L.S., 2011, Advanced mechanics of solids, Tata Mcgraw Hill.

Srivastav, S. and Redding, M., 1994, 3D Modelling of imperfect contact conditions between turbine blades and disk, Advances in Steam Turbine Technology for the Power Generation Industry: ASME International Joint Power Generation Conference, Vol. 26, pp. 197–204.

Vale, T.O, Villar G.d.C. and Menezes, J.C., 2012, Methodology for Structural Integrity Analysis of Gas Turbine Blades, Journal of Aerospace Technology and Management, 4(1), pp. 51–60.

Witek, L., 2006, "Failure analysis of turbine disc of an aero engine," Engineering Failure Analysis, Vol. 13(1), pp. 9–17.

Energy conservation and management

Multi-disciplinary Sustainable Engineering: Current and Future Trends – Tekwani, Bhavsar & Modi (Eds)
© 2016 Taylor & Francis Group, London, ISBN 978-1-138-02845-6

Economic analysis of heat pump integrated icecream making plant

Kalpan Patel & Dilip Sarda
Synergy Agro Tech Private Limited, Ahmedabad, Gujarat, India

Balkrushna Shah
*Department of Mechanical Engineering, Institute of Technology, Nirma University,
Ahmedabad, Gujarat, India*

ABSTRACT: Worldwide icecream is considered favorite dessert and its consumption is increasing day by day. Icecream manufacturing comprises of processes like pasteurization, homogenization, chilling of mix, and freezing. Around 65°–70°C of temperature is required for pasteurization which is achieved by using hot water generated by electricity or gas or other fuels. Around 1°C chilled water is required to be supplied by refrigeration for cooling the icecream mix. Aging of this mix is to be done for 5–6 h in aging tank at 4°C and then it is taken to the freezer to produce icecream. Also, icecream plant requires large amount of hot water for cleaning and sanitizing the equipments, piping, utensils and washing floors. As in icecream manufacturing process, both heating and cooling are required; the use of heat-pump may prove to be a better solution. In this paper, detailed economic analysis has been given for heat pump integrated icecream making plant. The aim of process integration is to reduce the energy consumption and, thereby, the operating cost of the icecream making. Heating and cooling requirement for 2,000 L of icecream production per day has been calculated. Suitable heat pump has been integrated to the existing plant in order to have benefit of heating utility for partial pasteurization upto temperature 58°C and cooling utility for the partial cooling of mix upto temperature 18°C. Saving of around 34% of energy consumption is made by the heat pump integrated systems. As per the calculations, simple payback period for modified icecream making plant with heat pump is just 2.2 years.

NOMENCLATURE

m	Mass flow rate, *(kg/s)*
ΔT	Temperature difference, *(°C)*
Cp	Specific heat, *(kJ/kg.K)*
Q	Load, *(kW)*
PHE	Plate heat exchanger
COP	Coefficient of performance

1 INTRODUCTION

Study of energy saving in an industrial process using heat pump integration begins with the collection of data, finding out the scope of its integration in existing system and finally, the modifying the system. Heat pump is widely used in domestic applications like space heating in cold countries. Use of heat pump in industry is very advantageous if both heating and cooling utilities are required in the process. A.H. Zaidi et al. (1979), in their study, explained the use of heat pump in dairy industry. They integrated heat pump system with pasteurizers, evaporators, spray dryers, refrigeration, and cold storage equipment which are used in the dairy plant.

Icecream making is one such process where simultaneous heating and cooling are required. After studying the flow diagram of the existing process, it was found that there are certain areas where integration of heat pump is possible to some extent for pasteurization, cooling, aging and in freezing. Also, icecream industry requires lot of hot water for cleaning and sanitizing utensils, and washing the shop floor. This requirement can also be fulfilled by heat pump. J. Ubbels and S. Bouman (1979) studied the opportunities of saving of energy in milk cooling and heating water in farms. They showed integration of heat pump in refrigerated farm tank with the use of precooler, and explained the conservation of energy analytically and experimentally where water is being heated with a heat pump. Pasteurization is a process which is used in dairy and icecream industry requires heat energy. Thermal and economic analysis of heat pump and auxiliary heater for the use of milk pasteurization process is carried out on heat transfer basis and the life cycle cost of system is evaluated. Omer omakli et al. [Soylemez M.S.] studied water source heat pump integrated in the existing system and experiments were carried out for variation in COP with time at certain temperatures. In this paper, an icecream mix plant of 500 L/h capacity is taken as a case study and energy conserved by integrating heat pump to the existing set up is calculated. A new process flow diagram is also generated. Economic analysis of the project is carried out to ensure its viability considering 2000 L/d of icecream production.

2 PROCESS INTEGRATION METHODOLOGY

2.1 Introduction to case study

The industrial process that is used for this analysis is an icecream mix manufacturing unit of 500 L/h capacity with total production 2000 L/d of icecream. Here, it may be noted that 1 L icecream mix can produce 2 L of icecream considering overrun of 100%.

2.2 Process operation and parameters

The process operation steps in a conventional ice-cream making plant are summarized in Figure 1.

For the case study, an existing icecream making plant with following details have been considered.

As the capacity of pasteurizer, homogenizer and P.H.E is 500 L, two batches of icecream mix are required for the production of 2000 L icecream per day.

2.3 Process flow diagrams

The actual process diagram of existing plant is shown in Figure 2(a). Heat pump is used to generate hot water of 60°C and cold water of 10°C. Hot and cold water are stored in its

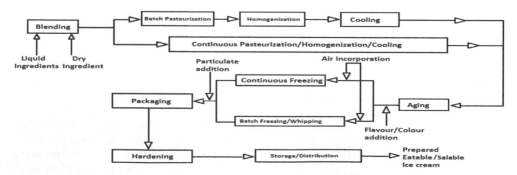

Figure 1. Process description.

446

respective insulated tanks. Hot water is used for partial heating of icecream mix upto 58°C for pasteurization process. Thus, the use of electric heater is reduced for pasteurization process. Cold water is used for the second stage cooling in P.H.E to cool icecream mix upto 18°C. The modified process flow diagram with heat pump is shown in Figure 2(b). The hot water required for cleaning purpose is drawn directly from hot water storage tanks.

Table 1. Details of existing icecream making plant.

Process	Equipment	Capacity (L/h)	Operating temperature (°C)
Pasteurization	Pasteurizer, Electric heater	500	20° to 70°
Homogenization	Homogenizer	500	70°
Cooling	P.H.E	500	70° to 8°
Aging	Aging tank	1000	8° to 4°
Storage	Aging tank	1000	4°
Icecream production	Continues freezer	400	4° to −5°
Cleaning	Pasteurizer	500	30° to 60°

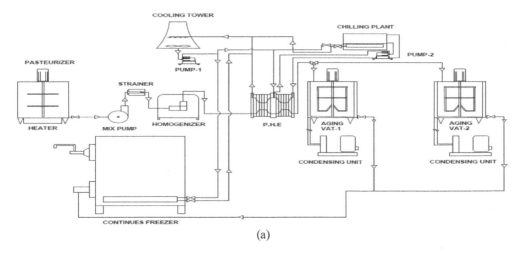

(a)

Figure 2(a). Existing icecream making plant.

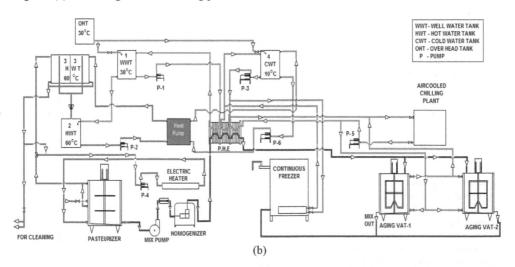

(b)

Figure 2(b). Heat pump integrated icecream making plant.

447

Table 2(a). Electric energy requirement of existing plant.

Equipment	Rating (kW)	Multiplication factor	Load factor	Load (kW)	Runtime (h)	Total load (kWh)
Pasteurization	16.70	1:1	–	16.70	3	50.1
Water heating (Cleaning)	17.41	1:1	–	17.41	4	69.64
Homogenizer	3.75	1:1	0.8	3	2	6
Glycol compressor	18.2*	1:2.64**	–	6.89	2	13.78
Aging compressor (Aging)	5.02*	1:1.92**	–	2.61	2	5.22
Aging compressor (Storage)	0.44*	1:1.92**	–	0.22	18	4.12
Continues freezer Compressor	12.15*	1:1.62**	–	7.5	5.5	41.25
Continues freezer dasher	3.75	1:1	0.8	3	5.5	16.5
Continues freezer mix pump	0.75	1:1	0.8	0.6	5.5	3.3
Mix Pump	0.75	1:1	0.8	0.6	2	1.2
Pump-1	0.75	1:1	0.8	0.6	7.5	4.5
Pump-2	0.75	1:1	0.8	0.6	2	1.2
Pasteurizer Agitator	0.37	1:1	0.8	0.3	4	1.2
Aging Vat Agitator	0.36	1:1	0.8	0.28	20	5.6
Cooling tower fan motor	0.27	1:1	0.8	0.21	7.5	1.62
Aging condenser fan	0.27	1:1	0.8	0.21	18	3.88
Exhaust fan	2.25	1:1	0.8	1.8	10	18
Fresh air fan	3.75	1:1	0.8	3	10	30
					Total	**277.11**

Table 2(b). Electric energy requirement of modified plant.

Equipment	Rating (kW)	Multiplication factor	Load factor	Load (kW)	Runtime (h)	Total load (kWh)
Electric heater	18.11	1:1	–	–	0.66	11.95
Heat pump	12.36*	1:2.32**	–	5.32	7.5	39.9
Homogenizer	3.75	1:1	0.8	3	2	6
Glycol PHE compressor	5.35*	1:1.73**	–	3.09	2	6.18
Glycol aging compressor (Aging)	5.48*	1:1.73**	–	3.16	2	6.32
Glycol aging compressor (Storage)	0.44*	1:1.73**	–	0.25	18	4.5
Continues freezer compressor	12.15*	1:1.62**	–	7.5	5.5	41.25
Continues freezer dasher	3.75	1:1	0.8	3	5.5	16.5
Continues freezer mix pump	0.75	1:1	0.8	0.6	5.5	3.3
Mix Pump	0.75	1:1	0.8	0.6	2	1.2
Pump-1	0.75	1:1	0.8	0.6	2	1.2
Pump-2	0.75	1:1	0.8	0.6	7.5	4.5
Pump-3	0.37	1:1	0.8	0.3	7.5	2.25
Pump-4	0.75	1:1	0.8	0.6	7.5	4.5
Pump-5	0.37	1:1	0.8	0.3	20	6
Pump-6	0.75	1:1	0.8	0.6	7.5	4.5
Pasteurizer Agitator	0.37	1:1	0.8	0.3	4	1.2
Aging Vat Agitator	0.36	1:1	0.8	0.28	20	5.6
Glycol condenser fan	0.27	1:1	0.8	0.21	20	4.32
Exhaust fan	1.5	1:1	0.8	1.2	10	12
					Total	**183.17**

*Refrigeration kW; **C.O.P of refrigeration system.

Table 3. Parameter for icecream plant.

Items	Rating
Capacity of icecream mix plant (L/h)	500
Capacity of icecream production (L/D)	2000
Electricity used in existing plant (kWh)	277.11
Electricity used in modified plant (kWh)	183.17
Electricity saved (kWh/D)	93.94
Rate of electricity (Rs/kWh)	8
Assumed working days	250
Total saving in electricity (Rs)	1,87,880
Cost of additional equipments (Rs)	4,12,000
Additional cost for maintenance (Rs)	20,600
Total saving (Rs)	1,67,280
Life of system (Year)	10
Discount rated (%)	20
Depreciation (%)	10

Table 4. Economic analysis of 500 L/h plant.

Sr no.	Factor	Value
1	% Saving in Energy	33.8
2	Simple payback period (Year)	2.46
3	Net present value (Rs)	2,89,316
4	Annual life cycle cost (Rs)	1,18,871
5	Life cycle cost (Rs)	4,98,364
6	Depreciation amount (Rs)	41,200
7	Revised investment cost (Rs)	3,70,800
8	Revised net present value (Rs)	4,95,316
9	Revised simple payback period (Year)	2.21
10	Benefit to cost ratio	2.12
11	Internal rate of return (%)	38

2.4 *Analytical calculation of energy usage*

Heating and cooling requirement for each process is worked out using following equation:

$$Q = m \times C_p \times \Delta T \tag{1}$$

The electricity required for heating is worked out using the (1). While electricity required for other equipment is worked out considering rated motor capacity with 0.8 as load factor. C.O.P of the whole system cannot be worked out but Danfoss compressor selection charts are used to work out C.O.P of each refrigeration system. Density of water, icecream mix and frozen icecream is taken as 1, 3.31 and 2.72 kJ/kg.K, respectively. Tables 2(a) and 2(b) show the calculation of existing and modified systems, respectively, to manufacture 2000 L/d icecream.

3 ECONOMIC ANALYSIS OF 500 L/H CAPACITY ICECREAM MIX PLANT—A CASE STUDY

By comparing Table 2(a) and Table 2(b), it has been found that total saving of energy in modified system is 93.94 kWh. By comparing Fig. 2(a) and Fig. 2(b) it is seen that the modified system has additional equipments like heat pump, 4 insulated tanks and additional 4 water pumps while cooling tower is eliminated. Parameters are considered in economic analysis has been given in Table 3.

4 RESULT AND DISCUSSION

Based on parameters discussed above, the following factors are calculated and shown in Table 4 [Motwani K.H., Jain S.V., Patel R.N]. Each factor plays an important role in judging the viability of the project. Internal rate of return is coming as 38% which is more than the assumed discount rate 20%. Hence, modified project is techno-economically viable.

REFERENCES

Danfoss commercial compressor selection chart, www.commercialcompressors.danfoss.com.

Motwani K.H., Jain S.V., Patel R.N., 2012, Cost Analysis of Pump As Turbine for Pico Hydropower Plants-a Case Study, NUiCONE.

Soylemez M.S., 2005, Optimum heat pump in milk pasteurizing for dairy, Journal of Food Engineering, Volume 74546–551.

Ubbels J., Bouman S., The saving of energy when cooling milk and heating water on farms, International Journal of Refrigeration, Volume 2, Issue 1, January 1979, 11–16.

Zaidi A.H., Rawal S.R., Sarma S.C., 1979, Applications of Heat Pump in Dairy Industries, Indian Dairyman, Volume 31, Issue 5303–309.

Multi-disciplinary Sustainable Engineering: Current and Future Trends – Tekwani, Bhavsar & Modi (Eds)
© 2016 Taylor & Francis Group, London, ISBN 978-1-138-02845-6

Surface roughness analysis of piston assembly of CI engine fuelled with diesel, bio-diesel and Di Ethyl Ether (DEE) during long term endurance test

Paresh D. Patel, Rajesh N. Patel, Sajan Chourasia & A.M. Lakdawala
Institute of Technology, Nirma University, Ahmedabad, Gujarat, India

ABSTRACT: The current work is extension of our earlier experimental work, based on which, a 20% (v/v) and a 4% (v/v) mixture of Jetropha biodiesel and di-ethyl ether in diesel (B20A4) was established as an optimum bio-diesel fuel blend. The Current experimental research is targeted to inspect the influence of bio-diesel and oxygenated fuel on wear of various components of diesel engine. A long term endurance test is carried out as per IS 10000 (part IX) on a DI diesel engine with B20A4. Test is done under scheduled loading cycles in two stages: engine fuelled with mineral diesel (B0) and engine fuelled with B20A4. After accomplishment of these tests, both the engines are disassembled for detecting the physical surface roughness of various key components of engine, e.g., piston, piston rings, cylinder liner. Surface roughness measurements of these key components are also done to evaluate the wear of these components. The wear of different component except compression ring are found to be lower in the case of B20A4 fuelled engine. To enumerate the wear of engine components, surface roughness at different positions are measured and compared. A qualitative analysis is also carried out by conducting surface profile test and metallurgical microscopy at the same locations.

Keywords: CI engine; bio diesel; Di-Ethyl Ether (DEE); life cycle analysis; wear analysis

1 INTRODUCTION

Although combustion-associated properties of plant oils are similar to that of mineral diesel oil, the pure plant oils or their mixtures with diesel pose many long-standing problems in CI engines, e.g., Poor atomization characteristics, ring-sticking, injector-choking, injector deposits, injector pump disappointment, and lubricating oil thinning by crank-case polymerization. Such difficulties do not rise with short-range engine operations. Occasionally, the engine fails terribly while operated on pure vegetable oils continuously for an extended period. Hence, the main motivation to carry out the life cycle analysis of CI engine is to analyze and compare wear of vital parts of CI engine like Cylinder Liner, Piston, Pistons Rings, Crank shaft, Crank shaft bearing and journals etc. for a long-term utilization of Bio-diesel or Diesel-Bio-diesel blend. Moreover, the comparison of long term wear of an engine fueled with diesel—biodiesel blended to that with an engine fueled with mineral diesel reveals modifications required to be made in existing CI engine for effective utilization of diesel—biodiesel blend.

According to IS 10000, 1980 for life cycle analysis of internal combustion engine the test engine should be completely dismantled and examined physically so that design features and also the condition of various parts is to be noted before the tests are commenced. After the physical examination, the dimensions of the main moving parts should be checked and recorded in the performance charts in terms of wear. Some of the vital parts are Cylinder Liner, Piston and Pistons Rings. After completion of the initial performance test and physical examination, the engine should be run for 32 cycles (each of 16 hours continuous running) at the rated speed. The test should be carried out for two different CI engines—one fuelled with

Diesel and second fuelled with optimize blend. At the completion of the recommended tests, the engine should be again dismantled. The condition should be noted and the dimensions of critical parts should be recorded in performance chart. The wear of critical components should be recorded in performance chart and should be compared with the declaration made by manufacturer [IS 10000 - V 3].

2 EXPERIMENTAL SETUP AND METHODOLOGY

The experimentations is carried out on computerize test rig equipped with single Cylinder, Kirlosker 4 Stroke, Water cooled, direct injection diesel engine. The test rig is coupled with eddy current dynamo-meter, crank angle encoder, rota meter and fuel tank with digital piezo sensor, temperature and pressure sensors etc. The setup enables study of VCR engine performance parameters like brake power, frictional power, BMEP, brake thermal efficiency, indicated thermal efficiency, specific fuel consumption, mechanical efficiency, volumetric efficiency, A/F ratio, heat balance and combustion analysis at different values of compression ratio, injection timing and injection pressure. The schematic diagram of experimental setup is shown in Fig. 1 and its specification is provided in Table 3.

Based on the earlier experimental work conducted for performance and emission, investigations at full fuel delivery settings and different engine load, a 20% blend of jatropha bio-diesel with 4% di-ethyl ether (B20A4) was found to be optimum blend (superior thermal efficiency and lower emission) and selected for long-term endurance test in the current

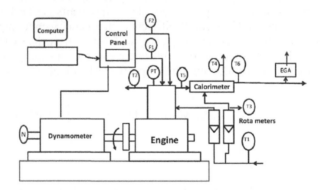

Figure 1. Line diagram of the experimental setup used in the present study. T1 – inlet (engine) water temperature (°C), T2 – outlet (engine) water temperature (°C), T3 – inlet (calorimeter) water temperature (°C), T4 – outlet (calorimeter) water temperature (°C), T5 – exhaust gas temperature before calorimeter (°C), T6 – exhaust gas temperature after calorimeter (°C), F1 & F2 – fuel consumption and air flow measurement, PT – Pressure transducer, EGA – exhaust gas analyser and N – engine speed measurement.

Table 1. Loading cycle for preliminary runs for constant speed diesel engine, showing load (% of rated load) VS running time (hour) provided by IS 10000 Part-V, 1980 [2]. The cycle is shown for 7 hours. Seven such cycle is carried out for 49 hours preliminary runs.

Load (% of rated load)	Running time (Hours)
25	1.5
50	2
75	1.5
100	2

Table 2. Loading cycle for long-term endurance test for constant speed diesel engine, showing load (% of rated load)VS running time (hour) including warm up period provided by IS 10000 Part-IX, 1980 [2].

Load (% of rated load)	Running time (Hours)
100 (Including warm up)	4
50	4
110	1
No load (Idling)	0.5
100	3
50	3.5

Table 3. Engine specifications.

Components	Specification
Engine	Research Engine test setup 1 cylinder, 4 stroke, Multi fuel (Computerized), Water cooled, STROKE 110 mm, BORE 87.5 mm
Diesel mode	Power 3.5 kW C.R. range 12.1–18.1 Speed 1500 rpm Injection Variation 0–250 BTDC
Compression ratio	18:1
Inlet valve open BTDC	15 Degree
Injection start BTDC	20 Degree
Inlet valve close ABDC	30 Degree
Exhaust Valve Open BBDC	30 Degree
Exhaust Valve close ATDC	15 Degree
Fuel injection Pressure	180 Bar
Dynamometer	Eddy current type, water cooled, with loading unit
Fuel tank	Capacity 15 lit, Type: Duel compartment, with fuel metering pipe of glass
Air box	M S fabricated with orifice meter and manometer
Calorimeter	Type Pipe in Pipe
Rota meter	Engine cooling 40–400 LPH, Calorimeter 25–250 LPH
Piezo sensor	Combustion: Range 5000 PSI, with low noise cable Diesel line: Range 5000 PSI, with low noise cable
Crank Angle Sensor	Resolution 1 Degree, speed 5500 rpm with TDC pulse
Temperature Sensor	Type RTD,PT100 and Thermocouple
Load Sensor	TYPE K Load cell, type strain gauge, range 0–50 Kg
Fuel Tank	Capacity 15 LIT
Fuel flow transmitter	DP transmitter, Range 0–500 mm WC
Software	ENGINESOFT, engine exhaust performance analysis software
Rota meter	Engine cooling 40–400 LPH; Calorimeter 25–250 LPH
Data acquisition device	NI USB-6210, 16-bit, 250kS/s
Digital voltmeter	Range 0–20V, panel mounted
Air flow transmitter	Pressure transmitter, Range (–) 250 mm WC
Pump	Type, Monoblock
Overall dimensions	W 2000 × D 2500 × H 1500 mm

work. The long-term engine endurance experiments is conducted in two phases. In the first phase, the engine is run with diesel (B0) and in the second phase, the engine is run with B20A4. The engines are initially dismantled for physical inspection. After physical examination, dimensions and weight of various parts such as cylinder bore/liners, pistons, and piston rings are recorded. Further, the engines are subjected to 49 hours duration preliminary run before commencing the endurance test. This is done according to test procedure given in IS:

10000 part V, 1980 [IS 10000-V]. The cycle for preliminary test is shown in Table 1. After a preliminary run, the lubricating oil from the sump is drained off and the engines are refilled with fresh lubricating oil (SAE 20W40), as specified by the manufacturer.

The endurance test is performed as per Indian standard code (IS: 10000 Part IX, 1980) [IS 10000-IX]. The test is conducted for a total of 512 hour duration and consists of 32 nonstop running cycles of 16 hours duration as shown in Table 2. The lubricating oil samples is collected from the engine after every 128 hour for wear analysis. After conducting the endurance test for the first phase, the engine is dismantled and the physical condition of various engine components is inspected carefully. Moreover, the surface roughness measurement and its microscopic surface images is also inspected. For the second phase of life cycle analysis, a new engine is installed on the same test rig and fuelled with B20A4. The above mentioned procedure, i.e., preliminary runs, endurance test, lubricating oil collection, and wear measurement are followed for second phase (B20A4) also.

Note that the surface roughness profiles of various engine components is measured using stylus based surface roughness meter (Maker: CARL-ZEISS, Specification: Minimum display value 0.01 µm, Accuracy: CLASS 1 (DIN 4777)) while for the wear analysis metallurgical microscopy is used. The magnification used in the said microscope is 50X and 1000X [Sinha Shailendra Agarwal, Kumar, 2010, A. K. Agarwal J. Bijwe L. M. Das Bora Dilip Kumar L M Das and M K G Babu].

3 RESULTS AND DISCUSSION

Figure 2 shows the comparison of surface roughness and close microscopic image analysis of two fresh engine components fuelled with diesel and optimum blend B20A4. The surface roughness is measured with ISO 97/JIS 01 standard at IGTR. In the case of piston top it

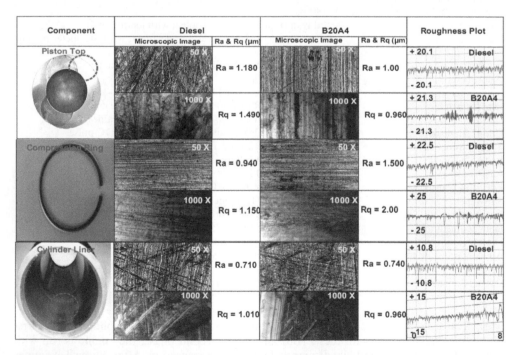

Figure 2. Microscopic and surface roughness analysis of CI engine components including piston, nozzle tip, inlet valve, exhaust valve, cylinder head, crank shaft bearing and cylinder liner. Figure shows the comparison between diesel and B20A4 fuelled engine components microscopic surface image taken at 50X and 1000X magnification along with that the surface roughness measurement data including Ra and Rq value were shown with their roughness plot.

shows the lower Ra and Rq value for B20A4 compare to diesel. Note that the Ra and Rq values are the average and root mean square values of surface roughness measured with ISO 97/JIS 01. Moreover, in microscopic image taken at 50X and 1000X magnification with metallurgical microscope, it is clearly visible that the wear found to be quit high with diesel as compared to B20A4. Further, in the case of compression ring it is found that the Ra and Rq values are higher and also the microscopic image for surface roughness are found to be slightly rough in B20A4 as compared to diesel. For the case of cylinder liner it is observed that the Ra and Rq valves are almost similar in both case of B20A4 and diesel.

From the qualitative assessment shown in Fig. 2 it is concluded that in over all cases the surface roughness of diesel fuelled engine is found to be more as compared to its counterpart of B20A4.

The reasons for more surface roughness and more wear in diesel compare to B20A4 is as follows: The compression ring belt region, with the top area and the piston ring, is straight thermally exposed with the combustion hot gases and hence it is subjected to temperature changes. The peak temperatures of gases inside the combustion chamber during combustion in a CI engine fuelled with diesel can increase to nearly 2500°C. Great heat fluctuation causes difficulties such as thermal stress with reduction of the lubricating oil film in between parts. Due to high combustion temperature difference on piston top area reduces the hardness of the piston. Furthermore, as the temperature rises, the yield strength and shear strength of material reduces. However, in the case of B20A4 the combustion temperate cannot reach to such a high value as in case of diesel. This is due to the fact that the calorific value of B20A4 is lower as compared to diesel. At the same time B20A4 contains 20% v/v of bio-diesel which has self-lubricating properties—in turn reduces the amount of wear and roughness [Andersson,2002].

4 CONCLUSION

An endurance test is carried out on the DI engine with Jatropha bio-diesel and additive di-ethyl ether to see its viability as a substitute for mineral diesel. The wear pattern of surfaces of various components is analysed. A low wear is observed for the piston for B20A4. Surface roughness profiles, various roughness parameters, and microscopic image show that wear is relatively higher at the crank shaft bearing and inlet valve with B20A4 but overall wear is lower than diesel fuelled engine.

A long-term endurance test conclusively proves that bio-diesel can be successfully used for partial substitution of mineral diesel. It can also be concluded that bio-diesel can readily be adopted as an alternative fuel in the existing DI engines without any major modifications in the engine hardware.

ACKNOWLEDGEMENT

The authors are sincerely acknowledge the financial support received from Gujarat Council on Science and technology (GUJCOST), Gandhinagar, Gujarat, India for the Minor Research Project entitled "An Experimental Investigation on Life cycle Analysis and Combustion Characteristics of CI Engine operating on Esterified oil and its blends with bio Additives" wide letter No. GUJCOST/MRP/2014–15/392, Dated 30/06/2014.

REFERENCES

Agarwal, A.K., J. Bijwe, L.M. Das. Effect, April 2003, of Bio-diesel utilization of wear of vital parts in Compression Ignititon Engine Transaaction of ASME vol 125603–611.
Bora, Dilip Kumar, L.M Das and M.K.G Babu.,, Aug 2010, Wear and tear analysis of a single cylinder diesel using karanja bio-diesel (B20) after 512 hours, Journal of Scientific & Industrial Research, Vol 69 pp. 639–642.

Bureau of Indian Standard: 10000: Part V.

Bureau of Indian Standard: 10000: Part IX.

Peter Andersson, Piston ring tribology A literature survey, Jaana Tamminen & Carl-Erik Sandström Helsinki University of Technology, 2002, Internal Combustion Engine Laboratory

Sinha, Shailendra, Agarwal, Avinash Kumar., 2010, Experimental Investigation of the Effect of Bio-diesel Utilization on Lubricating Oil Degradation and Wear of a Transportation CIDI Engine, Journal of Engineering for Gas Turbines and Power, ASME, Vol. 132/042801-1.

Multi-disciplinary Sustainable Engineering: Current and Future Trends – Tekwani, Bhavsar & Modi (Eds)
© 2016 Taylor & Francis Group, London, ISBN 978-1-138-02845-6

Numerical analysis on the effect of baffle inclination angle on flow and heat transfer characteristic of shell and tube heat exchanger with helical baffle

Unnat Prajapati
Heat Chem Engineers Pvt. Ltd., Ahmedabad, Gujarat, India

Absar Lakdawala
Department of Mechanical Engineering, Institute of Technology, Nirma University, Ahmedabad, Gujarat, India

ABSTRACT: In the present work numerical simulation is done to show the effect of baffle inclination angle on thermal and hydraulic performance of shell and tube heat exchanger. For this analysis a geometric similar model with 09 circular tubes in line arrangement is considered. The CAD model of the same is created using an unstructured Tetra/Hybrid mesh. Using finite volume method, steady flow momentum and energy equation are solved. Moreover, Renormalization Group (RNG) k–ε is selected for the turbulent modeling. Simulations are carried out for ten different Reynolds number ranging from $400 \leq Re \leq 2200$; three different Prandtl number between $0.7 \leq Pr \leq 118.2$, and four different baffle angles between $20° \leq \theta_b \leq 50°$. The methodology is validated with published experimental results and finds it a good agreement. The results shows that, shell and tube heat exchanger with helical baffle producing more vortex flow and hence have higher heat transfer co-efficient and lower pressure drop as compared with the segmental baffle. The simulation results shows that 40° helical baffle has 76.35% higher heat transfer co-efficient per unit pressure drop (Nu/f) as compared with the segmental baffle heat exchanger.

Keywords: helical baffle; numerical analysis; k–ε model; thermo-hydraulic performance

1 INTRODUCTION

In most of the industries, shell and tube heat exchanger is used with the segmental baffle plate on the shell side. However, it should be noted that the segmental baffle plate has few adverse effects—more pressure drop, higher fouling, and higher maintenance cost—on the thermo-hydraulic performance of heat exchanger. The problems can be minimized with the use of helical baffle instead of the segmental baffle. The effect of baffle inclination angle on fluid and heat transfer of heat exchanger using numerical methodology was carried out by (Lei et al. 2008). They showed that due to elimination of dead zone in continuous baffle heat exchanger, the pressure drop was found considerably lower as compared with the segmental baffle. Moreover, it was shown that although the heat transfers co-efficient is lower in case of helix heat exchanger, the heat transfer per pressure drop is found superior as compared with the segmental baffle. Chen et al. (2013) showed numerically the formation of circulation and rotating spiral flow—called as secondary flow on the shell side. The phenomenon enhances the heat transfer of heat exchanger (Chen et al. 2013). Through the numerical investigation (Lei et al. 2008) showed that 40° helical angle is found to be the best selection for improvement in thermo-hydraulic performance. Moreover, 40° helical baffle is no more the best selection if the Prandtl number of the fluid on the shell side is higher. When Prandtl number is large enough, heat exchanger with small helical angle reveals to be the optimal choice (Xiao et al. 2013). The experimental tests are performed on several shell-and-tube heat exchangers,

457

with the segmental baffles as well as with helical baffles at helix angles of $20° \leq \theta_b \leq 50°$. Nusselt number (Nu) and friction factor (f) correlations are obtained at helical angle 20°, 30°, 40°, and 50° (Zhang et al. 2009, 2013).

2 COMPUTATIONAL MODEL AND NUMERICAL METHOD

2.1 Computational model

The present study aims to numerically investigate flow in helical baffle heat exchanger. The computational domain with continues helical baffle at 40° of baffle angle is shown in Fig. 1 representing side view and front view. Geometry parameters considered in the present study are shown in Table 1.

2.2 Governing equation and boundary condition

Three-dimensional fluid motions in continues helical baffle shell and tube heat exchanger can be solved by mathematical modeling and numerical implementation for the fluid region.

Mass conservation (continuity) equation:

$$\frac{\partial}{\partial x_i}(\rho u_i) = 0$$

Momentum conservation equation:

$$\frac{\partial}{\partial x_i}(\rho u_i u_k) = \frac{\partial}{\partial x_j}\left(\mu \frac{\partial u_k}{\partial x_i}\right) - \left(\frac{\partial P}{\partial x_k}\right)$$

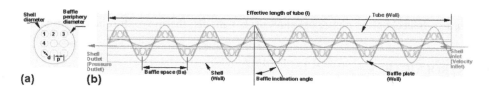

Figure 1. Schematic diagram (a) side view with flow parameter (b) front view.

Table 1. Geometry parameter.

Parameter	Value
Tube pitch	0.025
Number of tubes	09
Shell diameter	0.110
Tube outside diameter	0.019
Effective length of tube	1
Tube layout pattern	90°
Tube pass	1
Baffles angle	20°, 30°, 40°, 50°
Baffle spacing	0.02743–0.18535

Energy conservation equation:

$$\frac{\partial}{\partial x_i}(\rho u_i t) = \frac{\partial}{\partial x_i}\left(\frac{k}{C_p}\frac{\partial t}{\partial x_i}\right)$$

Inlet velocity of fluid is taken according to the Reynolds number range from $400 \le Re \le 2200$. Pressure boundary condition is used at the outlet and sets 0 gauge pressure. The fluid inlet temperature is set to be 335 K. The working fluids are air $(Pr = 0.74)$, water $(Pr = 6.99)$ and oil $(Pr = 118.2)$. Adiabatic boundary conditions are set at the shell wall and the baffle plate, and the no-slip boundary condition is applied at the impermeable heat exchanger tubes, shell wall, and baffle plate. Over the tube the dimensionless length y^+ is $0.30 < y^+ < 3.0$. The tube wall temperature is set to be 300 K and the properties of the fluid and the solid wall are constant.

2.3 Grid generation and numerical solution

The computational domain is discretized (grid generation) by the commercial code Ansys mesh. The domain is meshed with unstructured Tet/Hybrid with default scheme. Due to complexity of the present computational domain, the grid is generated with a boundary layer over each tube. Grid elements considered for simulation work at different baffle angles are 2831868 to 2750688. Side view of the geometry with grid generation of shell and tube heat exchanger with continues helical baffle at 40° of helix angle is shown in Fig. 2. The Fig. 2b$_1$ shows the exploded view of boundary layer generated over tubes.

The computer code FLUENT 6.3 is used to solve—steady state—fluid flow and heat transfer equations in the computational domain. Pressure base solver is selected. For the turbulence modeling k–ε is selected for the swirl flow on shell side to enhance the simulation accuracy. The governing equations are iteratively solved by the finite volume method using SIMPLE pressure–velocity decoupling algorithm. The convective terms in momentum, turbulent kinetic energy and turbulent dissipation governing equations is discredited by the QUICK scheme. The under relaxation factor are set to be 0.3, 0.7, and 1 for pressure, momentum and energy, respectively. The convergence criteria considered to be 10^{-3} for the flow equations and 10^{-6} for the energy equation.

2.4 Parameter reduction

For the calculation of various governing and engineering important parameters of helical heat exchanger, following methodology is adopted:

Transverse area: It is area from where the shell side fluid is flowing
$S = \frac{1}{2} - B\,D_s\left(1 - \frac{D_{to}}{P_t}\right)$ where, B (baffle spacing) in equation is given as: $B = \sqrt{2}D_s\tan\theta_b$
The Reynolds number of the shell side fluid flow is calculated as:
$Re = \frac{\rho D_e u}{\mu}$ D_e is equivalent diameter taken as tube outer diameter d_{to}.
LMTD of heat exchanger:

$$\Delta T_{LMTD} = \frac{(Th_i - T_w) - (Th_o - T_w)}{\ln\left(\dfrac{Th_i - T_w}{Th_o - T_w}\right)}$$

Frictional factor:

$$f = \frac{1}{4}\left(\frac{\Delta P}{\frac{1}{2}\rho u^2}\frac{D_e}{L}\right)$$

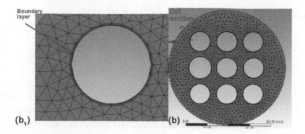

Figure 2. Grid of the configuration (b) side view (b₁) grid generated over tubes.

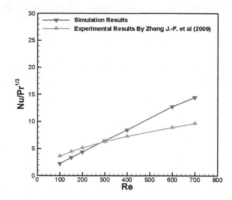

Figure 3. Comparison between experimental and simulation results represents at 30° inclination of baffle.

3 MODEL VALIDATION

In order to verify the results of simulation model, comparisons of present numerical results are done with the experimental results of (Zhang 2009). Simulation model is validated with experimental results for heat exchanger having 30° of baffles angle and $100 \leq Re \leq 700$ as shown in the Fig. 3. As per the Bell–Delaware methodology, the shell side flow can be divided into five different streams (Kakac 2002). In CFD simulations the bypass streams in the tube bundle and leakages are neglected. Moreover, in experiment setup due to manufacturing limitations there exist tolerance and fits. Due to this results are deviating. The minimum and maximum deviation between present numerical and published experimental (Zhang 2013) values of $Nu/Pr^{1/3}$ is found to be 2.83–33.50%.

4 RESULT AND DISCUSSION

In order to understand the effect of baffle angle on thermal and hydraulic performance of shell and tube heat exchanger, 130 cases are simulated.

4.1 Shell side velocity and vorticity distribution

In Fig. 4(a) and (d) it illustrates the contours of the velocity magnitude in longitudinal section plane with $Re = 2200$ and $Pr = 6.99$ of segmental baffle and 40° baffle angle heat exchanger. The velocity magnitude is higher and uniform in the active zone as compared with the dead zone. The velocity magnitude is nearly equal to 0.4 for segmental baffle on the other hand the velocity magnitude is decreasing to 0.17 for 40° of helical baffle. For segmental baffle the velocity distribution is not uniform near the baffle plate whereas the velocity distribution is uniform near the baffle for helical baffle plate. The entering and exit velocities of segmental baffle are low as compared with the flow velocity whereas velocity remains uniform through for helical baffle.

In Fig. 4(a) four transverse plane a_1 to a_4 represents vorticity magnitude at distance of 0.01 m, 0.35 m, 0.70 m, and 0.99 m, respectively. At $Z = 0.01$ m vorticity magnitude is higher over the tube segment and low and uniform at other places. As the distance increases from $Z = 0.01$ to $Z = 0.35$ m the vorticity is higher near the periphery of shell wall and non-uniform at other places. Now as the distance increases, the vorticity decreases non-uniformly. For 40° helical baffle vorticity is uniform over the periphery of shell wall. Moreover, it should be noted that 40° helical baffle vorticity is uniform along the distance.

4.2 Shell side temperature distribution

Fig. 5 shows the contour of temperature distribution along the longitudinal direction with $Re = 2200$ and $Pr = 6.99$ for segmental baffle and helical baffle. The shell-side flowing fluid temperature rises steadily and uniformly along the shell. For helical baffle heat exchanger as the baffle angle increases the temperature is consistently rises in three meridian planar sections. The temperature is higher near periphery than near the axis throughout the shell. The temperature gradient is apparently sharper in the area around the tube wall than in other regions.

4.3 Comprehensive performance analysis

The ratio of Nusselt number and frictional factor (Nu/f) is illustrated in Fig. 6. The analysis results show that the ratio of non-dimensional parameter has increased with helical angle.

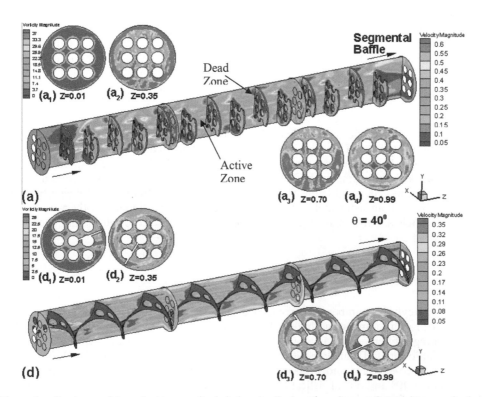

Figure 4. Contours of the velocity magnitude in longitudinal section plane and vorticity magnitude in transvers plan: (a) segmental baffle (d) $\theta_b = 40°$.

461

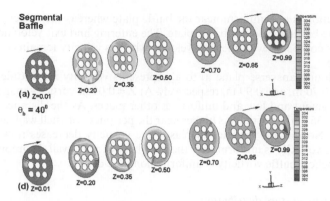

Figure 5. Contour of temperature distribution in longitudinal section plane along the distance: (a) segmental baffle, (d) $\theta_b = 40°$.

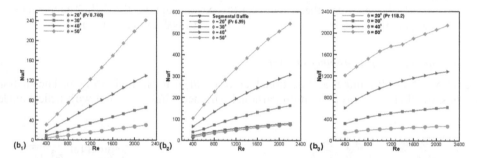

Figure 6. Reynolds number vs Nusselt number/Frictional Factor (b₁) Air ($Pr = 0.74$) (b₂) Water ($Pr = 6.99$) (b₃) Oil ($Pr = 118.2$).

Baffle angle of 40° and 50° have higher value as compared with the segmental baffle at same Reynolds number. From Figs. 6b₁, b₂, and b₃, it is evident that the ratio Nu/f on shell side for 40° and 50° of helical angles are 66.34–76.35% and 79.72–87.34%, is respectively higher as compared with the segmental baffle. For $Pr = 6.99$ the comparison made between the segmental baffle and helical baffle for Nu/f ratio shows that the difference is not significant for 20° helical baffle than segmental baffle. It is observed that the difference in the ratio of Nu/f between 20° helical baffle than segmental baffle is not significant for all Pr simulated in the present work.

5 CONCLUSION

For varying Reynolds number the percentage difference of heat transfer co-efficient increases with lower Prandtl number and higher Baffle angle. Similarly for varying Reynolds number the percentage difference of pressure drop per unit length increases with higher Prandtl number and lower Baffle angle. Shell and tube heat exchanger with helical baffle have lower heat transfer co-efficient and pressure drop than segmental baffle. The non-dimensional parameters at 40° and 50° of helical baffle have higher performance than segmental baffle. However, the effects of inclination angle on pressure drop and frictional factor on shell side are small when $\theta_b = 40°$.

REFERENCES

Chen Y.P., Sheng Y.J., Dong C. & Wu J.F., 2013, Numerical simulation on flow field in circumferential overlap trisection helical baffle heat exchanger. App. Therm. Engg. Vol. 50, 1035–1043.

Heat Exchangers by Sadik Kakac, 2002, Selection Rating and Thermal Design.

Lei Y.G. et al., 2008, Effects of baffle inclination angle on flow and heat transfer of a heat exchanger with helical baffles. Chem. Engg. and Pro. Vol. 47, 2336–2342.

Xiao X. et al., 2013, Numerical investigation of helical baffles heat exchanger with different Prandtl number fluids, International Journal of Heat and Mass Transfer Vol. 63, 434–444.

Zhang J.F. et al., 2009, Experimental performance comparison of shell-side heat transfer for shell-and-tube heat exchangers with middle-overlapped helical baffles and segmental baffles. Chem. Engg. Sci. Vol. 64, 1643–1653.

Zhang J.F. et al., 2013, Experimental performance comparison of shell-and-tube oil coolers with overlapped helical baffles and segmental baffles. App. Therm. Engg. Vol. 58, 336–343.

Multi-disciplinary Sustainable Engineering: Current and Future Trends – Tekwani, Bhavsar & Modi (Eds)
© *2016 Taylor & Francis Group, London, ISBN 978-1-138-02845-6*

Numerical simulation and modeling of a downdraft gasifier

Jagdish R. Kanjariya, Rajesh N. Patel & Absar M. Lakdawala
*Department of Mechanical Engineering, Institute of Technology, Nirma University,
Ahmedabad, Gujarat, India*

ABSTRACT: In the gasification process, raw materials react with air, oxygen, stem, or their mixture to produce syngas. The syngas produced has useful heating values and can be used as fuel for generating heat and power. The gasifier considered in this study is a downdraft throat type, fixed bed, and 10 kW downdraft gasifier that is located at Nirma University's gasifier laboratory. Design data and material grades for modeling were taken from the same gasifier. The objective of this study is to develop a numerical model to investigate the thermal-hydraulic and gasification process inside a downdraft gasifier using the commercial CFD solver ANSYS/FLUENT 14.5. Turbulence model selected is standard k–ε model and combustion process has been displayed by species transport model to calculate global gasification reaction. The chemical reactions that are considered in CFD gasification are combustion, gasification, methanation, partial combustion, and water gas shift reaction. Radiation model P1 is considered for radiation heat transfer. Discrete phase model is used to consider heat, mass transfer, and momentum between solid phase and gas phase. Stochastic particle tracking method is used for considering turbulent dispersion. The influence of variation in fuel composition, calorific value, and Equivalence Ratio (ER) on temperature distribution in various zones is evaluated. It is observed that with ER 0.25 the outlet temperature is 960 K while that with 0.35 is 1024 K.

Keywords: downdraft gasifier; equivalence ratio; species transport

1 INTRODUCTION

Gasification is a very efficient method for extracting energy from many different types of organic materials. Considering environmental pollution from energy conservation system gasification is near zero pollution technology with high thermal efficiency for syngas production. Gasification is a thermochemical process that converts carbon based materials like coal, petroleum, and biomass into gaseous fuel. In gasification, raw materials react with air, oxygen, steam, or their mixture at a temperature higher than 700°C to produce syngas. The syngas produced has useful heating values and can be used as fuel for generating heat and power. The calorific value of syngas produced is determined by gasifying medium. In the gasification process, oxygen molecules are less than carbon molecules which indicate that the gasification process operates at a limited amount of oxygen. Oxygen is deliberately supplied in a less amount so that carbon and hydrogen present in feed stock does not get converted to CO_2 and H_2O, respectively. The syngas is a mixture CO, H_2, CH_2, CO_2, N_2, and H_2O out of which CO, H_2, and CH_4 are combustible rest are noncombustible components (Reed & Das 1988). There are four types of gasifiers: 1. fixed bed gasifier, 2. fluidized bed gasifier, 3. entrained flow gasifier, and 4. transport gasifier. The main benefits of fluidized bed gasifiers are, it can handle any coal practically, and the produced synthetic gas is free of oils and tars, with a very high carbon conversion ratio (Chen, Horio, & Kojima, 1999).

2 DESIGN AND BOUNDARY CONDITION

A downdraft gasifier is studied in this research. The geometry of the gasifier used in the simulation is shown in Fig. 1. The geometry and the operating conditions are based on Nirma

University laboratory scale gasifier. The gasifier is divided into four regions: drying zone and pyrolysis zone are upper stage of gasifier and combustion zone and reduction zone are lower stage of gasifier. The gasifier has two symmetrical injectors of the air in combustion region, which is the lower stage of gasifier. All the wood and biomass wastes are injected in upper section of the gasifier. Dimension and other parameters are in use similar to Patel

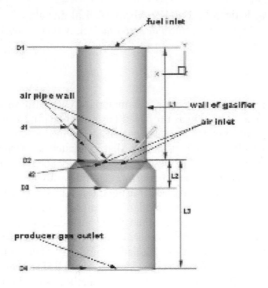

Figure 1.   Downdraft gasifier model.

Table 1.   Dimensions of downdraft gasifier.

	Name	Dimension (mm)
D1	Inlet diameter	500
D2	Throat diameter	500
D3	Throat outlet diameter	155
D4	Ash outlet diameter	630
d1	Air inlet diameter	20
d2	Air outlet diameter	20
l	Air pipe length	350
L1	Hopper length	765
L2	Throat length	185
L3	Reduction cell length	661
L	Total length	1426
θ	Throat angle	45

Table 2.   Boundary condition table at 0.25 equivalence ratio (ER).

Boundary name	Mass flow specification	Mass flow rate (kg/s)	Temperature (K)	Turbulence method	Turbulence intensity	Hyd. Dia	Em-Isivity
Air inlet	Mass flow rate	0.0046	430	Intensity & hyd. dia	10%	20	–
Air pipe wall	No slip	–	300	–	–		0.77
Fuel inlet	Mass flow rate	0.00278	300	Intensity & hyd. dia	10%	500	–
Producer gas outlet	Out flow	–	–	Intensity & hyd. dia	10%	630	–
Wall of gasifier	No slip	–	300	–	–		0.77

et al. (2014). This model of gasifier is discretized using unstructured hybrid mesh in ANSYS meshing modeler. The total number of elements used in the present study is nearly 5 lakhs.

3 GOVERNING EQUATION AND CHEMICAL REACTION

The governing equations for continuity, momentum, energy, and species transport equation for the fluid region are as follows (Silaen & Wang 2010):

3.1 *Mass conservation equation*

$$\frac{\partial}{\partial x_i}\left(\rho u_i\right) = S_m \tag{1}$$

where ρ is density, S_m is source term.

3.2 *Momentum conservation equation*

$$\frac{\partial}{\partial x_i}\left(\rho u_i u_j\right) = \rho \overline{g_j} - \frac{\partial p}{\partial x_i} - \frac{\partial p}{\partial x_j} + \frac{\partial}{\partial x_i}\left(\tau_{ij} - \rho \overline{u_i' u_j'}\right) + S_j \tag{2}$$

where τ is stress tensor.

3.3 *Energy conservation equation*

$$\frac{\partial}{\partial x_i}\left(\rho C_p u_i T\right) = \frac{\partial}{\partial x_i}\left(\lambda \frac{\partial T}{\partial x_i} - \rho C_p \overline{u_i' T'}\right) + \mu \Phi + S_h \tag{3}$$

where C_p is specific heat, λ is heat conductivity, and Φ is viscous dissipation.

3.4 *Species transport equation*

$$\rho \frac{\partial}{\partial x_i}\left(u_i Y_j\right) = \frac{\partial}{\partial x_i}\left(D_t \frac{\partial Y_j}{\partial x_i} - u' Y'\right) \tag{4}$$

where D_t diffusion coefficient and Y is mass fraction of species.

For CFD analysis of gasification process, along with steady state mass, momentum, and energy standard k–ε model for turbulence modeling and species transport model for combustion modeling is selected. Moreover, P1 model for radiation modeling, discrete phase model with stochastic particle tracking for coupling between solid phase and gaseous phase are selected.

This study deals with global chemical reaction of wood gasification. The generalized equations are listed as Eq. (5) to (11), In this study, the water gas shift reaction and methanation chemical reaction are considered.

$$C(s) + 0.5\ O_2 = CO \text{ Partial combustion} \tag{5}$$
$$C(s) + CO_2 = 2\ CO \text{ Gasification, Boudouard reaction} \tag{6}$$
$$C(s) + H_2O = CO + H_2 \text{ Gasification} \tag{7}$$
$$CO + 0.5\ O_2 = CO_2 \text{ Combustion} \tag{8}$$
$$CO + H_2O\ (g) = CO_2 + H_2 \text{ Water gas shift reaction} \tag{9}$$
$$CO + 3\ H_2 = CH_4 \text{ Methanation} \tag{10}$$
$$H_2 + 0.5\ O_2 = H_2O \text{ Oxidation} \tag{11}$$

4 RESULTS AND DISCUSSION

It has been noted that the different ERs are obtained by varying mass flow rate of air for fixed mass flow rate of fuel. The results obtained from the CFD simulation are discussed here. For the species transport model, the proximate and ultimate analysis of fuel used in the gasifier is required to be given as an input parameter. The proximate and ultimate analysis is shown in Table 3 for wood biomass.

4.1 *Flow distribution in gasifier*

Fig. 2 shows variation of velocity magnitude in a downdraft gasifier for different ER. It can be seen in Fig. 2 that the velocity field can be divided into two parts: (1) Passive region (upper part of the gasifier): In this region the velocity magnitude is small and the flow is diverted towards the downward direction. The passive velocity field is a first qualitative indication of pyrolysis and drying zone with lower temperature field. (2) Active region (lower part of gasifier): The active region is seen between air inlet and outlet of gasifier (see Fig. 2 the velocity contours indicated with red color). The formation of vortex is seen in the active region. The formation of vortex is indication of combustion and reduction zone. Moreover, the production of vortex is responsible for the transport of H_2 and CO to the lower part of the gasifier. The comparison of vortex at different ERs (Fig. 2 a, b, c, and d) clearly shows that the intensity of vortex increases with the ER. The increase in vortex strength can be correlated with the increase in temperature of combustion zone as ER increases.

4.2 *Temperature distribution in gasifier*

Fig. 3 shows temperature distribution into perpendicular planes at mid-section. It can be seen in Fig. 3 and that the upper zone (passive region as discussed in previous section) experiences lower temperature while lower zone (active region) experiences higher temperature indicating pyrolysis for upper and combustion and reduction for lower zone, respectively. This can be explained as follows: The heat is absorbed during pyrolysis reaction leading to reduction in temperature while heat release takes place during oxidation reaction which increases the temperature of combustion zone. The maximum temperature observed is around 1280 K near the throat section of gasifier for ER = 0.35. (See Fig. 3 (d)). The higher temperature near the throat section is observed due to higher concentration of oxygen leading to combustion. Finally slightly lower temperature as compared with the throat section is seen in reduction zone of gasifier. This is because of hot gases are reduced by unreacted char in this zone. Moreover, due to increase in availability of oxygen with increasing ER, the gas temperature is found to be higher (compare temperature distribution in Fig. 3 at different ERs).

4.3 *Mass fraction concentration*

The zone-wise temperature variation in different zone of gasifier at different ERs is shown in Fig. 4. The variation of temperature is found to be in good agreement with the theory.

Table 3. Ultimate and proximate analysis of wood as well as boundary condition for various mass fraction.

Proximate		Ultimate		Oxide	
Volatile matter	0.8285	Carbon	0.4960	Nitrogen	0.7666
Ash	0.0079	Hydrogen	0.063	Oxygen	0.2333
Moisture	0.0669	Nitrogen	0.004		
Fixed carbon	0.0967	Sulfur	0.0002		
		Oxygen	0.04368		

Fig. 5 illustrates syngas composition for different ERs of 0.2, 0.25, 0.3, and 0.35. With increasing ER, it is observed that the concentration of H_2 decreases, while N_2 increases. This is due to infiltration of more air in the gasifier which brings more nitrogen and oxygen. It is also observed that the mass fraction of hydrogen decreases due to dilution effect. Fig. 5 also shows that CO concentration increases first and then decreases viz a viz for CO_2. The results lead to a conclusion that there is an optimal value of ER about 0.3 producing best quality of syngas. Due to availability of higher oxygen for complete combustion CH_4 found to be decreased with the increase of ER.

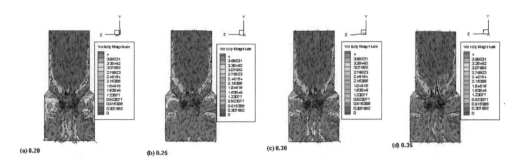

Figure 2.   Velocity magnitude in axial direction for downdraft gasifier with different equivalent ratio. Here, (a) ER = 0.2, (b) ER = 0.25, (c) ER = 0.30, and (d) ER = 0.35.

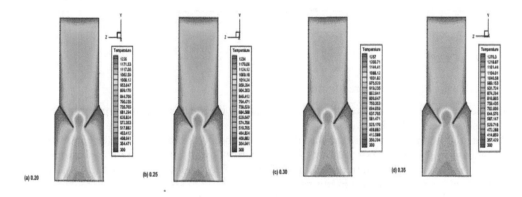

Figure 3.   Temperature profile across (X = C) mid plane of downdraft gasifier with different equivalence ratio. Here, (a) ER = 0.20, (b) ER = 0.25, (c) ER = 0.30, and (d) ER = 0.35.

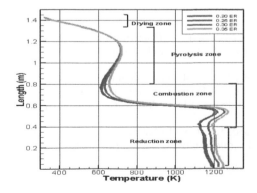

Figure 4.   Temperature variations at different equivalence ratios in different zones.

469

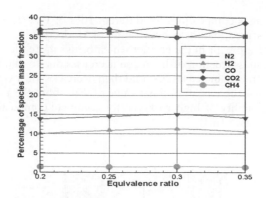

Figure 5.   Mass fraction of species at outlet of gasifier.

Table 4.   Summary.

Case No.	Equivalence ratio	Inlet temp. of air (K)	Outlet temp. of syngas (K)	N_2	H_2	CO	CO_2	CH_4
1	0.20	430	928	36	9.97	13.78	36.8	1.47
2	0.25	430	960	36.17	10.85	14.47	37.04	1.45
3	0.30	430	996	37.41	11.22	14.96	34.88	1.5
4	0.40	430	1024	35.24	10.57	14.10	38.68	1.4

5   CONCLUSIONS

The following conclusions are obtained:

1. The increase in vortex strength can be correlated with the increase in temperature of combustion zone as ER increases.
2. The high temperature near the throat section is observed due to the higher concentration of oxygen leading to combustion. Finally slightly lower temperature as compared with the throat section is seen in reduction zone of gasifier. This is because of the hot gases are reduced by unreacted char in this zone.
3. The results lead to a conclusion that there is an optimal value of ER about 0.3 producing the best quality of syngas.

REFERENCES

Chen, C., Horio, M. & Kojima, T. 1999, Numerical Simulation of Entrained Flow Coal Gasifier part-1 Modeling of Coal Gasification in an Entrained Flow Gasifier, *Chemical Engineering Science*, 35(3861–3874).

Patel, V., Upadhya, D. & Patel, R. 2014, Gasification of Lignite in a Fixed bed Reactor: Influence of Particle Size on Performance of Downdraft Gasifier, *Energy*, 78(323–332).

Reed, T. & Das, A. 1988, Handbook of Downdraft Gasifier Engine System, 1st edition, *Biomass Energy Foundation Press*.

Silaen, A. & Wang, T. 2010, Effect of Turbulence and Devolatilization Model on Coal Gasification Simulation in an Entrained Flow Gasifier, *International Journal of Heat and Mass Transfer*, 53(2074–2091).

Multi-disciplinary Sustainable Engineering: Current and Future Trends – Tekwani, Bhavsar & Modi (Eds)
© 2016 Taylor & Francis Group, London, ISBN 978-1-138-02845-6

Design and analysis of heat pipe heat exchanger for stenter machine

H.V. Mardhekar
*Department of Mechanical Engineering, Marwadi Education Foundation
Group of Institutions, Rajkot, Gujarat, India*

S.V. Jain, V.J. Lakhera & R.N. Patel
*Department of Mechanical Engineering, Institute of Technology, Nirma University,
Ahmedabad Gujarat, India*

ABSTRACT: Stenter machine uses air to air heat exchanger as heat recovery unit from the high-temperature exhaust gases. In the present study, the design of heat pipe heat exchanger is presented for the Stenter machine. The materials of the heat pipe, wick and fin were proposed as Copper, Stainless Steel and Aluminum respectively. Depending on requirements, 4 different options of heat pipe heat exchangers are proposed. To predict the performance of the heat pipe heat exchanger, the heat pipe is fabricated and experiments were performed at different flow rates and inclinations of the heat pipe. The analysis was done by recirculating the water through the heat exchanger, passing the water once through heat exchanger as well as with stagnant water. The collection efficiency and instantaneous efficiency of heat pipe were obtained in the range of 47–49% and 85–94% respectively for different flow rates.

1 INTRODUCTION

Stenter machine is used for stretching of fabrics to bring the length and width to predefined dimensions. It is also used for heat setting of fabric and for applying chemicals for shade variation. Heat exchangers are used for waste heat recovery in the Stenter machine. The heat pipe is an element for heat transfer in which wick and working fluid are provided under vacuum. Heat pipes can have several hundred times more thermal conductivity than the pipe of the same material and hence it is also sometimes referred as a superconductor.

Srimuang and Amatachaya (2012) reviewed the applications of heat pipe heat exchangers for the heat recovery in air preheater for energy saving in industrial applications. Some other applications of heat pipe heat exchanger are also shown. Borges et al. (2011) described the use of heat pipe heat exchanger for Stenter machine's exhaust maintained at 100–200°C. Heat pipe heat exchanger was designed for the heat recovery from the exhaust gases. For calculation of heat and mass balance in Stenter machine and hot-air generator, a routine was developed and implemented in EES. For designing of heat pipe, TROCATER software was used. Shah and Giovannelli (1988) described characteristics of different types of heat pipes. For complete design procedure of heat pipe heat exchanger NTU method was explained. The details of material selection, working fluid, types of wick etc. are also discussed.

In this paper, the design and optimization of heat pipe heat exchanger are presented. The optimized HPHE was fabricated and parametric studies were done by varying flow rates, power inputs and inclinations of the heat pipe.

Table 1. Design specifications for heat pipe heat exchanger.

Parameter	Description
Heat transfer medium	Air to air
Temperature range in exhaust gas	80°C (min) to 200°C (max)
Flow rate of exhaust air	3000 m³/hr (Design)
Flow rate of fresh air	1000 m³/hr (min)
Weight of Heat transfer unit	upto 450 kg
Pressure drop across HRU	1 mm H₂O (max)
Imported parts	Less imported parts.
Corrosion effect	Use of corrosion resistant material.
Oil condensation	Provision for managing condensed oil.

2 DESIGN OF HEAT PIPE HEAT EXCHANGER

2.1 *Major specifications and design methodology*

The major design specifications and constraints in the design are given in Table 1.

Designing of heat pipe heat exchanger involves designing and analysis of air side of heat exchanger followed by the designing of the heat pipe. The steps to design heat pipe heat exchanger are given below (Shah & Giovannelli, 1988):

- To find out the effectiveness and NTU of the heat exchanger.
- To fix the pipe dimensions and pitch of heat exchanger.
- To decide the pressure drop and to find out the mass velocity of air and hot gases entering the heat exchanger.
- To find Reynolds number of the heat exchanger.
- To find heat transfer coefficient and overall heat transfer coefficient of the heat exchanger.
- To find out dimensions of heat exchanger and number of heat pipes required for heat transfer.
- To find out actual pressure drop and iterating until decided pressure drop is achieved by varying the Reynolds number of the heat exchanger.
- To design the heat pipe.

3 GOVERNING EQUATIONS

3.1 *Design of heat pipe heat exchanger (Shah & Giovannelli, 1988)*

The number of transfer units (NTU) was found out using

$$NTU = \frac{1}{1-c} * ln\left[\frac{1-c^* * \varepsilon}{1-\varepsilon}\right] \tag{1}$$

where, C* = Heat capacity ratio, C = Heat capacity (kJ/kg.K), ε = Effectiveness.

The mass velocity was found out using

$$G = \left[\frac{2g_c \, \Delta p}{Deno}\right]^{1/2} \tag{2}$$

$$Deno = \frac{f}{f} ntu \, pr^{2/3}\left(\frac{1}{\rho m}\right) + 2\left[\frac{1}{\rho o} - \frac{1}{\rho i}\right] \tag{3}$$

where, G = Mass velocity (kg/m².s), g_c = Proportionality constant, ρ = Density (kg/m³), f = Friction factor, J = Colburn factor, Pr = Prandtl number.

Reynolds number was found out using

$$\mathrm{Re} = \frac{G D_h}{\mu} \tag{4}$$

where Dh = Hydraulic diameter (m), μ = Viscosity (kg/ms).

Heat transfer coefficient was found out by

$$h = \frac{J G C_p}{Pr^{2/3}} \tag{5}$$

Overall heat transfer coefficient was found out by

$$\frac{1}{U} = RA = \frac{1}{\eta_o h} + \frac{R_s}{\eta_o} + \frac{r_o - r_i}{k_p} \frac{A}{A_{b_p}} + \frac{r_i - r_v}{k_e} \frac{A}{A_{ws}} \tag{6}$$

where, A = Area (m²), η = Efficiency, R_s = Scaling resistance, K = Conductivity (W/m°C).

Number of heat pipes were found out using

$$q_{HP} = q_{HX} / N_t \tag{7}$$

Minimum free flow area of one side of fluid is given by

$$\mathrm{Ao} = (m/G) \tag{8}$$

where, m = Mass flow rate (kg/s), A_o = Minimum free flow area of one fluid (m²).

Frontal area of heat exchanger is given by

$$A_{fr} = A_o/2 \tag{9}$$

where, A_{fr} = Frontal flow area of the heat exchanger (m²).

Geometrical dimension L_2 is given by

$$L_2 = \frac{D_h A}{4 A_o} \tag{10}$$

where, A = Total area of heat transfer required (m²).

Geometrical dimension L_3 is given by

$$L_3 = \frac{A_{fr,e}}{L_{1,e}} \tag{11}$$

Minimum thickness of heat pipe is given by

$$\delta \min = \Delta P d_o / 2 S_u \tag{12}$$

where, d_o = Outer diameter of the heat pipe, S_u = Ultimate strength of heat pipe material.

3.2 Design of heat pipe (Reay & Kew. 2006)

For correct working of heat pipe the following condition should be satisfied

$$\Delta P_c \geq \Delta P_l + \Delta P_v + \Delta P_g \tag{13}$$

where, Maximum capillary pressure

$$\Delta Pc = \frac{2\sigma l \cos\theta}{r_e} \tag{14}$$

Pressure drop of liquid in the wick

$$\Delta P_l = \frac{\mu l\, Q\, l_{eff}}{\rho l\, L\, A_w K} \tag{15}$$

Pressure drop due to gravitation

$$\Delta P_g = \rho g l \sin\varphi \tag{16}$$

Pressure drop required for vapor flow

$$\Delta P_v = \frac{8\mu_v m l_{eff}}{\rho \pi r_v^4} \tag{17}$$

Sonic limit check for heat pipe is given by

$$Q_s = A_v L \rho_v \left[\frac{1+\gamma_v}{2+\gamma_v}\right] (\gamma_v RT_v)^{0.5} \tag{18}$$

Boiling limit check for heat pipe is given by

$$Q_b = 2\pi * le * Ke * Tv * Nt \left[2 * \frac{\sigma_l}{r_n} - pc\right] L * \rho_v ln(r_i / rv) \tag{19}$$

Entrainment limit check for heat pipe is given by

$$Q_{ent} = \pi r_v^2 L \sqrt{\frac{2\pi\rho_v\sigma_l}{z}} \tag{20}$$

Table 2. Options of heat pipe heat exchangers.

Parameters	Option 1	Option 2	Option 3	Option 4
Fins/Inch	Hot side: 8.8 Cold side: 8.8	Hot side: 4 Cold side: 4	Hot side: 4 Cold side: 4	Hot side: 0 Cold side: 8.8
No. of heat pipes	60	135	112	210
Outer dia. of heat pipe	0.0254 m	0.0254 m	0.0254 m	0.0254 m
Diameter of Fin	0.04411 m	0.04411 m	0.04411 m	0.04411 m
Thickness of Fin	3.04* 10^{-4} m	3.04* 10^{-4} m	3.04* 10^{-4} m	3.04* 10^{-4} m
Reynolds No	Hot Side: 272 Cold Side: 400	Hot Side: 570.6 Cold Side: 839	Hot Side: 987.6 Cold Side: 1507	Hot Side: 2731 Cold Side: 400
Overall heat transfer coefficient	9.749 W/m² K	8.819 W/m² K	10.907 W/m² K	2.85 W/m² K
Pressure drop	3.323 Pa	3.274 Pa	16.93 Pa	0.4353 Pa
Width L_1	1.2095 m	1.2095 m	1.2095 m	1.2095 m
Depth L_2	0.1025 m	0.2377 m	0.3367 m	0.3429 m
Height L_3	1.4663 m	1.33 m	0.7542 m	1.5 m

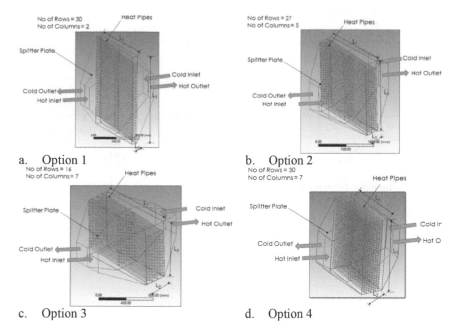

a. Option 1

b. Option 2

c. Option 3

d. Option 4

Figure 1. Models for different options of heat pipe heat exchanger.

where, L = Latent heat of vaporization (kJ/kg), σ = Surface tension on liquid surface (N/m), z = Characteristic number, Q = Power (kW), φ = Inclination of heat pipe, r_v = Vapor core radius(m).

Based on above equations, four different heat pipe heat exchangers were designed. The materials of the heat pipe, wick and fin were taken as Copper, Stainless-steel and Aluminum respectively. The major dimensions of these options are given in Table 2 and the three—dimensional models of these options are shown in Fig. 1 (Mardhekar, 2015).

4 EXPERIMENTAL INVESTIGATIONS ON HEAT PIPE

After the design optimization, 2 numbers of heat pipes (outer dia. 0.0254 m, 4 fins/inch) were fabricated. The parametric studies on heat pipe were carried out as under:

- by recirculating the water in closed loop test rig
- by passing the water once through the shell
- with stagnant water at different power inputs and inclinations of the heat pipe

The schematic diagram and experimental setup are shown in the Fig. 2.
The collection efficiency of heat pipe was worked out using Eq. (21).

$$\eta = \frac{Heat\ gained\ by\ water}{Electric\ heat\ supplied} = \frac{m c_p \Delta T}{V\ I\ cos\phi} \tag{21}$$

where, m = mass flow rate of water (kg/s), C_p = specific heat of water (J/kgK), ΔT = rise in water temperature, V = voltage (V), I = current (A), cos ϕ = power factor.

From the experiments, the time required to reach to steady state for heat pipe was found as 135 minutes. The instantaneous efficiency of heat pipe was found in the range of 85–94% at different flow rates (0.0070–0.016 kg/s). The efficiency of the heat pipe at different inclinations (90°, 70°, 4.5°) and power input (155–425 W) was found in the range of 87–97%.

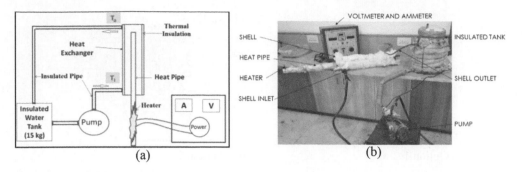

Figure 2.   (a) Schematic diagram and (b) experimental setup.

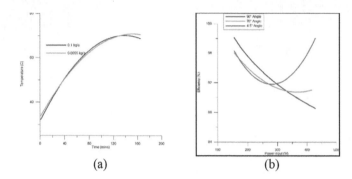

Figure 3.   Variation of (a) temperature versus time (b) efficiency versus power input.

The variation of water temperature against time is shown in Fig. 3(a) and variation of efficiency with different heat inputs is shown in Fig. 3(b) (Mardhekar, 2015).

5   CONCLUSIONS

Based on the design, Copper heat pipes with 1 layer of Stainless Steel wick and Aluminum fins are recommended. It was concluded that the Option 2 (135 heat pipes with 4 fins/inch on hot and cold sides) can be used in the applications where low fluff and lint are expected in the exhaust gases going inside the heat exchanger. However, for the Stenter machine subjected to high fluff and lint Option 4 (210 heat pipes with 8.8 fins/inch on cold side and bare hot side) can be used to avoid clogging on hot side. The instantaneous and collection efficiency of heat pipe was found in the range of 85–94% and 47–49% and at different flow rates and time required to reach steady state is 135 minutes.

REFERENCES

Borges, T., Pickler, G. & Henriques, M. 2011, Heat pipe air heaters in stenters for textile industry, *21st Brazilian congress of mechanical engineering*: 24–28.
Mardhekar, H.V. 2015, Design and analysis of heat pipe heat exchanger for stenter machine, M Tech Thesis, Nirma University.
Reay, D. & Kew, P. 2006, *Heat pipes theory design and applications*. Burlington: USA.
Shah, R. & Giovannelli, A. 1988. Heat pipe heat exchanger design theory, Heat transfer equipment design: 609–653.
Srimuang, W. & Amatachaya. P. 2012, A review of the applications of heat pipe heat exchangers for heat recovery, *Renewable and Sustainable Energy Reviews* 16: 4303–4315.

Multi-disciplinary Sustainable Engineering: Current and Future Trends – Tekwani, Bhavsar & Modi (Eds)
© *2016 Taylor & Francis Group, London, ISBN 978-1-138-02845-6*

Identification of energy saving potential of different thermal insulation through mathematical modelling

Somakshi Mattoo, B.A. Shah & A.M. Lakdawala
Institute of Technology, Nirma University, Ahmedabad, Gujarat, India

ABSTRACT: In the present work, a mathematical model is developed to demonstrate energy saving potential of a thermally insulated, air conditioned room and identify the best insulation. Solar radiation and cooling capacity of air conditioner are applied as boundary conditions of the room. The room is discretised using Finite Volume Method and governing equations are solved simultaneously. A code is developed to calculate the mean temperature inside the room and AC working time. Simulations have been carried out for four different conductivities and seven different thicknesses of insulations namely Polyurethane Foam, Extruded Polystyrene Foam, Fiberglass and Rockwool. AC on-off cycle for different insulations and thicknesses, 2-D temperature contour variations and velocity profiles are generated as outcome of the code. The amount of energy saved is calculated on the basis of how long the AC works. A graph is plotted for Cost of electrical energy saved per day vs thickness of insulated wall for different insulation and optimum thickness of insulated wall is obtained from it.

Keywords: energy saving potential; conjugate heat transfer; ac on-off cycle; cost of electrical energy saved/day; optimum thickness of insulation

Nomenclature	
X	Non-dimensional horizontal coordinate
Y	Non-dimensional vertical coordinate
Z	Non-dimensional transverse coordinate
U	Non-dimensional velocity of fluid in x-direction
V	Non-dimensional velocity of fluid in y-direction
W	Non-dimensional velocity of fluid in z-direction
U	Non-dimensional velocity vector
P	Non-dimensional pressure
g	Acceleration due to gravity
T_s	Hot wall temperature
T_∞	Free stream temperature inside room
L	Characteristic length of enclosure
Pr	Prandtl number
Gr	Grashof number
Ra	Rayleigh number
Greek symbols	
θ	Non-dimensional temperature
ρ	Density of fluid (air)
τ	Non-dimensional time
α	Thermal diffusivity
β	Thermal expansion coefficient
ν	Kinematic viscosity

1 INTRODUCTION

The world's energy resources are fast depleting with the rapidly increasing consumption of the growing population, thus creating a need to develop methods for saving or optimization of these resources. Use of thermal insulation is an efficient way of conserving energy to bridge the gap between energy demand and supply. Insulation material is important tool for constructing energy conserving buildings. The better the insulation, the better resistance it will provide to the heat transfer in and out of the insulated room.

In an air conditioned room heat gain through walls, roofs, ceilings etc. are major source of heat gain including heat gain through glass and others. The conduction heat transfer through wall or roof depend on wall thickness and thermal conductivity of material used. Thus there is a large scope of energy saving is possible by providing thermal insulation on wall and roof of an air conditioned room. Conjugate heat transfer is fundamental for many devices and processes, including heat exchange between building and environment, heat exchangers, material processing and many others (Zhao 2006). A large number of studies have been reported on conjugate heat transfer in square, triangular, rectangular or open cavities with different fluids and thermal boundary conditions, square and open cavities being more common (Ha 2000, Serrano-Arellano 2014). This study aims to develop code to find AC on/off cycle in a close insulated room and thereby calculate energy saving potential of different thermal insulation of varying thickness.

2 MATHEMATICAL MODELLING

In the present work a cubic room with dimensions 3 m × 3 m × 3 m without any doors or windows, which is maintains temperature at 25°C using an air conditioner, have been considered. The inside of the room is thermally insulated on all sides. Aim of the project is to identify energy saving potential of different thermal insulation by mathematical modelling. Different thermal insulation have different value of thermal conductivity, k. The inside wall temperature of the room is determined by solar radiation falling on the walls and roof of the room which varies with time and place. A code is prepared to calculate the mean temperature of the different points (nodes) inside the room using governing equations. When the temperature reaches a temperature of 25°C, the AC gets cut-off and switches on when the temperature reaches 27°C. Figure 1 shows a three-dimensional view of the insulated room described in the problem.

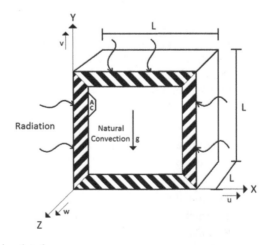

Figure 1.   3D view of insulated room.

2.1 Governing equations in non-dimensional form

In the present work non-dimensional form of governing equations are solved. Governing equations are non-dimensionalised (normalised) by using scaling constants for length as L, for velocity as α/L, for pressure as $\rho\alpha^2/L^2$ temperature as $(T_s - T_\perp)$ and time as L^2/α.

$$X = \frac{x}{L}, Y = \frac{y}{L}, Z = \frac{z}{L}, U = \frac{uL}{\alpha}, V = \frac{vL}{\alpha}, W = \frac{wL}{\alpha}, P = \frac{pL^2}{\rho\alpha^2}, \theta = \frac{T-T_\infty}{T_s-T_\infty}, \tau = \frac{t\alpha}{L^2}$$

Continuity equation in non-dimensional form for an incompressible fluid can be written as

$$\frac{\partial U}{\partial X} + \frac{\partial V}{\partial Y} + \frac{\partial W}{\partial Z} = 0 \tag{1}$$

Momentum equation in non dimensional form for an incompressible unsteady fluid can be written as

$$\frac{\partial \mathbf{U}}{\partial \tau} + U\frac{\partial \mathbf{U}}{\partial X} + V\frac{\partial \mathbf{U}}{\partial Y} + W\frac{\partial \mathbf{U}}{\partial Z} = \Pr\left(\frac{\partial^2 \mathbf{U}}{\partial X^2} + \frac{\partial^2 \mathbf{U}}{\partial Y^2} + \frac{\partial^2 \mathbf{U}}{\partial Z^2}\right) - \frac{\partial P}{\partial X}\hat{i} - \frac{\partial P}{\partial Y}\hat{j} - \frac{\partial P}{\partial Z}\hat{k} + Ra\Pr\theta\hat{j} \tag{2}$$

where $\Pr = \frac{v}{\alpha}$, $Gr = \frac{g\beta(T_s-T_\infty)L^3}{v^2}$, $Ra = Gr \cdot \Pr = \frac{g\beta(T_s-T_\infty)L^3}{v\alpha}$

Pr, Gr and ***Ra*** are the governing non-dimensional parameters. Boussinesq approximation is used in the Navier-Stokes (momentum) equation in the pressure gradient term. Variation in density due to temperature variation generates the buoyancy force causing flow of air inside the room resulting in heat transfer by natural convection.

Energy Equation in non dimensional form for an incompressible unsteady fluid can be written as

$$\frac{\partial \theta}{\partial \tau} + U\frac{\partial \theta}{\partial X} + V\frac{\partial \theta}{\partial Y} + W\frac{\partial \theta}{\partial Z} = \left(\frac{\partial^2 \theta}{\partial X^2} + \frac{\partial^2 \theta}{\partial Y^2} + \frac{\partial^2 \theta}{\partial Z^2}\right) \tag{3}$$

Three boundary conditions are applied on the problem at hand to calculate the temperature at different points (nodes) inside the thermally insulated room. First is the **solar radiation** falling on the walls and roof of the room, which is used to calculate inside wall temperature. Second is the **no-slip condition** at the walls. Velocity of the fluid at the walls (boundary) of the room is zero, i.e. $U = 0$, $V = 0$, $W = 0$. But there is movement of fluid (air) in the y-direction (v-velocity) inside the room arising due to buoyancy force (due to temperature variation) which is taken into account in the Y-Momentum (Navier-Stokes) Equation. The third is the **constant wall heat flux** dissipated by the AC.

2.2 Numerical methodology

The governing equations are discretised using finite volume method on a staggered grid where the pressure is located at the cell center whereas the velocity components u and v are staggered and located at the cell face centers. The finite volume discretised form of navier-stokes equation is

$$\frac{\Delta V\left(\mathbf{U}^{n+1} - \mathbf{U}^n\right)}{\Delta\tau} = F^D - F^C - \nabla P^{n+1}\Delta S + F_g \tag{4}$$

where **Convective Flux** $F^C = \nabla\mathbf{U}(\mathbf{U}\cdot\Delta S)$, **Diffusive Flux** $F^D = \Pr\nabla(\nabla\mathbf{U}\cdot\Delta S)$ and **Gravitational Force** $F_g = \Delta V\left(Ra\cdot\Pr\cdot\theta\right)\hat{j}$

In the present work, **U** in the convective flux is calculated using the Total Variation Diminishing (TVD) scheme. The gradient of velocity across a cell face in diffusion flux term is calculated using central difference scheme.

The projection method is used to solve time-dependent Navier-Stokes equation for incompressible fluids. The main advantage of this technique is that the pressure and velocity field computations are decoupled.

Navier-Stokes equation (dimensionless form) for an incompressible fluid can be written as:

$$\frac{\partial \mathbf{U}}{\partial \tau} + (\mathbf{U} \cdot \nabla)\mathbf{U} = -\nabla P + Pr\nabla^2\mathbf{U} + Ra\,Pr\,\theta\hat{j} \tag{5}$$

where vector velocity, $\mathbf{U} = U\hat{i} + V\hat{j} + W\hat{k}$ and the term $Ra\,Pr\,\theta\hat{j} = 0$ for Navier-Stokes equation in x- and z-direction (U and W velocities respectively).

$$\frac{\mathbf{U}^{n+1} - \mathbf{U}^n}{\Delta\tau} = -(\mathbf{U}^n \cdot \nabla)\mathbf{U}^n - \nabla P^{n+1} + Pr\nabla^2\mathbf{U}^n \tag{6}$$

Computation of the intermediate velocity, U*, is done explicitly by ignoring the pressure gradient term in the momentum equation.

Velocities at new time level, n+1 can be calculated by substituting the Pressure of new time level. Thus, the system has updated values for the new time level. Values of U, V, W and P at different nodes inside the room can be calculated by this method. These values can be substituted in the Non-dimensional Energy Equation and value of θ can be computed which further gives the value of temperature T at different nodes.

3 RESULTS AND DISCUSSION

In the present study only solar radiation data for a day (per day) in summer in Ahmedabad is considered and all the results obtained below are for Ahmedabad only. 2 D heat transfer is considered in the room. Thickness of bare (uninsulated) walls & roof is considered to be 0.1 m, to which the insulation thickness is added. Conductivity k and thickness of wall with insulation Δx is varied. *Polyurethane Foam (PUF), Extruded Polystyrene Foam (XPS), Fiberglass and Rockwool insulations* having conductivities *0.021 W/mK, 0.033 W/mK, 0.04 W/mK and 0.045 W/mK* respectively are used by varying thickness as *0.125, 0.15, 0.175, 0.2, 0.225, 0.25 & 0.275 m* for each insulation.

3.1 *Variation of AC on-off cycle with insulation*

In the present work, it is considered that an AC of cooling capacity 1 ton is installed on the west wall of the insulated room. Accordingly the west side wall acts as a sink for the hot air inside the room and dissipates it to the surroundings. The AC switches on when the mean temperature in the room reaches 27°C and the AC switches off when the mean temperature

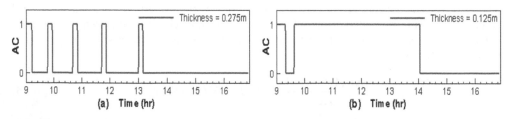

Figure 2.   AC On-Off Cycles for (a) Polyurethane Foam (PUF) insulation at insulated wall thickness of 0.275 m and (b) Rockwool insulation at insulated wall thickness of 0.125 m.

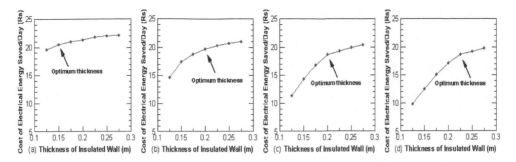

Figure 3. Cost vs Thickness Variation for (a) Polyurethane Foam, (b) Extruded Polystyrene Foam, (c) Fiberglass and (d) Rockwool.

in the room reaches 25°C. AC on-off cycle (time step) is generated by the code showing the duration for which the AC is on (working) and the time for which it is off. In the cycle *0* and *1* represent the off and on condition (time) of the AC respectively.

The AC working (on) time increases with the increase in conductivity of insulation material and decreases with the increase in thickness of the material. The mean temperature inside the room rises quickly as a result of increase in conductivity and decrease in thickness of the insulation material and attains the AC switch on temperature. Thus, the AC works for a longer time to achieve the required (switch off) temperature.

3.2 *Calculation of optimum thickness of insulation*

The AC installed at the wall in the insulated room under study is of capacity 1 ton (3.517 kW) and 3-star rating. Compressor work (power input) of the AC is calculated by dividing the refrigeration capacity (output) of the AC with Energy Efficiency Ratio (EER) which comes out to be 0.983 kW (Data of Carrier 3-star AC is used). Energy is consumed in the form of electricity. Cost of electricity (energy) for Ahmedabad is taken as Rs 3.2 per unit (kWh). Data from the AC on-off cycles is used for calculating the values in the graphs in Figure 5 which are used to determine the optimum thickness for each insulation.

4 CONCLUSION

- A graph is plotted for Cost of electrical energy saved/day (Rs) vs Thickness of insulated wall (m) for different insulations and optimum thickness of insulated wall is obtained from it. Energy saved is directly proportional to the thickness of insulation used and inversely proportional to the conductivity of insulation. This means the AC working time and energy consumption can be reduced by decreasing the conductivity of the insulated wall or increasing the thickness of the insulation.
- Optimum thicknesses for Polyurethane Foam (PUF), Extruded Polystyrene Foam (XPS), Fiberglass and Rockwool insulation are found to be 0.15 m, 0.2 m, 0.2 m and 0.225 m respectively.
- Saving in cost of electricity is found to be maximum for Polyurethane Foam insulation with conductivity, k = 0.021 W/mK at insulated wall thickness of 0.275 m and minimum for Rockwool insulation with conductivity, k = 0.045 W/mK at insulated wall thickness of 0.125 m.

REFERENCES

Ha, M.Y. & Jung, M.J. 2000. A numerical study on three-dimensional conjugate heat transfer of natural convection and conduction in a differentially heated cubic enclosure with a heat-generating cubic conducting body. *Int. J of Heat & Mass Transfer* 43: 4229–4248.

Incropera, F.P. & DeWitt, D.P. (5th ed.), 2002. *Fundamentals of Heat and Mass Transfer*. New York: Wiley.

Papadopoulos, A.M. 2005. State of the art in thermal insulation materials and aims for future developments. *Energy and Buildings* 37: 77–86.

Serrano-Arellano, J. & Gijón-Rivera, M. 2014. Conjugate heat and mass transfer by natural convection in a square cavity filled with a mixture of Air–CO2. *Int. J. of Heat & Mass Transfer* 70: 103–113.

Sukhatme, S.P. & Nayak, J.K. (3rd ed.), 2008. *Solar Energy Principles of Thermal Collection and Storage*. Tata McGraw-Hill Companies.

Versteeg, H. & Malalasekra, W. (2nd ed.), 2007. *An Introduction to Computational Fluid Dynamics: The Finite Volume Method Pearson*.

Zhao, F.Y., Tang, G.F. & Liu, Di. 2006. Conjugate natural convection in enclosures with external and internal heat sources. *Int. J. of Engineering Science* 44: 148–165.

Author index